BMW

DAVID HODGES

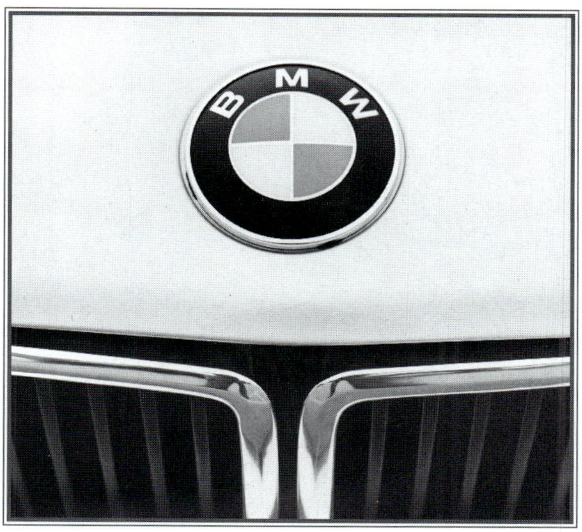

SUTTON PUBLISHING LIMITED

Sutton Publishing Limited
Phoenix Mill · Thrupp · Stroud
Gloucestershire · GL5 2BU

First published 1999

Title page: *The badge is a symbolic reminder of BMW's origins as an aero engine manufacturer. The design represents a spinning propeller; it was first seen in a more explicit form in 1917, and the familiar blue and white quartered stylized version was introduced in 1920. It has been subtly changed at intervals. The* nierenförmig *'kidney grille' is almost as familiar as the badge.*

British Library Cataloguing in Publication Data
A catalogue record for this book is available from the British Library.

ISBN 0-7509-2144-7

Typeset in 10.5/13.5 Photina.
Typesetting and origination by
Sutton Publishing Limited.
Printed in Great Britain by
Ebenezer Baylis, Worcester.

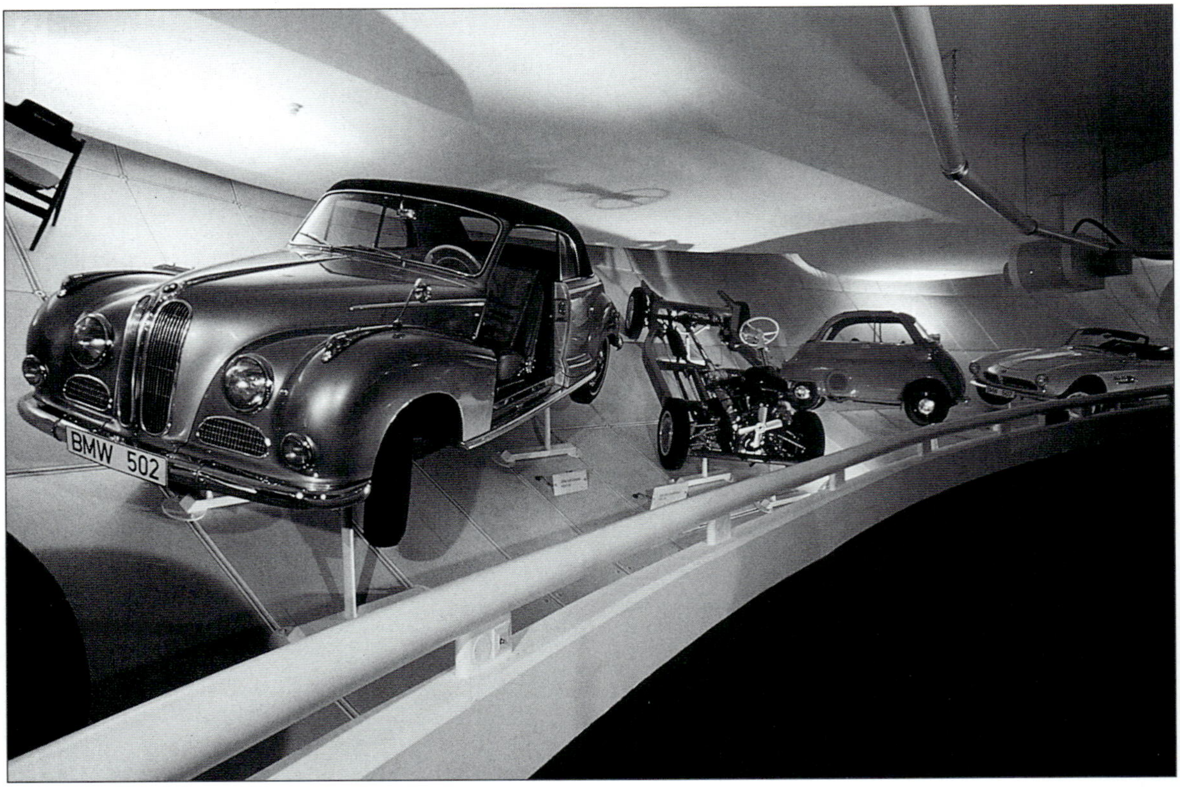

BMW's museum is a 'must-visit' building in Munich. Landmark machines from the company's past are imaginatively displayed, although seldom changed – one shortcoming in the modern architecture is difficult car movements! The complete vehicles shown here are a 502, an Isetta and a 507.

CONTENTS

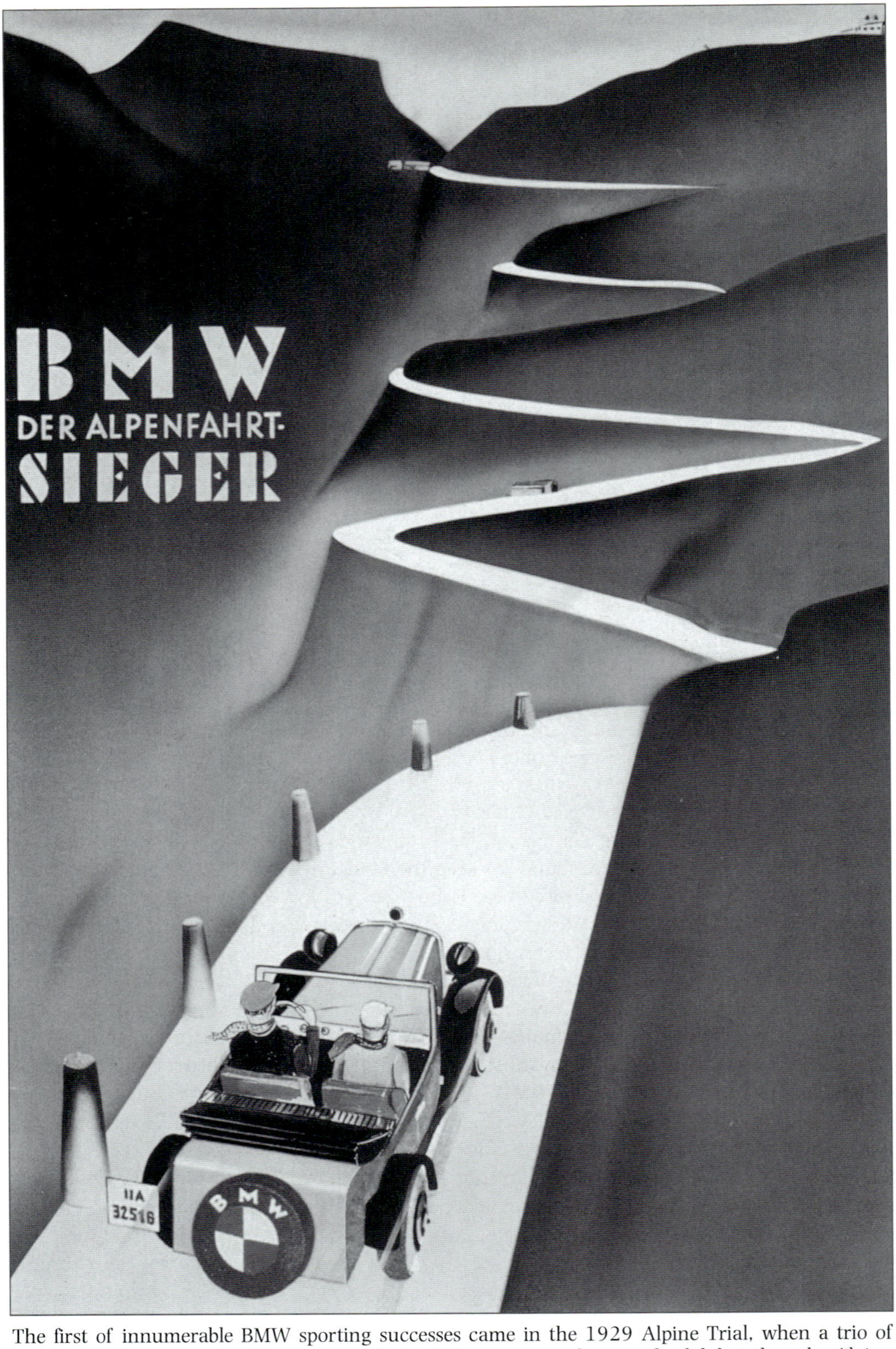

The first of innumerable BMW sporting successes came in the 1929 Alpine Trial, when a trio of 3/15s won the team prize. The artist made the little car appear large and solid, but then the Alpine pass is also stylized.

INTRODUCTION

BMW is renowned for its stylish and sophisticated cars, and the success that has made it one of the leading companies in the European motor industry. Its formative period as a car maker began with a simple light car, and it quickly developed an advanced range, only for car-building to cease for a dozen years. Through the 1950s and into the 1960s BMW cars tended to be at the extremes of the motoring spectrum, company policies were uncertain, and BMW's future was sometimes in doubt. Then in the 1960s the foundations for lasting success were laid, and within a quarter of a century BMW was positioned to challenge the oldest and greatest of the German car manufacturers.

Two small aero engine companies merged to form the company that was titled Bayerische Flugzeugwerke in 1916 and a year later became the Bayerische Moteren Werke. Three men central to the new company were Austrian financier Camillo Castiglioni, engineer Franz-Josef Popp and Max Friz, whose design experience had been honed under Paul Daimler.

Towards the end of the First World War the first BMW aero engine was put into production at Munich. This straight-six IIIa powered the most successful variant of the finest scout (fighter) aircraft of the first German Air Force, the Fokker D VIIIf. The BMW engine was preferred to the Mercedes 'six' used in earlier versions of the D VIII, and it was built by Opel while other companies looked to produce it under licence. Aero engine production was banned under the disarmament conditions imposed on Germany, but quiet development work continued at BMW and one of its engines was used in a successful but officially unrecognized altitude record attempt in 1919.

In common with other German companies, BMW was permitted to build aero engines again from 1923, and the company was to produce some outstanding power units. Meanwhile, other outlets were found, to keep the company in business. Some aero engine components were illicitly used elsewhere, while engines for commercial vehicles and the 'little engine' for the Victoria motorcycle company were developed. The horizontally opposed twin-cylinder unit designed by technical director Friz and first supplied to Victoria was to become a classic, and it was also used in the Helios built by BMW before its own first production motorcycle, the R32, was introduced in 1923. This was an advanced machine, with the transverse boxer engine in 494 cc form, a three-speed gearbox, and shaft drive.

The engine was also used by obscure German car manufacturers, or would-be manufacturers in the 1920s, as BMW looked to this market. Its first four-wheeler was little more than a Tatra 11 with a BMW engine in place of the Tatra flat twin, driving the front wheels. The company then undertook the completion of prototypes of a car designed by engineer and aerodynamicist Wunibald Kamm; this front-wheel-drive car with quasi-unitary construction was too far ahead of its time to be a realistic production project.

Towards the end of the 1920s Popp saw a sensible opportunity at the popular level where he felt BMW should enter the car market. (He was also a Daimler-Benz director, and in any case there would have been commercial risks in going head-on into competition with the likes of Mercedes-Benz.) The long-established Dixi-Werke company of Eisenach was slipping ever deeper into debt, and its modest car production had peaked at 867 vehicles in 1925. The large cars in its ranges had sold poorly, and it desperately needed the small model that came in 1927, when the first 3/15 was completed. This was the Austin Seven built under licence, using German materials. It sold well in 1928, its first full year of Eisenach production – the 9307 completed as the DA 1 exceeded the total of other Dixi cars, 1904–26!

Castiglioni negotiated his last major deal for BMW in 1928, whereby the Munich company took over the Eisenach factory, and of course the proven 3/15. Early in 1929 the first cars carrying the BMW badge appeared, and within months the modified DA 2 was introduced. The little car was progressively developed, and served BMW well through a harsh period, until the licence agreement with Austin was rescinded in 1932. The Dixi made way for the first true BMW car, the 3/20. That was flawed, but in 1933 the true progenitor of the BMW car line, the 303, was introduced. It had a six-cylinder engine, and generally attractive body lines with the detail of the *nierenförmig*, the 'kidney grille', appearing for the first time. The 303 set a pattern for the 1930s, and sales success meant that minor revenue-earning oddities such as three-wheel delivery vehicles could be set aside. More than that: in the 1930s a six-cylinder engine was often taken to show a manufacturer's intentions, in this case BMW's move towards the quality car market.

Fritz Fiedler took on the car design role, and the development of the straight six, where Rudolf Schleicher played a significant part. By the late 1930s BMW models ranged from elegant coachbuilt saloons and cabriolets to handsome production four-seaters and sports cars, from the mid-decade 315/1 that led to the Frazer Nash association and the outstanding 328. Car production then petered out in 1941.

Motorcycles were made again from 1946, but major difficulties had to be overcome before car manufacture was resumed, and then there were false starts. Germany was divided in 1945, and Eisenach fell in the Russian occupation zone. Cars were produced in the old plant, but after some 321s had been completed for the Soviet Autovelo organization BMW was successful in a legal action, and the use of its blue and white badge was denied to EMW, the company that took on the factory. Eisenach-built cars then carried red and white EMW badges until production ceased in 1955.

By that time BMW had made a fresh start. The attractive little 531 was taken to the prototype stage, before management policy decreed that prestige cars should be given priority and an expensive large saloon, the 501, was introduced. It was to be no more than modestly successful; the V8-engined 502 did little better, and sales of the much praised 507 two-seater were disappointing. Motorcycle business largely carried the company through to the mid-1950s, before sales fell away and financial losses became threatening.

The BMW management once again looked to small cars. Alongside the big models, the Isetta built and developed under licence made for an unbalanced range, and perhaps provided turnover rather than the profit or development resources needed for the vital models for the 1960s.

The late 1950s were in any case uncertain years, with management changes, rumours of takeovers or mergers, and a Deutsche Bank-inspired Board recommendation that BMW should become part of the Daimler-Benz empire, turned down by shareholders in 1959. A positive approach came as the Quandt brothers, Herbert and Harold, took a substantial shareholding, and soon a controlling majority. As financiers and industrialists, they could see the weaknesses; their first nominee joined the Board, and in the early 1960s the company began to take the shape it has today, with products that are widely respected. The Quandt family still has a controlling interest.

The new generation cars that were to reposition BMW in the market (roughly where it had been in 1939) were designed quickly and economically, although they did not come as quickly as some might have hoped; but once they did these '*Neue Klasse*' models gave the marque a modern image.

In 1967 BMW took over Glas, largely for its Dingolfing production facility, although two Glas models were continued and some tooling was passed to BMW's South African subsidiary, where production was extended into the mid-1970s. As BMW fortunes improved, the company entered motor sport again. The first cars had provided a modest platform in the 1930s, and the 328 had gained an excellent international reputation late in that decade. These cars were run again after the Second World War, by independent entrants, BMW engines were used by minor German constructors in the 1950s, and the Bristol derivatives were used in British Formula 2 cars as well as Bristol's Le Mans coupés. Then the little 700 showed well in small-capacity classes, in rallies as well as races into the 1960s. Independent entrants were encouraged, and direct company involvement gathered pace with the later '*Neue Klasse*' cars. The European Touring Car Championship was to become a happy hunting ground for BMW, while Munich engines were again used in other constructors' sports-racing cars and single-seaters. BMW's own Formula 2 venture brought modest results, and after that its International single-seater commitments tended to be in partnership with British constructors. Then there was a notable entry into endurance racing with McLaren in the 1990s, before BMW launched its own open sports car programme late in that decade.

The record of success in touring car racing through to the end of 1998 was remarkably consistent, and the connection between production road cars and competition cars was always strong. Most of the M engines designed by the talented Paul Rosche and his team have been based on production units, even the turbocharged Formula 1 engines of the 1980s.

The Quandts appointed one of their own men chairman in 1970. Eberhard von Kuenheim was an outsider, but proved to be absolutely right for this top automotive industry position. Through to the 1990s he refashioned BMW, maintained its independence, controlled its investment and directed some to important areas such as the FIZ R&D centre, oversaw production efficiencies that were essential to offset high German labour costs, ensured that it countered rivals (especially German and Japanese rivals), and save for a hesitation through a mid-1970s fuel crisis saw its sales increase year on year.

During his reign the BMW Motorsport subsidiary was set up, survived a difficult period, and then flourished. The M-designated cars have been distinctive high-performance versions of production models since the first M, the M1 supercar for road and circuit use. In two decades BMW M GmbH produced 100,000 cars and had its own 'branches', such as M-Technic which was responsible for M products (other than cars). Incidentally, it had no connection with BMW Technik, an R&D department.

BMW chassis had attracted leading coachbuilders for many years, but available design studies show that the company's own designers were never stifled, at least until late stages, when requirements to conform to exterior and interior 'family characteristics' in the front-line series were inevitably paramount. Until 1974 BMW did not have a test track, and its proving grounds were public roads and race circuits.

The 1970s saw four of the modern lines introduced. The 5 Series, the first mid-size BMWs for many years, came in 1972; the 3 Series launched in 1975 were the smallest models and were to be built in large numbers, with many variants; the 6 Series coupés from 1976 were based on the 5 Series and borrowed something from the 7 Series that was still to come; the 7 Series in 1977 was the large BMW, and a competitor for Mercedes-Benz in a prestige sector. The 8 Series coupé was to supersede the 6 Series, and with the others to continue through to the end of the century, albeit with revisions. These Series have determined the arrangement of this book.

There were of course engines. New four- and six-cylinder types were introduced in the 1960s and in the 1970s, none perhaps exciting as far as basic specifications were concerned, but carefully conceived in important areas such as combustion chambers, and with development potential. A V12 in the early 1980s did not progress beyond the prototype stages, but a larger one did reach production later in that decade, and so did a V8 early in the 1990s.

Many BMWs have had strong sporting qualities, and in 1987 the imaginative little Z1 was true to the broad sports car image of an open two-seater. There has been a tendency to regard the next Z production car as a roadster rather than a sports car. Either way, this product of BMW's Spartanburg factory in South Carolina has been another sales success – by the end of 1998 more than 150,000 had been built. In addition to that Z3, the M coupé and the X5 'sports activity vehicle' were to be built at Spartanburg.

The success of this green field plant reflected the productivity of the German factories. Through the 1990s BMW's flexible working methods were refined to a degree still rare in the automotive industry. Each of the three main factories built more than one model, even more than one on the same assembly line, and this meant that output could be closely matched to demand.

Towards the end of the twentieth century BMW attracted attention as much for its business activities as for its cars. Bernd Pischetsrieder, an engineering graduate and well known as a production expert, succeeded von Kuenheim in 1993. (Significantly, von Kuenheim was still very influential in the background, as the principal supervisor.) Pischetsrieder had played a major role in persuading the Board to sanction the US plant, and in the Rover takeover (when the plum for BMW was reckoned to be Land Rover; there was also a ring of irony, for the first production BMW car was really an Austin, and Rover was the successor to Austin). Then there was BMW Rolls-Royce GmbH, an aero engine company, the agreement to supply engines for Rolls-Royce cars, followed by an unseemly contest for the ownership of the great marque. BMW Motorsport Limited to be set up in Britain was launched in 1995 and functioned until 1999; then a BMW Group engine factory in Britain was announced. And there were more examples of BMW's spread as a major automotive company.

But however essential expansion was to preserve BMW's position, it was not achieved without problems, particularly with Rover's enormous losses, and these led to widely discussed boardroom disputes which came to a head as Bernd Pischetsrieder resigned in February 1999. The Supervisory Board appointed Engineering and Production Director Joachim Milberg to be the new Chairman of BMW AG.

Meanwhile, the mainstream lines flourished: with the broad pattern established, the versatility of some of the basic models could be exploited and BMW could confidently enter niche markets. Evolution rather than revolution often seemed to govern technical aspects, but the real engineering strength behind the cars could never be denied. There had been lapses in the company's past but there were few in the last decades of the twentieth century, when Bee Emms, Bimmers, BMWs – call them what you will – inspired enviable marque loyalty.

* * *

In the 1960s the E numbers system that is sometimes used as a model reference came into use. It superseded a three-figure number sequence that had run from the first Isetta (100) to a 1600 roadster project (124) in the mid-1960s. The E numbers were to become useful labels, but at first sight owe little to logic, largely because a complete list has never been available, as some E-designated projects remain confidential. These *Entwicklungs*, or project/development numbers, were issued as work was authorized and retained for vehicles that entered production (sometimes applying to revised versions, such as the E6 updated two-door saloons in 1973). The 2500 in 1968 was the first E-designated type to reach production.

E has also been used for electric cars, confusingly as E1 did not come until 1991, and then did not have an *Entwicklungs* number. (Logically, this would have been an early-30s number.)

Then there is Z, normally associated with production sports cars such as the Z3. However, *zunkunft* (future) numbers have been applied to other projects, some such as city cars that are at the other end of the motoring spectrum, as well as some assumed to be stillborn that remain confidential.

CHAPTER ONE

FORMATIVE YEARS

The Dixi 3/15 gave BMW an instant entry into the car market, but the acquisition also depleted capital resources, for BMW took on debts almost equal to the Eisenach purchase price. Development work on the little car was consequently limited, and Austin influences were to be found in its successor, the 3/20. The 3/15 DA 1 gave way to the DA 2 in 1929 (BMW retained the DA designations – *Deutsches Ausfuhrung*, or simply German Version). Four-wheel brakes were introduced – DA 1 foot brakes worked on the rear wheels – and there were three body styles, while a van was also listed and a few chassis had coachbuilt bodies. The Wartburg name from the earliest Eisenach days was revived for the DA 3 sports two-seater (1930–31); it then disappeared again, until a Wartburg two-stroke of dubious repute was built at Eisenach for three decades from 1956.

The four-cylinder line continued into the second half of the 1930s. The DA 4 in 1931 had a slightly more generous body (from the rear it seemed to overwhelm the under-pinnings) and the transverse spring suspension that apparently gave quirky handling, and a reputation that meant disappointing sales. The 3/20 (1932) and 309 extended the life of the small four-cylinder BMWs, and the 309 outsold the first of the new six-cylinder cars in 1934–36.

Eisenach's name for high-quality cars stems from the 303, launched early in 1933. It was powered by a straight six, to all intents and purposes the four with two cylinders added – an economic approach, as existing tooling could be used. It was preferred to a new four-cylinder engine designed by Friz, and BMW's original technical supremo turned away from automobile engines. Fritz Fiedler and Rudolf Schleicher were responsible for the six, and its development. The 303 had a welded tubular chassis, quite unlike the Austin types, and after an initial batch had been made by Daimler-Benz its standard bodies were made by Ambi-Budd.

The adaptable straight six was used in varying capacities in the 315, BMW's first sports car with this type of engine, and the 319. The 315s in the 1934 Alpine Trial impressed the Aldington brothers, who bid for the British concession in the name of Frazer Nash. The 319 was sold as a Frazer Nash-BMW, and the relationship did much to establish BMW in Great Britain. H.J. Aldington 'rescued' one of the famous 'Mille Miglia' 328s at the end of the Second World War, and Frazer Nash was to use the straight six in its own cars through to the 1950s, albeit the Bristol version of the engine.

The 326 announced early in 1936 was BMW's first four-door model; there were also some coachbuilt cabriolets. It was seen as a challenge to Mercedes-Benz, and in sales terms was the most popular BMW of the 1930s, remaining in production until 1941. (15,936 were built.) The 320 and 321 were essentially smaller and cheaper two-door derivatives.

A race victory at the Nürburgring in June 1936 gave the 328 a dramatic public debut, although this outstanding car was not on sale until the following year. It was to influence a generation of European sports cars, and showed that the type could be sophisticated in technical respects and appearance. Here and there it picked up from earlier cars, under the skin having a developed and very rigid version of the 315 chassis, while there were clues to the flowing body lines in Peter Schimanowski's roadsters.

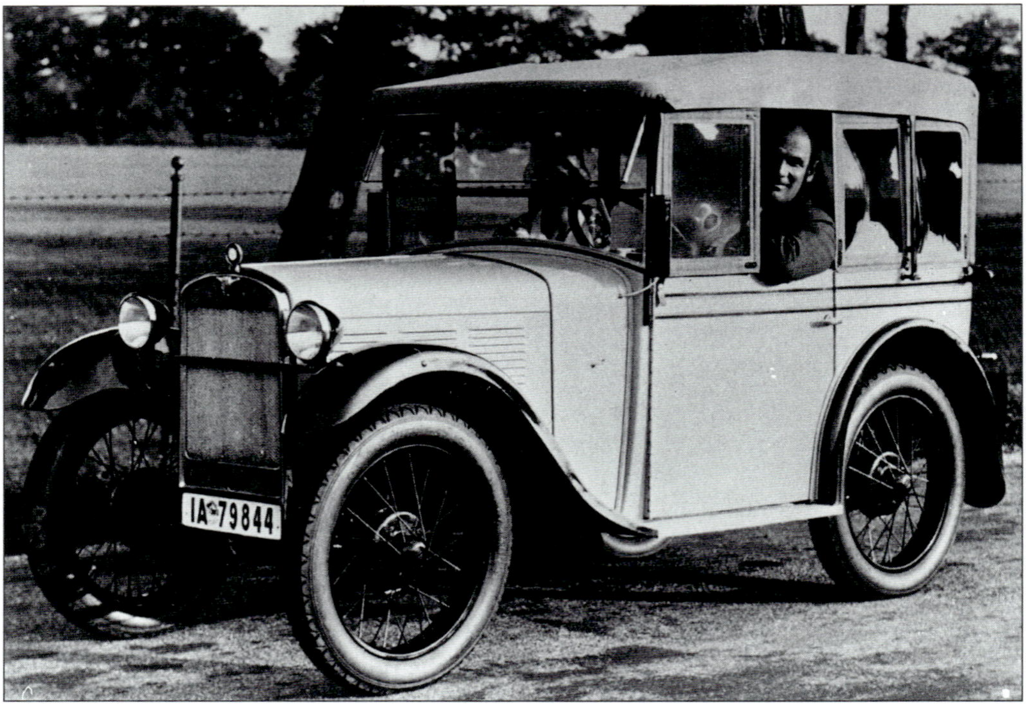

BMW continued to build the Dixi DA 1 until mid-1929, then began its first programme of model improvement, a policy usually followed for the rest of the century. The 3/15 DA 2, here in four-seat tourer form, shows the Austin origins very clearly.

Fiedler outlined an expedient means to increase the power of the straight six without the complication and cost of twin overhead camshafts, using inclined inlet valves directly operated by pushrods and exhaust valves operated by rockers and secondary pushrods. This engine gave competitive power from the outset, was reliable, and had development potential. In hindsight it came to be regarded as a masterpiece. The 328 was a genuine 100 mph (160 kmh) car, and an outstanding competitions car.

The 328 overshadowed the 327, or 327/28 when the 328 engine was fitted, which was usually seen in coupé or roadster forms. The last significant BMW of the era was the 335, a stretched 326 in most respects, and a big car with a 3.5-litre six-cylinder engine.

By the end of the 1930s the BMW range was impressive, although the 1929–41 production total was modest, just under 76,000 cars. But the marque was established, with a high reputation that would help in its eventual recovery after the Second World War.

There had been sporting successes from the first year – trials class victories for 3/15s in 1929, a class win in the 1930 Monte Carlo Rally, and seventh place overall in that winter classic in 1931. The mid-1930s saw more good results in rally-type events, and then there was the significant 328 victory in a sports car race at the 1936 Eifelrennen meeting, followed by a third place overall in the Tourist Trophy at the Ards circuit (a result repeated in 1937, when the race was run at Donington Park).

The 328s went on to dominate 2-litre sports car racing, and in 1939 a coupé was fifth overall at Le Mans. BMW's greatest triumph of the era came in the 1940 'Mille Miglia', run that year on a 102-mile (165-km) closed course. The 328s were far from standard, and the winning car had an attractive coupé body by Touring of Milan. Almost incidentally, 328s were raced independently in the later years of the 1940s.

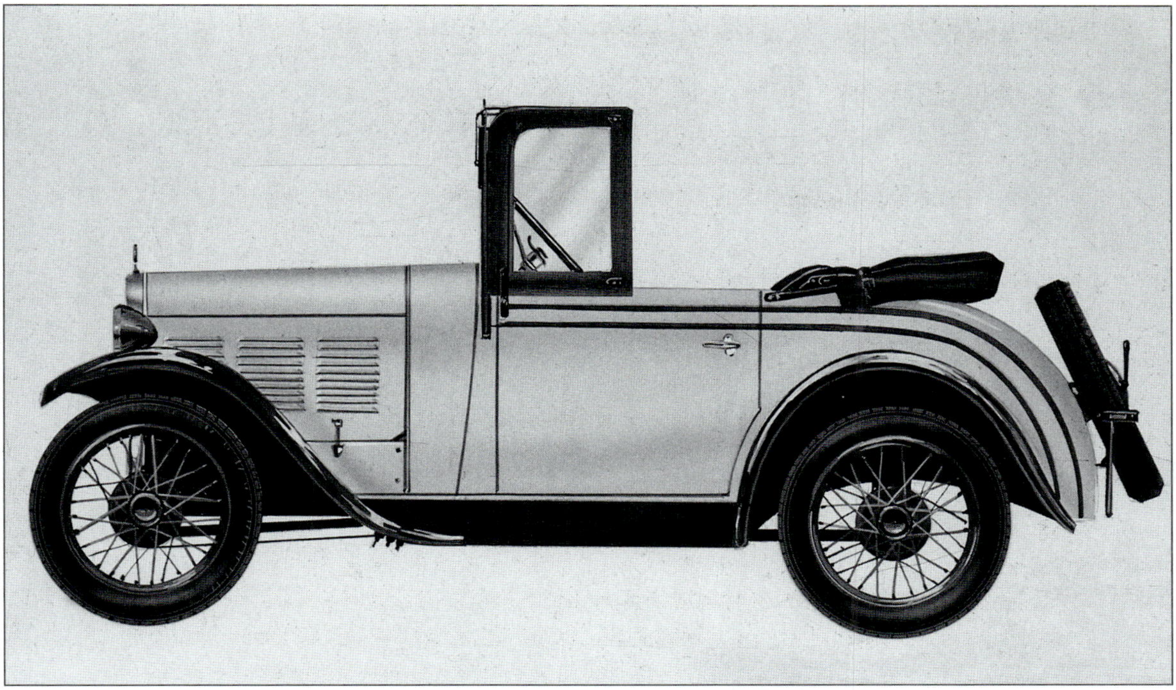

Spindly but sporty. Two-seater styles were popular, but misleading in that only the DA3 Wartburg variant could reach 50 mph (80 kmh). The DA 2 cabriolet (above) had lines that were flawed by its tall upright windscreen. The photograph below shows a DA 2 'cabrio-limousine'. Its wheel trims were presumably stylish. . . .

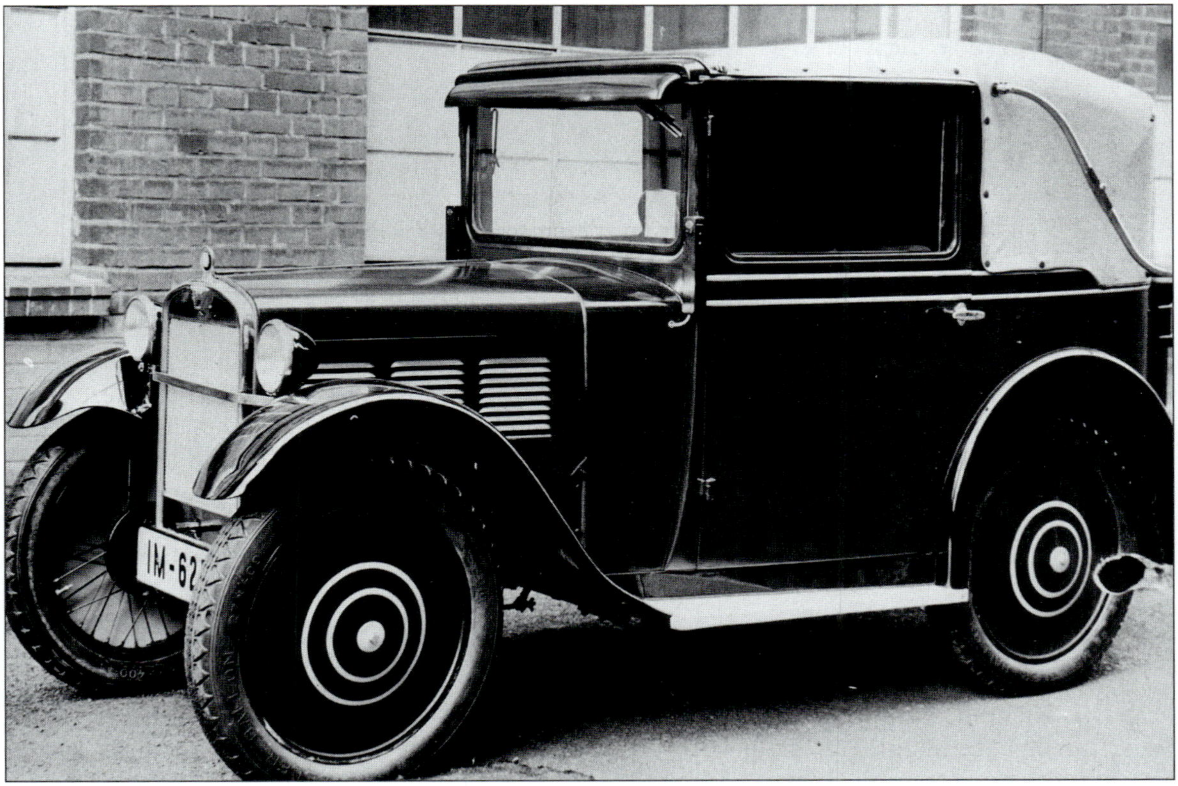

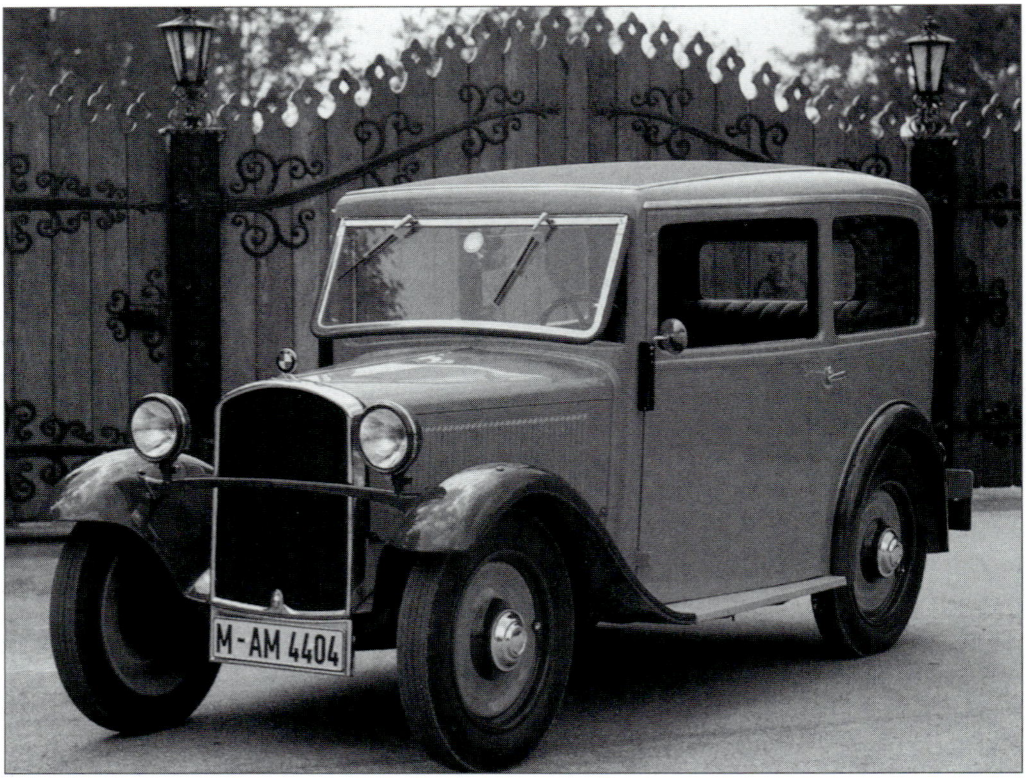

The 3/20, here in 1933–34 AM 4 form, marked the break from Austin/Dixi origins, although similarities remained. It was not a happy car, as transverse leaf suspension front and rear endowed it with curious handling qualities, and the heavy Daimler-Benz-built bodies meant that performance with a 782 cc engine was hardly sparkling. This car has the optional roll-back fabric roof.

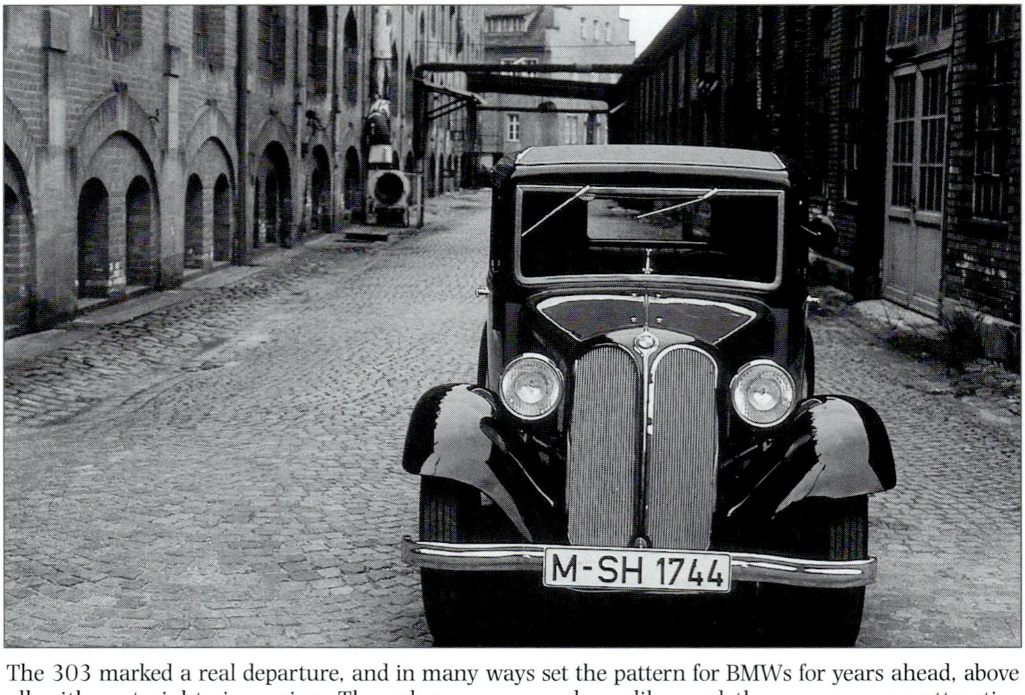

The 303 marked a real departure, and in many ways set the pattern for BMWs for years ahead, above all with a straight six engine. The saloons were workmanlike, and there were some attractive cabriolets. All were distinguished by the 'kidney grille', and its shape is echoed on modern BMW cars.

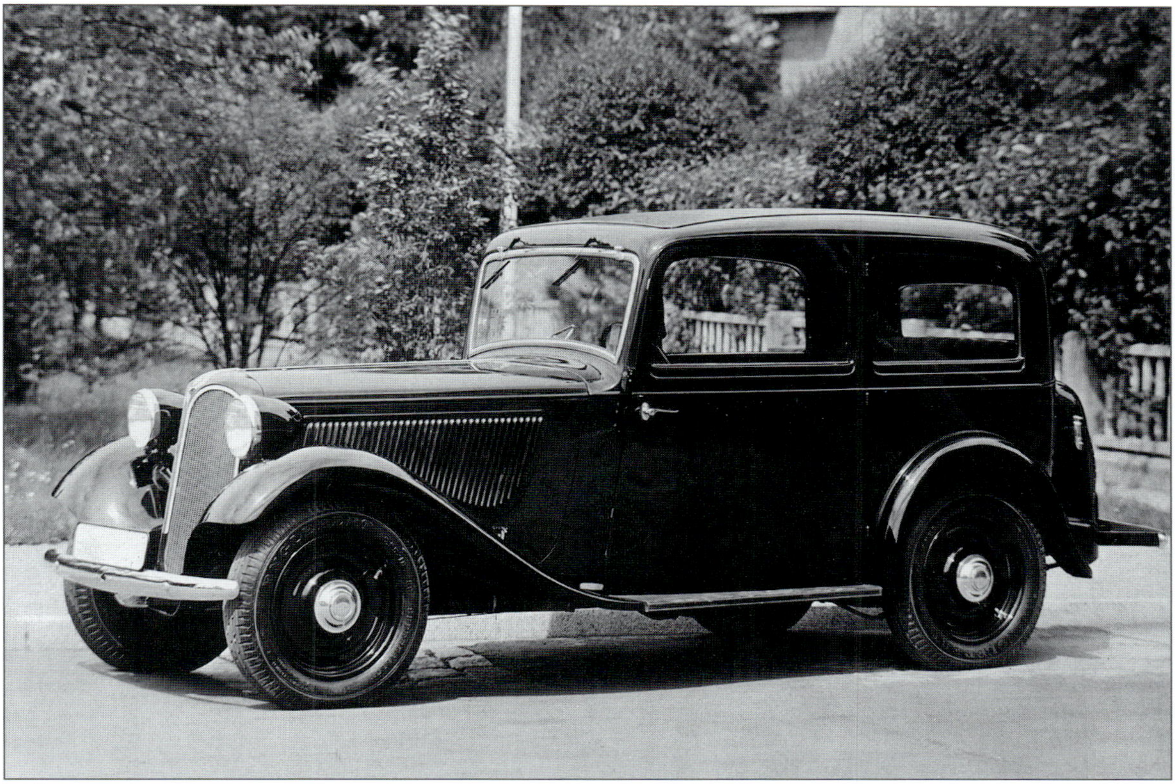

The 309 used the chassis of the 303, with an 845 cc four-cylinder engine. It succeeded the 3/20 in 1934 and some 6,000 were sold before production ended in 1936. Most were two-door saloons or cabriolets. Top speed was no more than 50 mph (80 kmh).

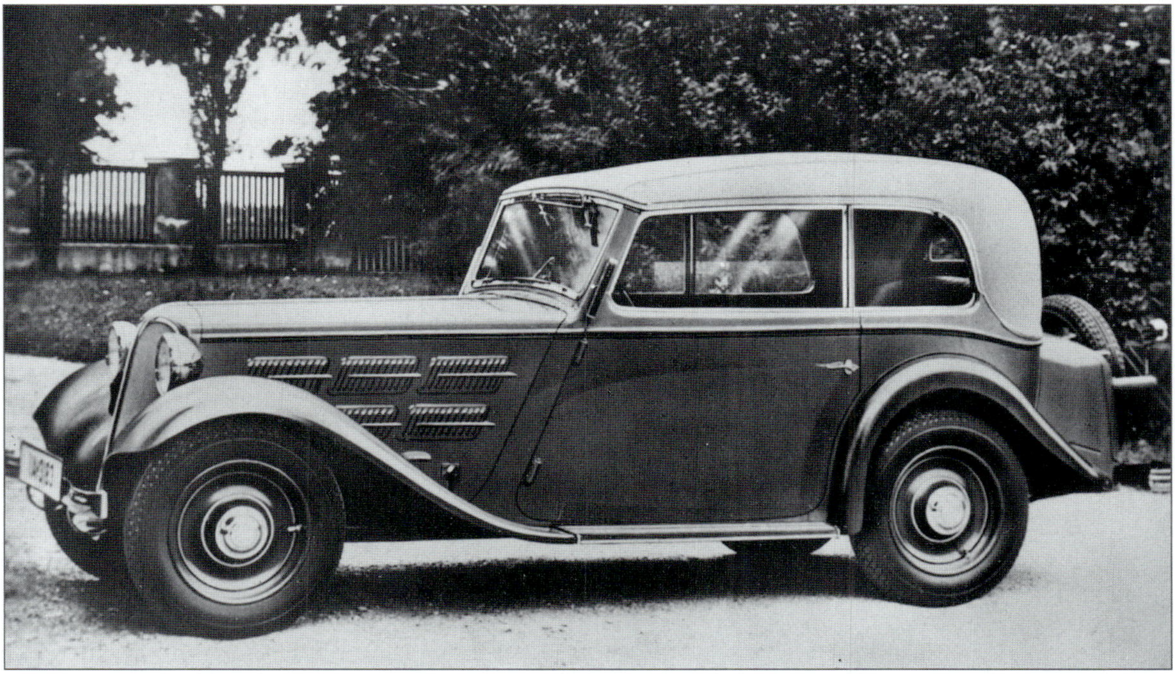

The 315 marked another step forward, and incidentally had a logical designation – Type 3, 1.5 litres. This 1934–35 four-window cabriolet has solid, sensible lines. The little boot carrying the externally mounted spare wheel appears to be an afterthought.

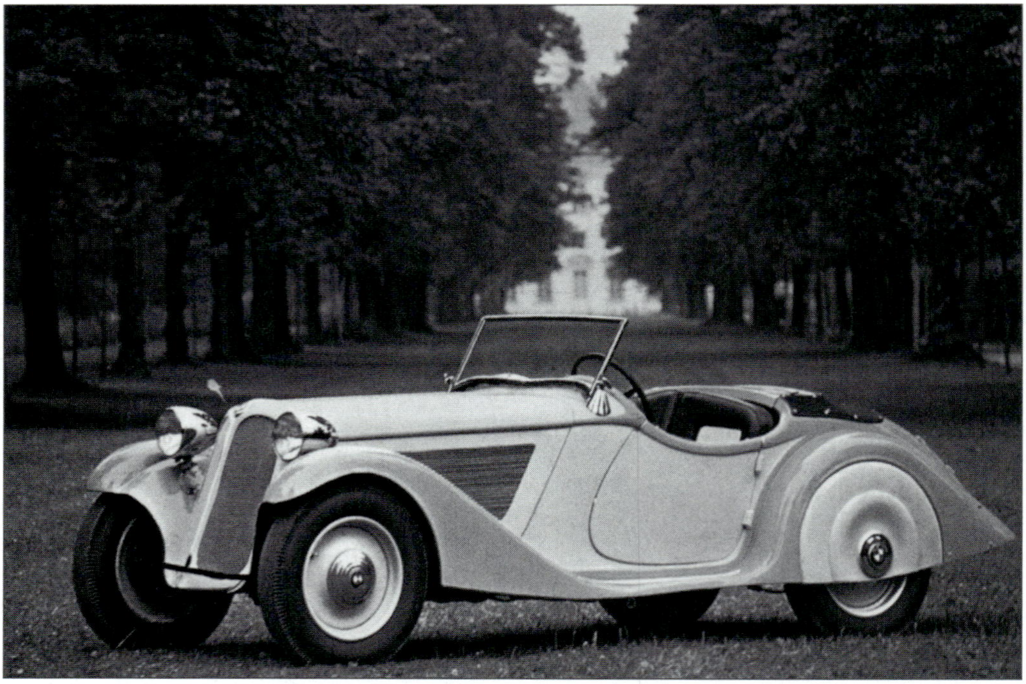

The 315/1 Sport needs only bonnet straps to be an archetypical rakish 1930s sports car. These cars sometimes had minimal bumpers. The badge is repeated on the rear wheel cover, aligned with the front wheel badge, the spare wheel is neatly stowed (at the expense of luggage space), and the soft top is tucked away.

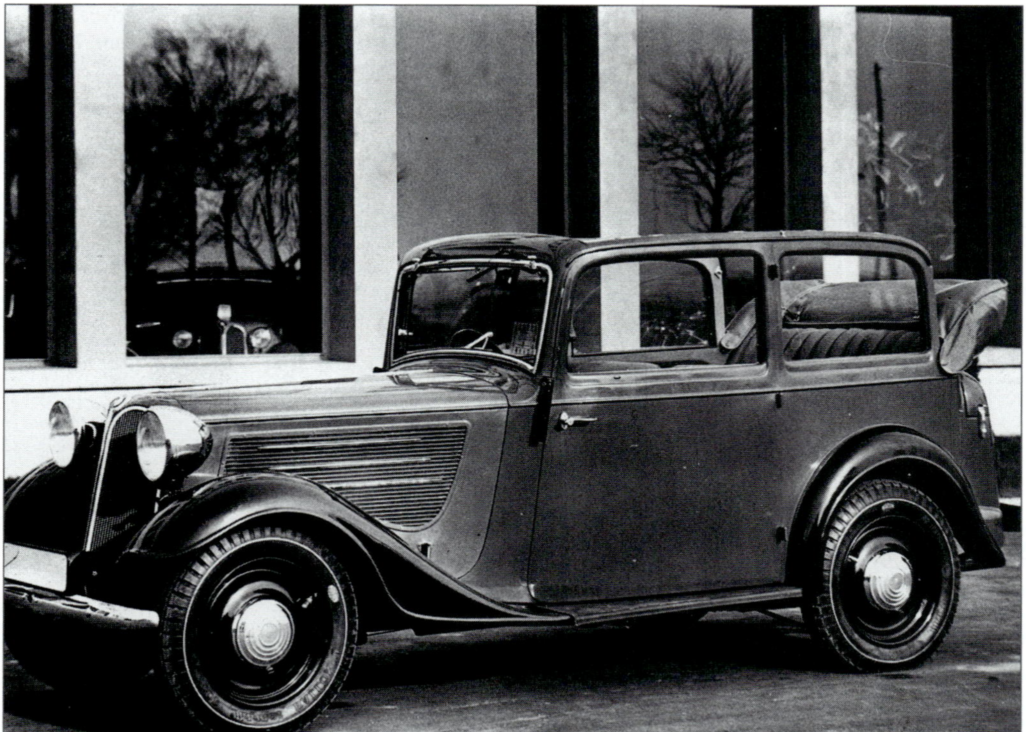

The largest BMW in 1935 was the 319 with a 1911 cc straight six, giving 45 bhp. It first appeared in 'Sport' form in 1934, but few of these two-seaters were built. Just over 2,000 319s were completed in this two-door cabriolet form, and these were 70 mph (114 kmh) cars.

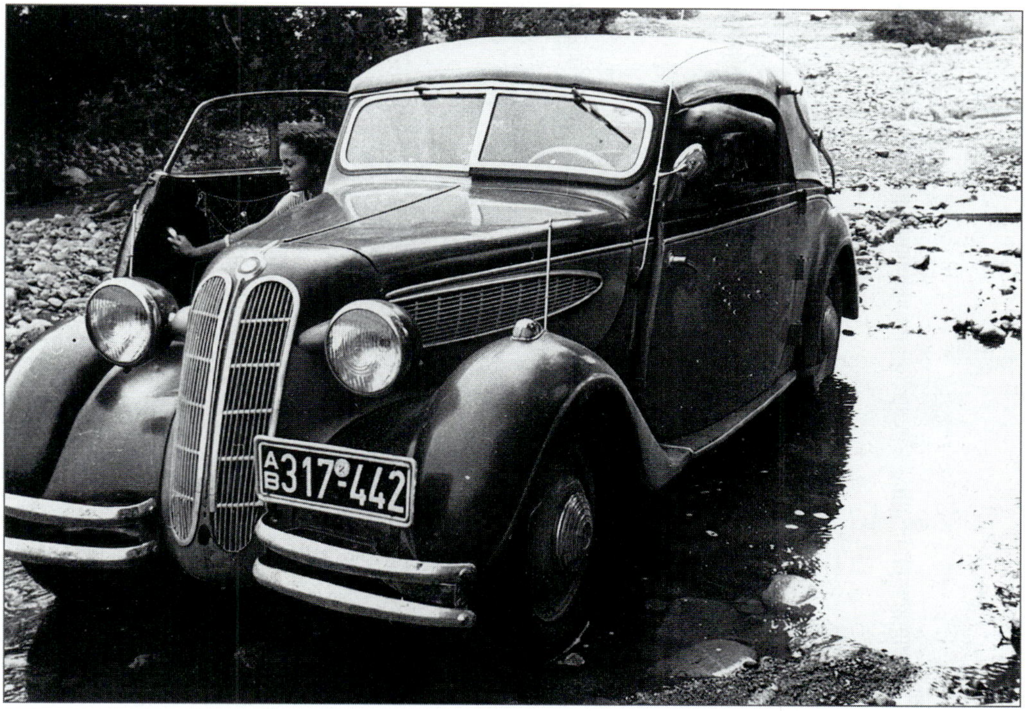

The 326 was a comfortable medium-size touring car, heavy in its saloon form (2,480 lb/1,125 kg) for the 2-litre 50 bhp engine, although the claimed top speed was 70 mph (114 kmh). A substantial box-section chassis was needed to maintain rigidity. Saloons and cabriolets had rear-hinged doors, with front-hinged doors on four-door types. All 1930s BMWs had opening windscreens, and the driver of this solid-looking cabriolet has his open. His passenger appears to be abandoning him in mid-stream. . . .

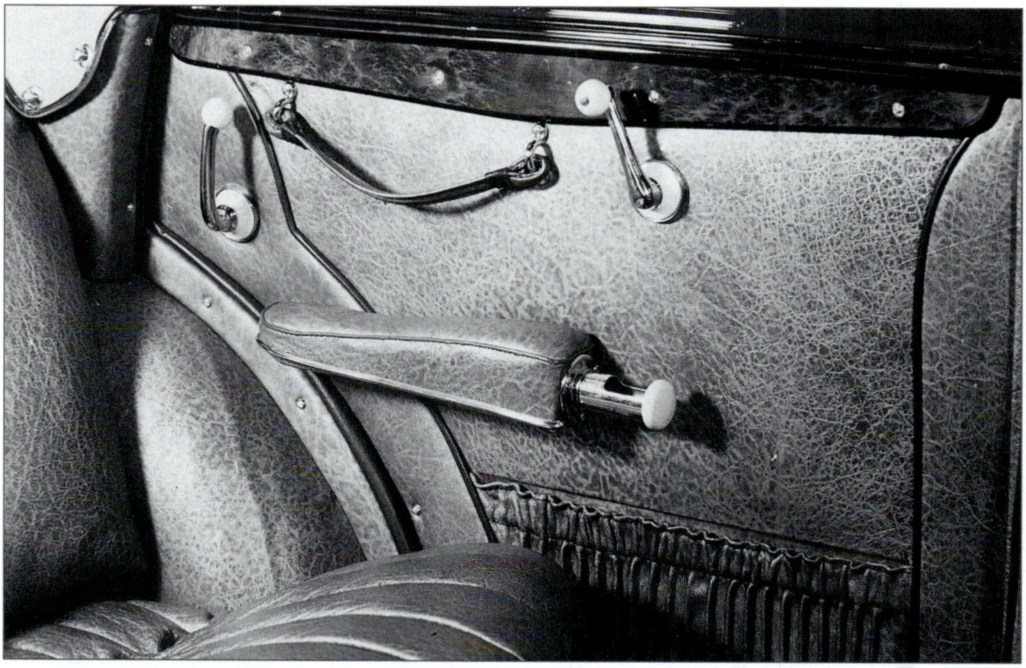

These were reasonably spacious cars and, as this rear-door detail suggests, they were well equipped. The seat does not promise great comfort or support by modern standards – that squab looks far from firm – but there were grab handles on the backs of the front seats and that door pull could be a useful support. The arm rest incorporates a pull-out ashtray.

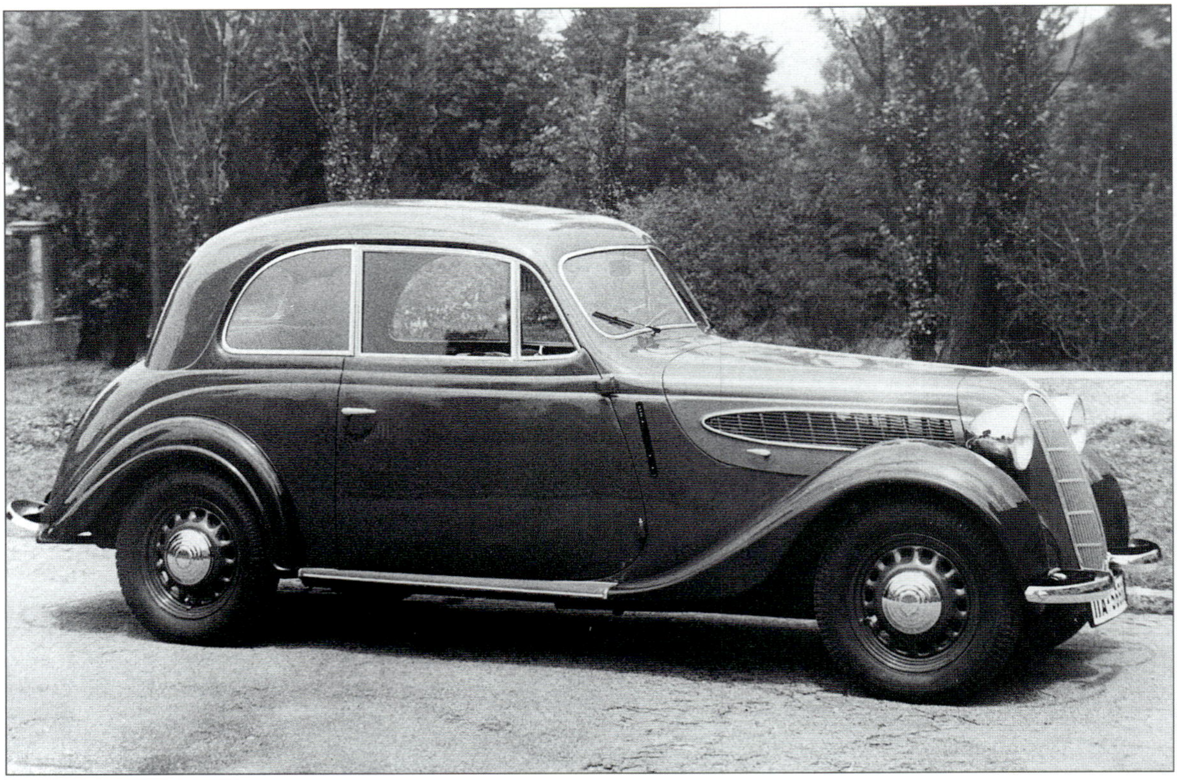

The 320 followed the 326, and indeed the transitional 329, and was listed in two-door saloon and cabriolet forms. They appeared in 1937 with 1911 cc engines, but the 1971 cc straight six was fitted before the end of that year, giving a top speed approaching 70 mph. The 321 which succeeded them in 1938 was essentially similar.

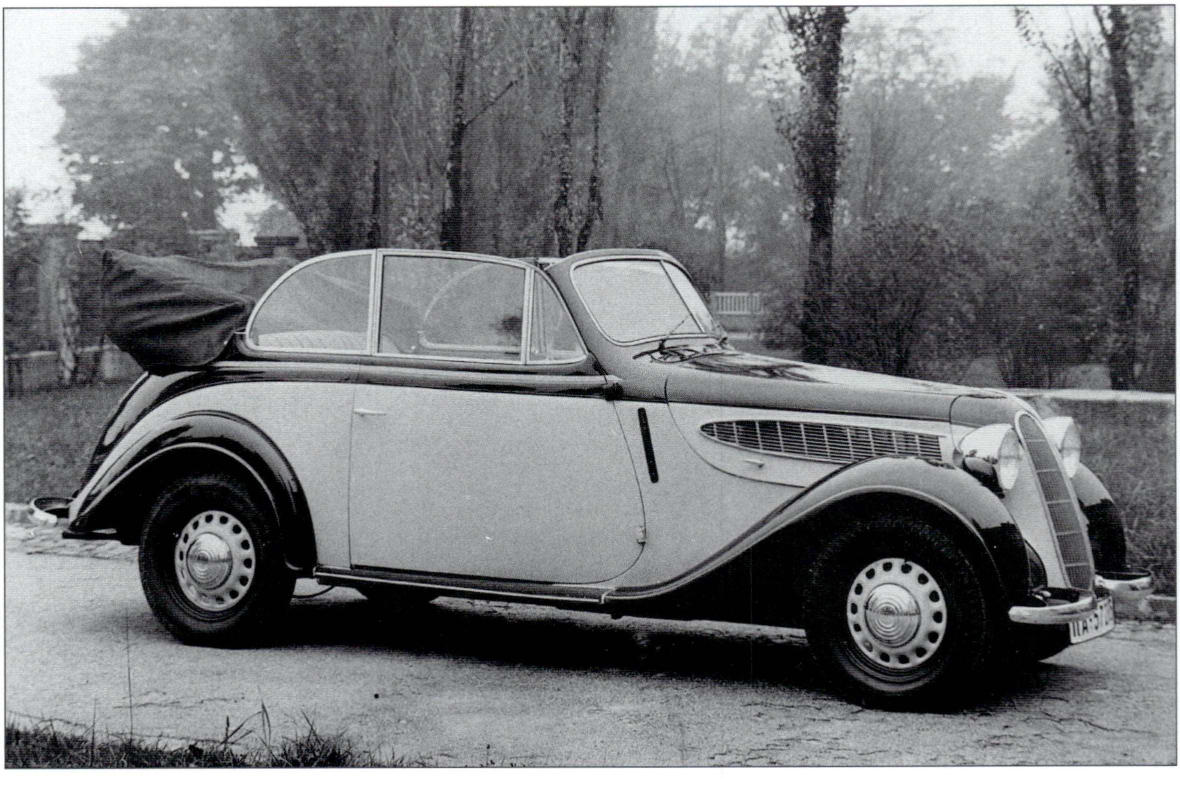

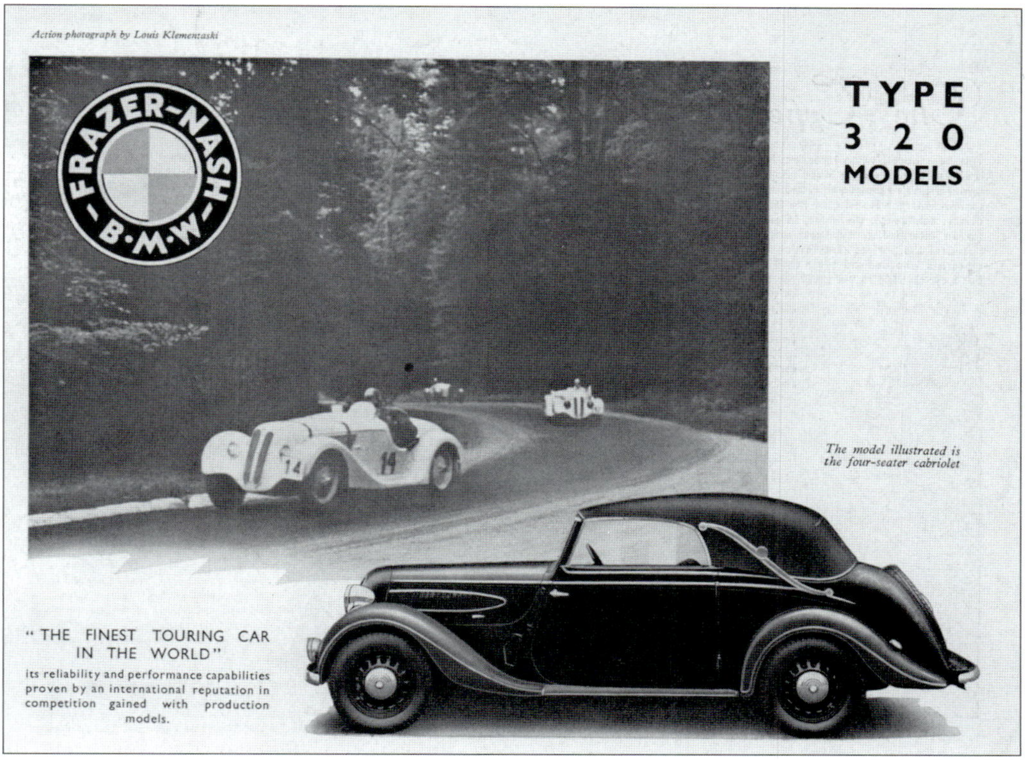

The Frazer Nash association was important to BMW as well as to the British company, and in view of later debates it is interesting to see that AFN chose to hyphenate the name in this 1937 brochure. The cabriolet looks much neater with its hood erected.

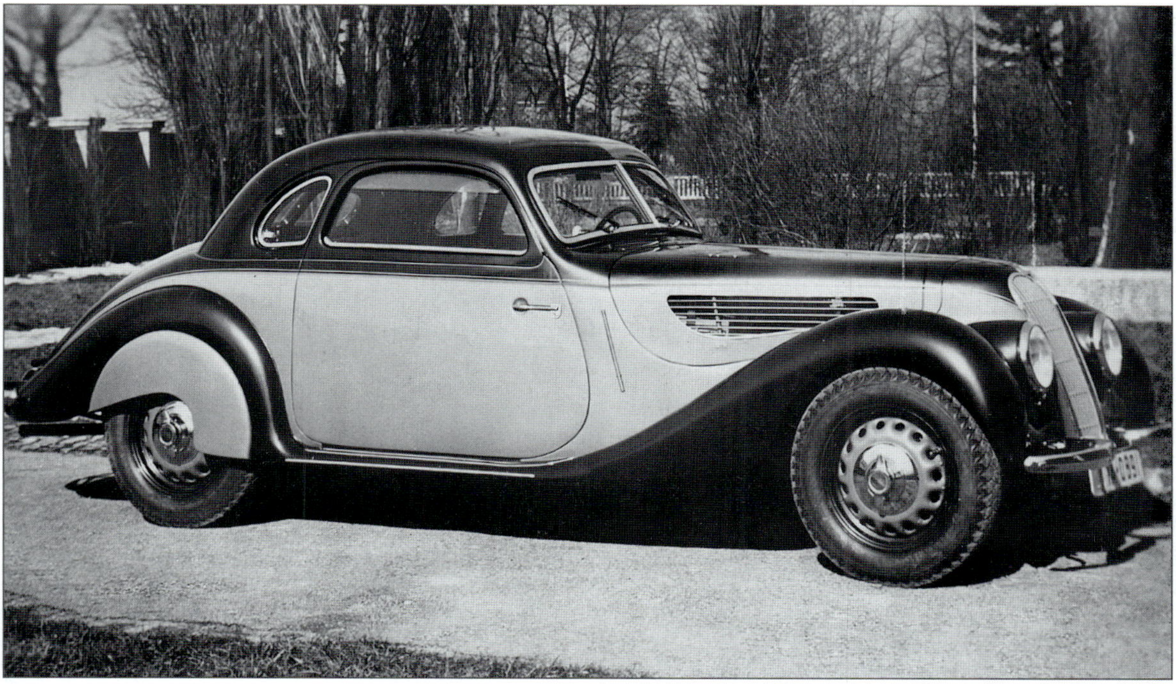

Although fewer than 2,000 were produced, the 327 was another important model, and particularly elegant with this Sport Coupé body on a 327/28. The 1971 cc engine was rated at 55 bhp in the first cars, but the 327/28 was generally 'sportier', with 80 bhp and an 87 mph (140 kmh) top speed. Chassis and running gear were conventional.

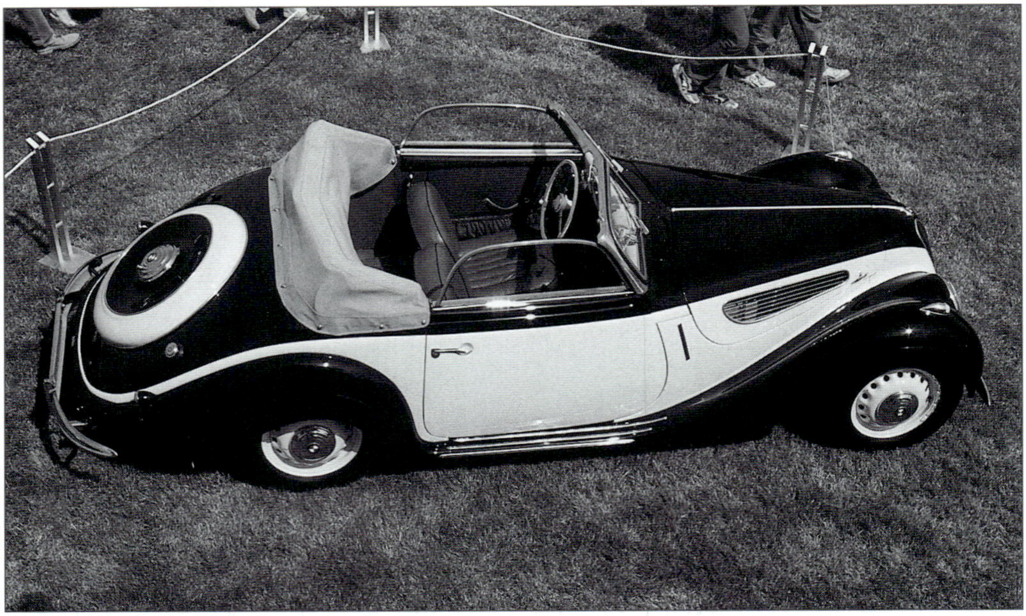

The open 327 was built in greater numbers than the coupé, although the total of just over 1,600 in 327 and 327/28 forms meant that this was still a rare model. This immaculate car is one of many preserved by BMW's Mobile Tradition division, which displays cars at important international events (here at the Goodwood Festival of Speed in 1994).

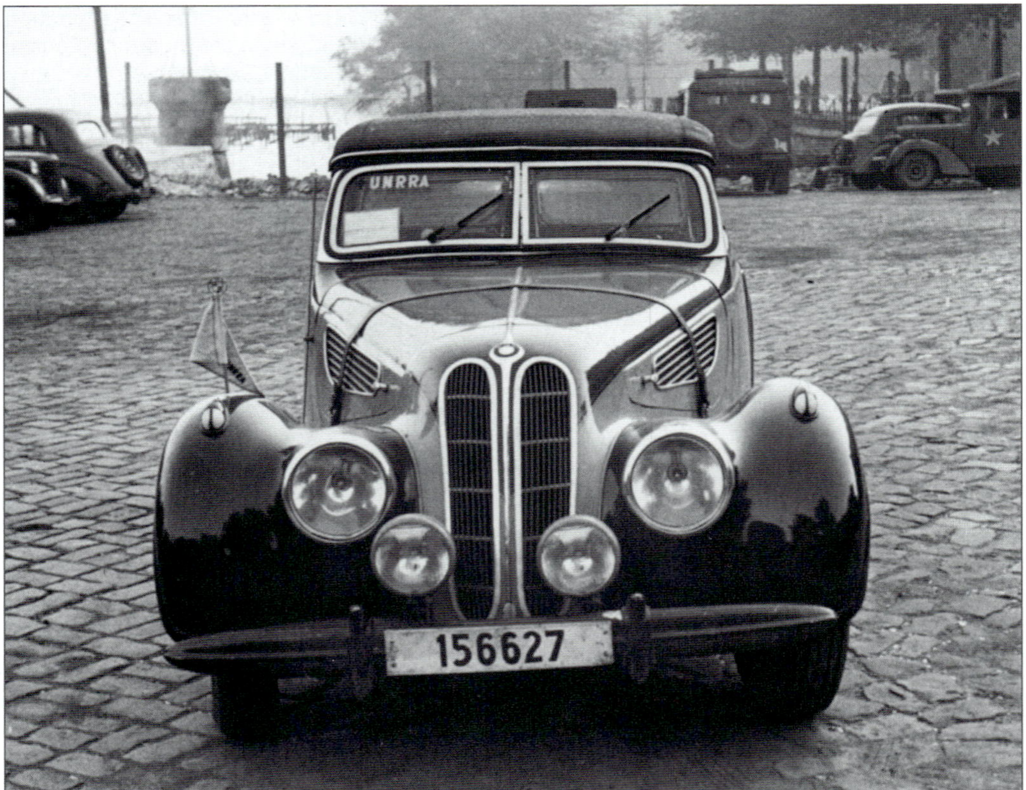

An enterprising – or lucky – official seemingly acquired this 327 for personal transport around Hamburg in 1945, and perhaps added the bonnet strap for effect. By no means all the cars had sidelights on the wings.

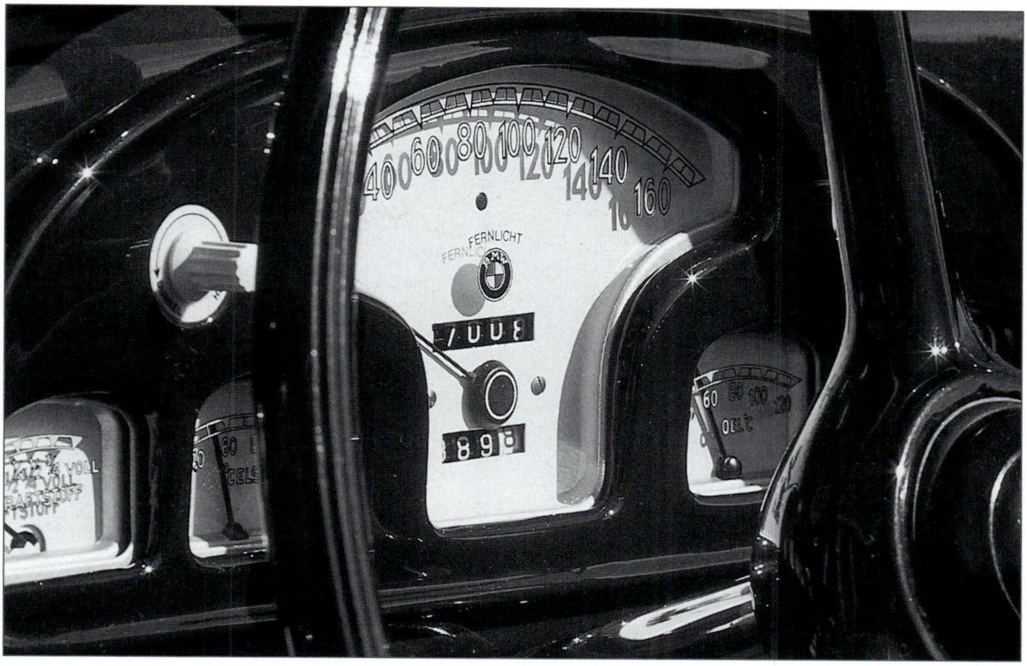

Period piece: the dashboard of a 327, with Bakelite-type material much in evidence. The speedometer of this Fast Tourer is optimistically calibrated to 160 kmh; the Sports also had a rev counter. Secondary instruments normally included an oil gauge, water temperature gauge, fuel gauge and clock, and this car appears to have a voltmeter. There was an 'easy start' control, provision for a Telefunken radio, and a combined ignition switch and steering lock on the column.

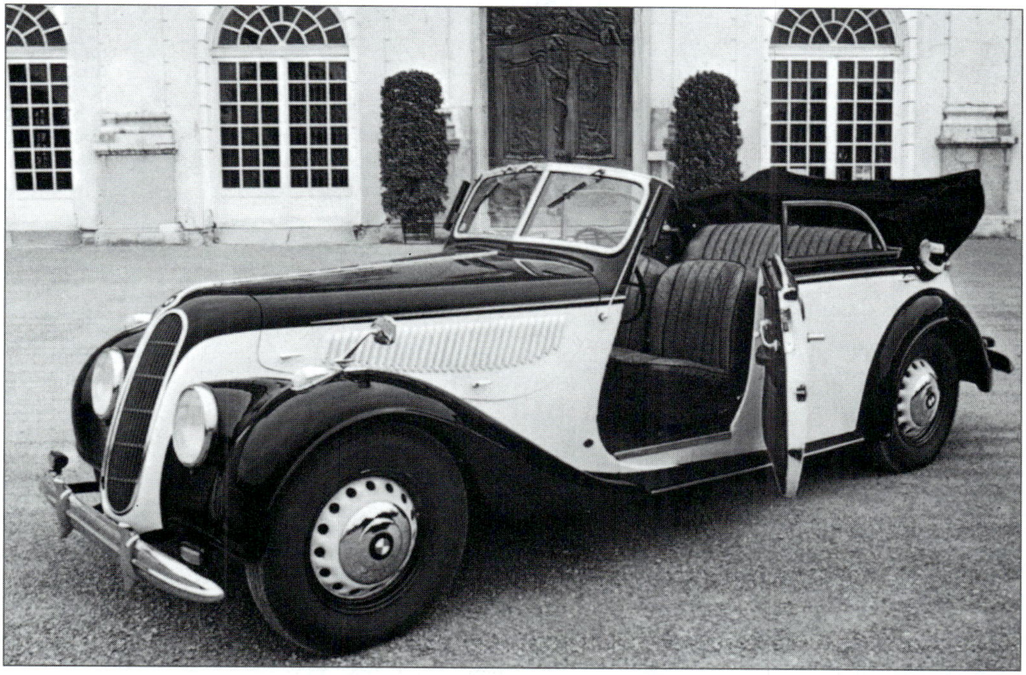

Largest BMW of the 1930s was the 335, introduced late in 1938 with a 3.5-litre development of the established straight six, giving 90 bhp. This four-door cabriolet looks a little ponderous, and coachbuilders (notably Graber) devised neater methods of stowing the hood. The maximum speed claimed for the cabriolet was 90 mph (145 kmh), while the four-door saloon reached 95 mph (152 kmh). The 335 was introduced late in 1938 and built until 1941, when a modest total of 410 had been delivered.

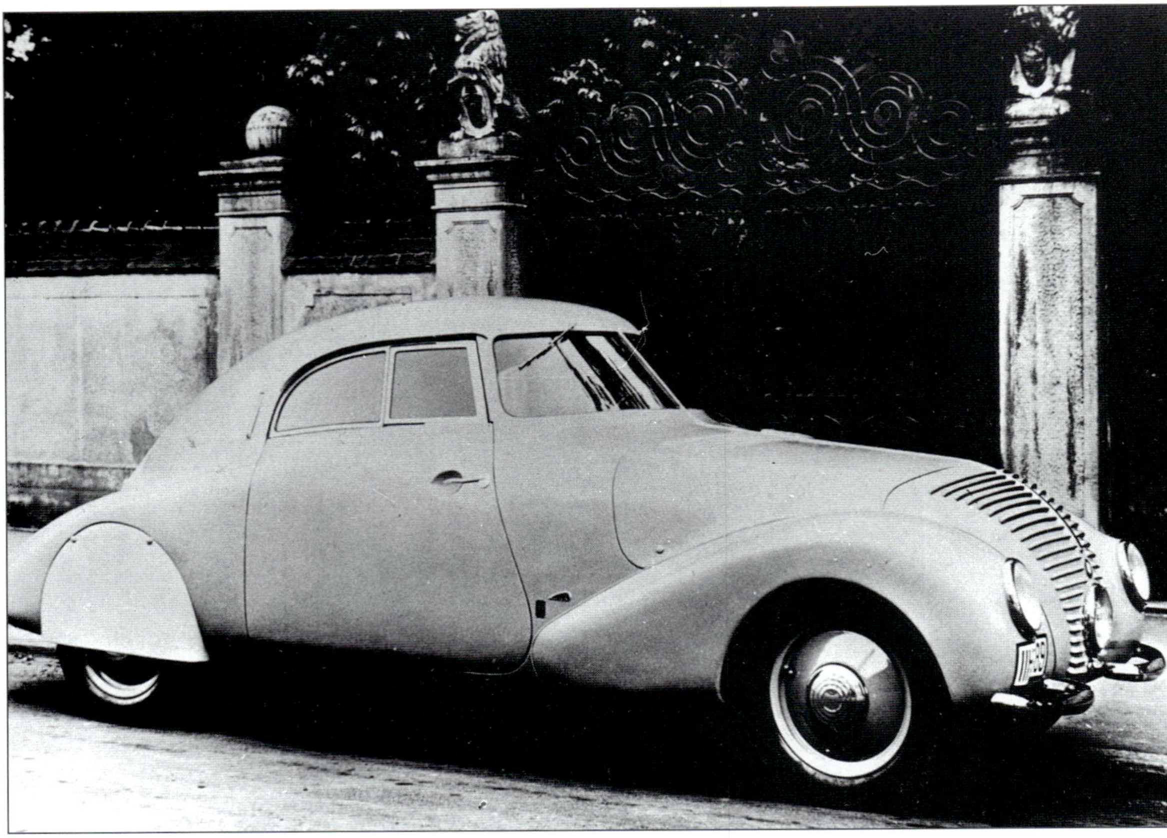

Streamlining was in vogue in the 1930s. Paul Jaray turned from his Fiats of the mid-1930s to this 326-based coupé (above), exhibited at the Berlin Motor Show early in 1938. Like other contemporary coupés, the cockpit promised to be claustrophobic, and view of the outside world limited: there was a small horizontally divided rear window.

Wunibald Kamm was responsible for the 335K, below, in 1939. It was built with high-speed autobahn cruising in mind, and in later decades might have fallen into an 'executive express' category. A more extreme version in 1940 had exaggerated rear fins.

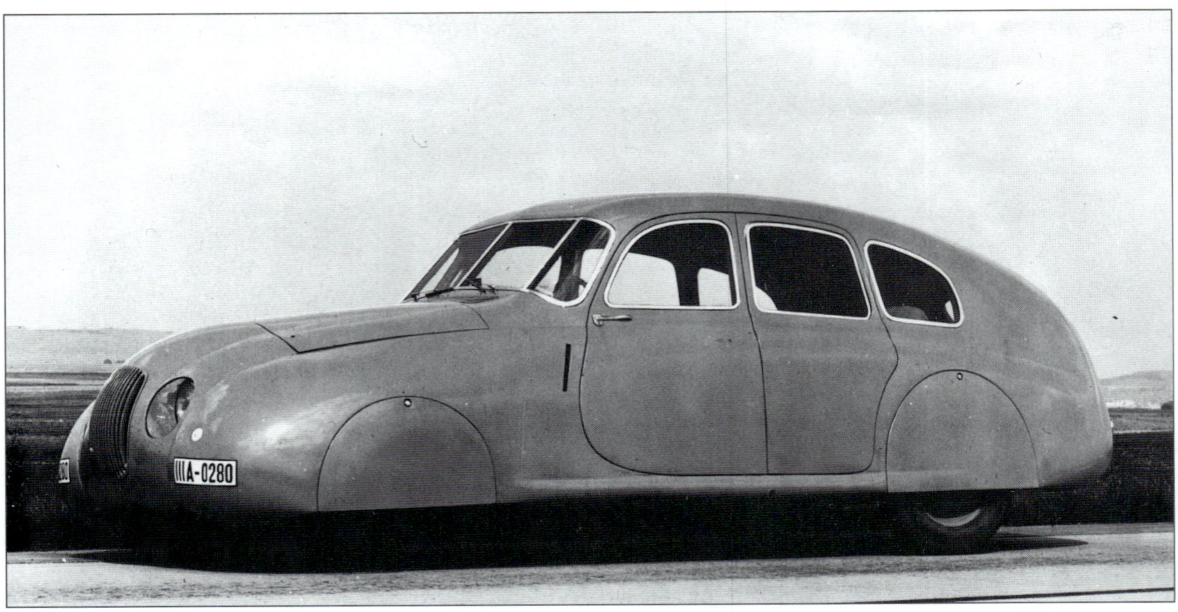

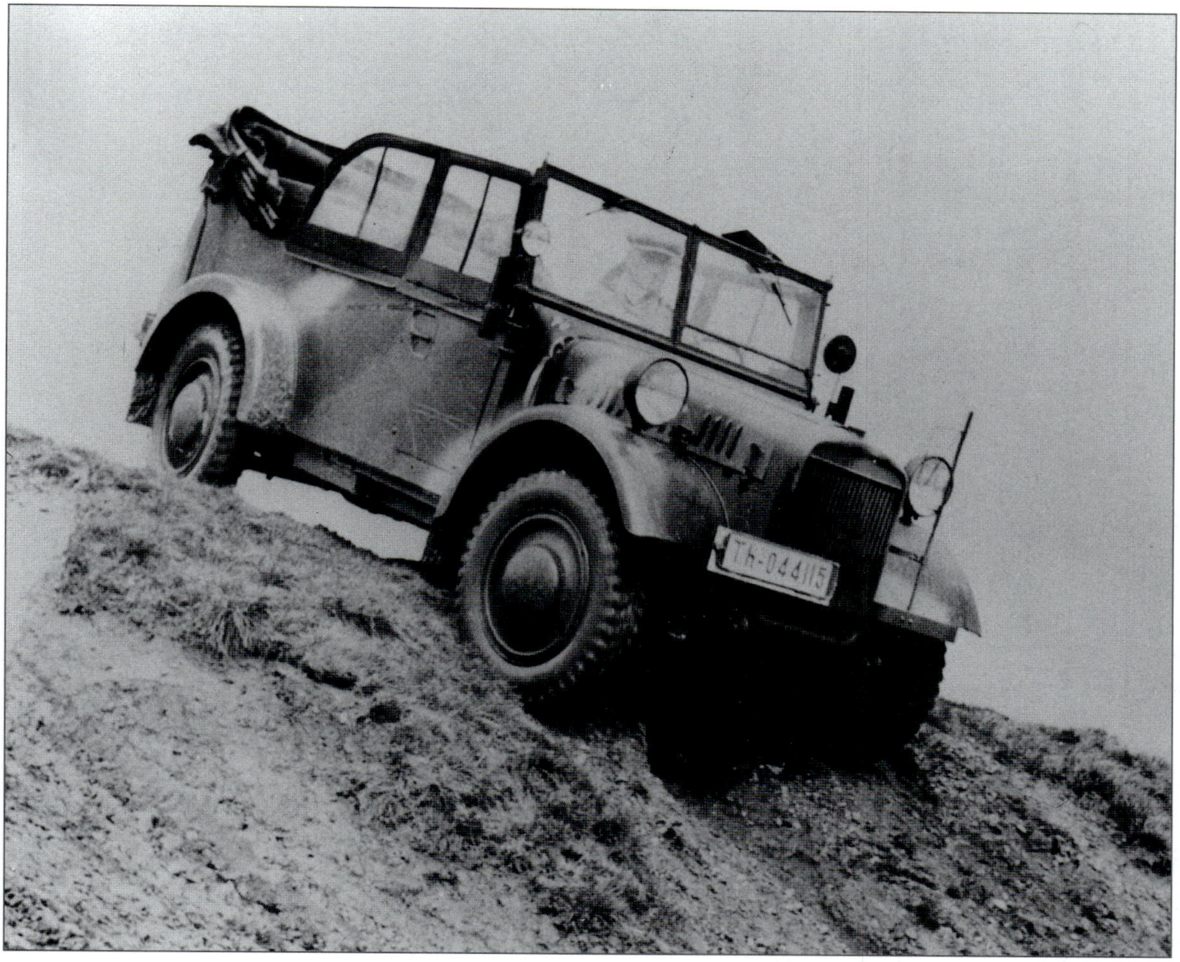

This hefty military vehicle was first seen in 1937, and did not carry BMW badges, although it was built with the type number 325 until 1940. It was also produced by Hanomag and Stoewer to a common design (albeit with each company's power units, the 1971 cc six in the BMW). It had full off-road capabilities, with four-wheel drive and four-wheel steering, transverse wishbones and double coil springs front and rear, and a five-speed gearbox – quite a specification for 1937!

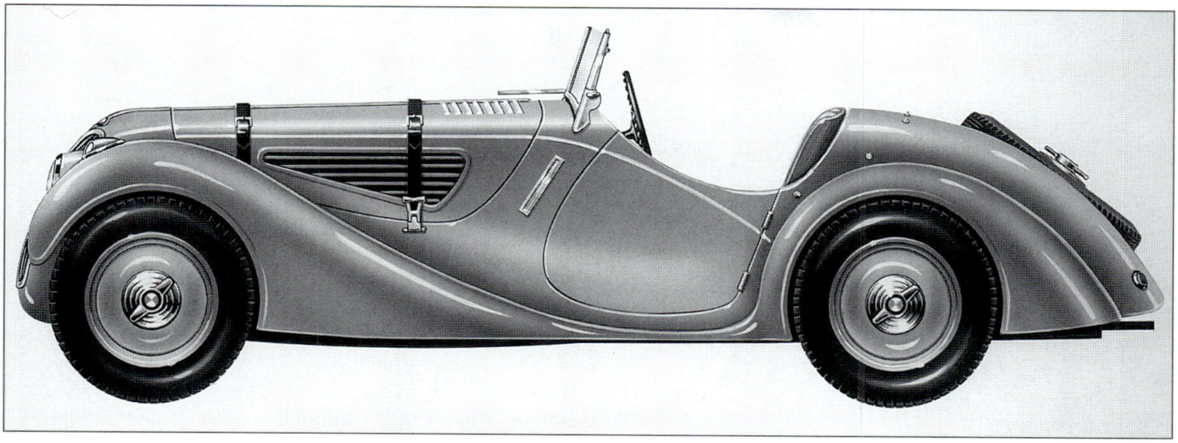

The 328 provides the abiding image of BMW in the 1930s, and this was a sports car that deserved the praise it has attracted – 'trend-setting', 'ahead of its time' and so on are not exaggerations. The main features have been recalled in the introduction to this chapter; its overall lines do not call for comment.

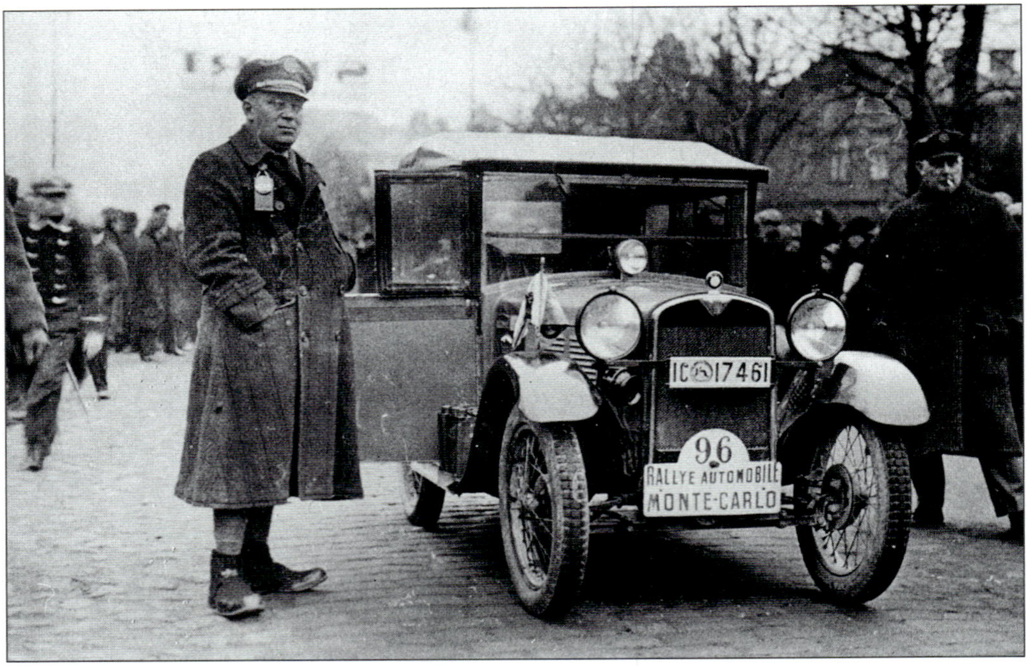

Two heroes about to set out on a great mid-winter adventure in the 1930s. From their Tallinn start they faced a 2,159-mile (3,454-km) journey to Monte Carlo, and they completed it successfully. Their 3/15 DA 2 has a BMW badge above the Dixi emblem on the nose. Rally equipment includes the tyres, large headlights and a spotlight, a large horn rather casually slung under one of the headlights, fuel cans on the running board, and a pair of flags on the nearest wing.

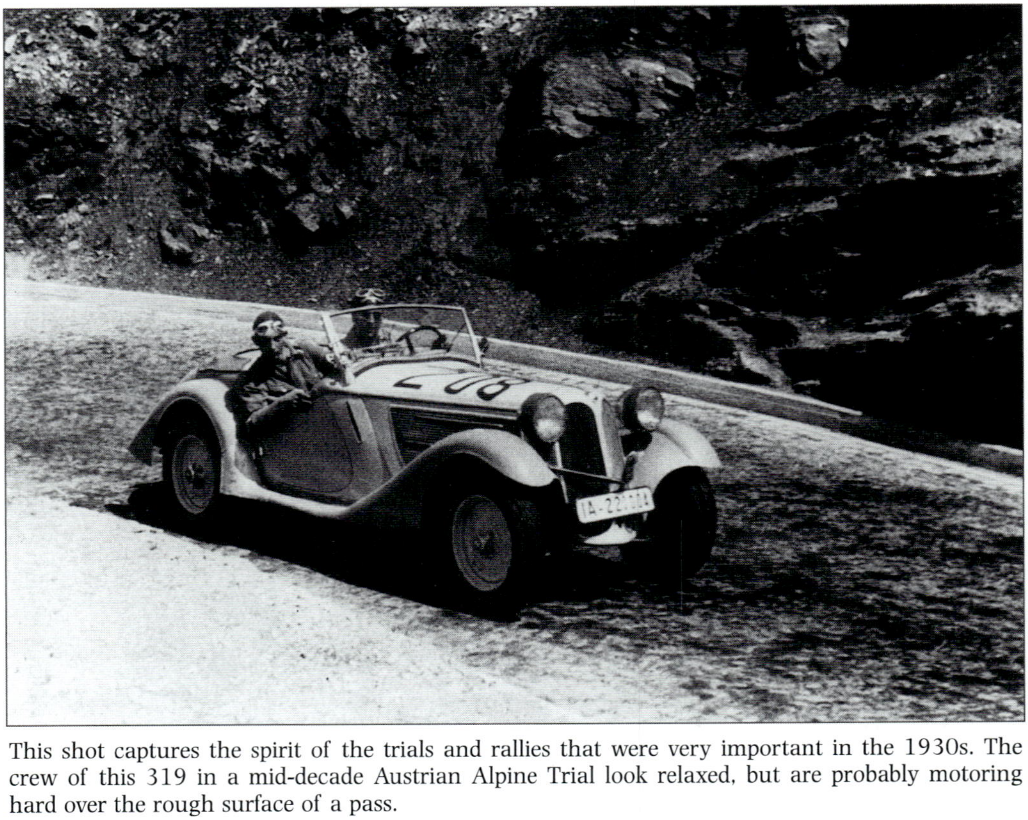

This shot captures the spirit of the trials and rallies that were very important in the 1930s. The crew of this 319 in a mid-decade Austrian Alpine Trial look relaxed, but are probably motoring hard over the rough surface of a pass.

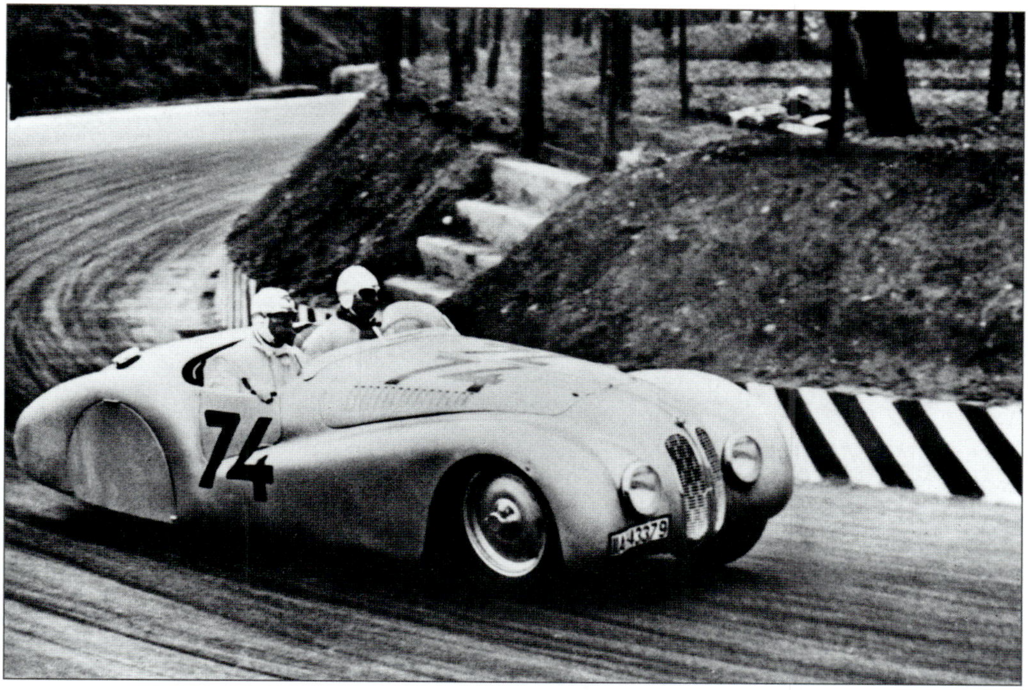

Three open works cars were built for the 1940 'Mille Miglia' (Gran Premio di Brescia) and known by BMW as the 'Mille Miglia Roadster'. The slippery body inspired later sports cars, and a 1940 prototype with very similar lines might have been a successor to the 328. As it was, Veritas Meteors followed the lines of this car. In that 1940 race the best of the open cars was driven into third place by Brudes and Roese. One of the trio was converted to right-hand drive in 1946, and for a while carried a Frazer Nash badge.

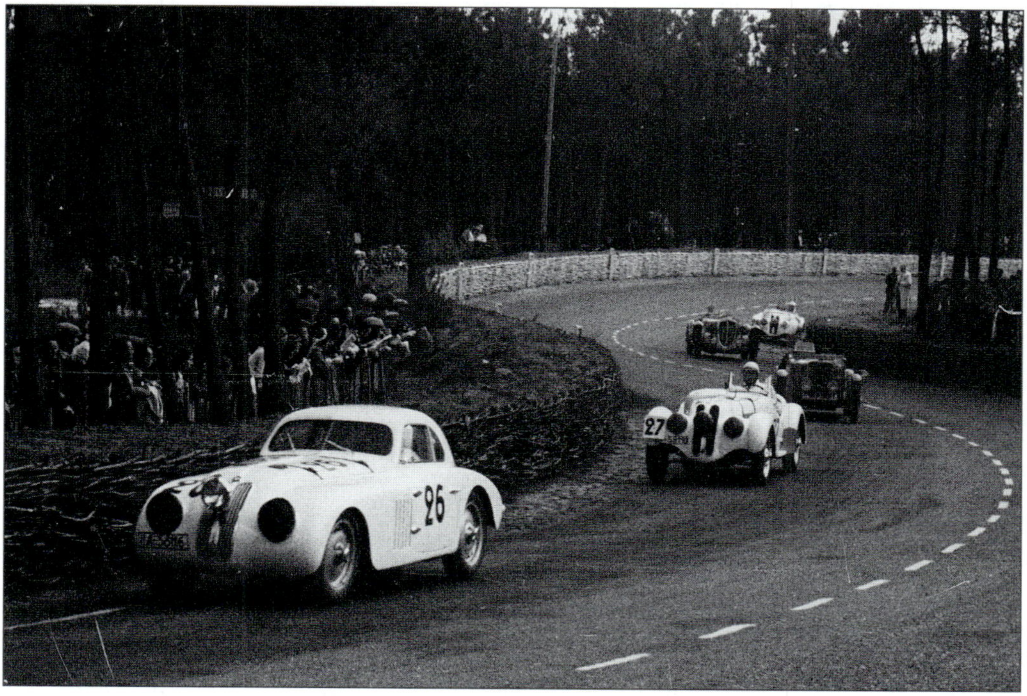

Touring of Milan built an attractive one-off coupé body for a 328 in 1939. This was first raced at Le Mans, where Prince Schaumberg-Lippe and Wenscher finished fifth overall, covering 5 miles more than the outright winners in 1938, at 82.55 mph (132.85 kmh). Here it is leading a standard 328.

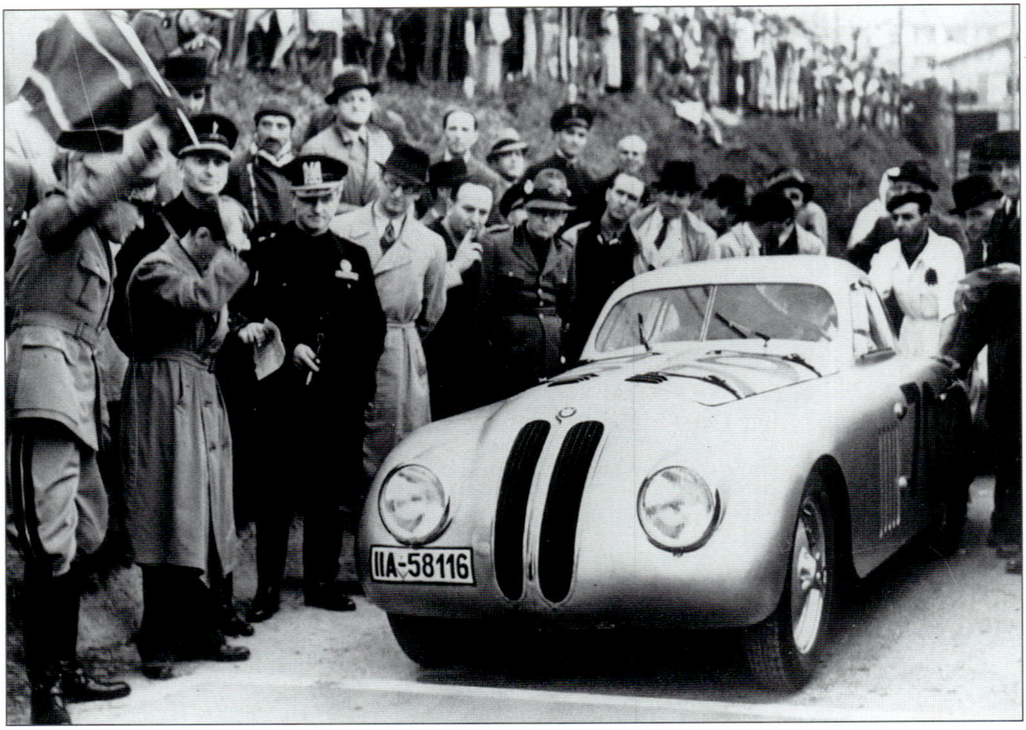

Quite sensationally, von Hanstein and Baumer then drove the Touring coupé to win the 1940 Mille Miglia. A second coupé, built by BMW, was quicker in back-to-back tests but had less attractive lines. It failed to finish in the 1940 Italian race.

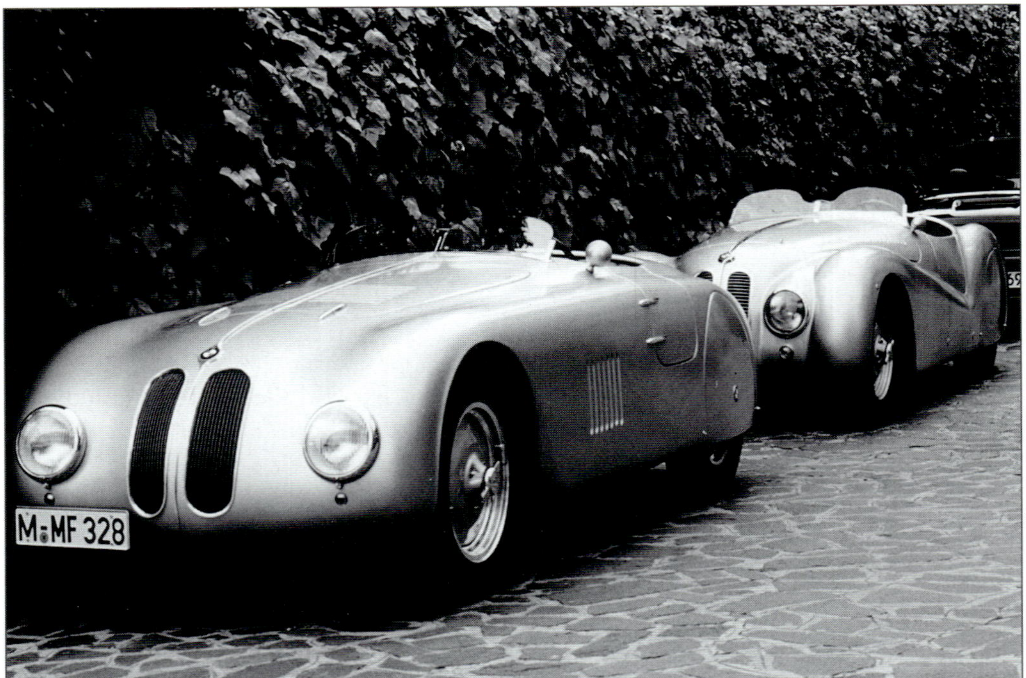

Touring was also commissioned to build an open sports-racing car on the basis of the 328, and three were completed too late to appear before wartime developments put an end to all motor sport. These inspirational cars looked more slippery than the coupé and were certainly light (kerb weight 1411 lb/6401 kg). One of these Touring roadsters survived.

CHAPTER TWO

DIFFICULT YEARS

There was a will to revive that part of BMW that fell in the western zone of divided Germany in 1945, but the means hardly existed. There was considerable damage to most factories, and then the US military government ordered that some of the production facilities that did survive were to be dismantled and removed as reparations. This applied to the Allach and Milbertshofen plants in the Munich suburbs; the Spandau and Eisenach factories in the Soviet zone were expropriated.

The US military order meant that the Aldington brothers were able to acquire a quantity of technical drawings, which were to be passed to the Bristol Aeroplane Co. (for its nascent car division). A projected Frazer Nash-Bristol fell through, but Fritz Fiedler worked on an anglicized version of the straight six for Bristol. This was used in the first Bristol 400 and was to power Bristol cars for a decade, AC and Frazer Nash models, and other cars.

Some buildings at Munich were repaired and pressed into use, and BMW manufactured an assortment of products, including bicycles. (Recent research suggests that only a handful were made, and none actually sold.) The 1948 prototype of the R24 motorcycle led to its production in 1949. Cars were little more than an ambition, although at the little-damaged Eisenach plant some were being built for Autovelo as early as 1945.

In the west, Veritas sports cars following 328 lines were built by a group of ex-BMW personnel from 1948 – half a century on, there is a feeling that at the time few BMW people approved of Veritas – and Alexander von Falkenhausen built a few AFMs very much in the spirit of BMWs; later he was to run the reborn BMW competitions department.

BMW's faltering approach followed small-car and large-car themes. The little 531 was a two-seater with a 600 cc twin-cylinder engine, designed by Fiedler with an attractive body styled by Schimanowski. The prototype has been likened to the Fiat Topolino, but it was sensibly larger than the Italian car, and in one of those nice little coincidences of history it was more like the near-contemporary 700 cc Rover M1 prototype. The 531 was stillborn: it would have made business sense only if it was produced in quantity, and that would have entailed tooling costs BMW could ill afford. In any case, company executives apparently felt that a small car was not appropriate for BMW. The idea that a pre-war design could be brushed down was little more realistic, but eventually the 501 took shape on lines that would not have been out of place in the 1930s, and indeed it did owe something to the 332. It was not a mass-produced car, and that was a positive advantage as BMW was re-equipping factories. (Limited facilities meant that bodies were made by Baur.) The pushrod straight six had to be used.

Sales were modest, but a version with an enlarged engine extended the life of the 501 – the last was built in 1958, and in seven years production reached almost 9,000 units; the 501 V8 was built in parallel from 1955, and lasted until 1962. Other V8-engined models included the closely similar 502, which was actually in production a little earlier than the 501 V8, and the better-known 507. That first V8 thread ran through to the 3200CS coupé of 1962–65, which outlived the saloons; its styling was to be followed in later CS coupés.

Meanwhile, low-volume car production and falling motorcycle sales led BMW to look again at small vehicles, and to an agreement to build the Isetta 'motocoupé' (bubble car) under licence. The design was modified, and BMW four-stroke engines were used in place

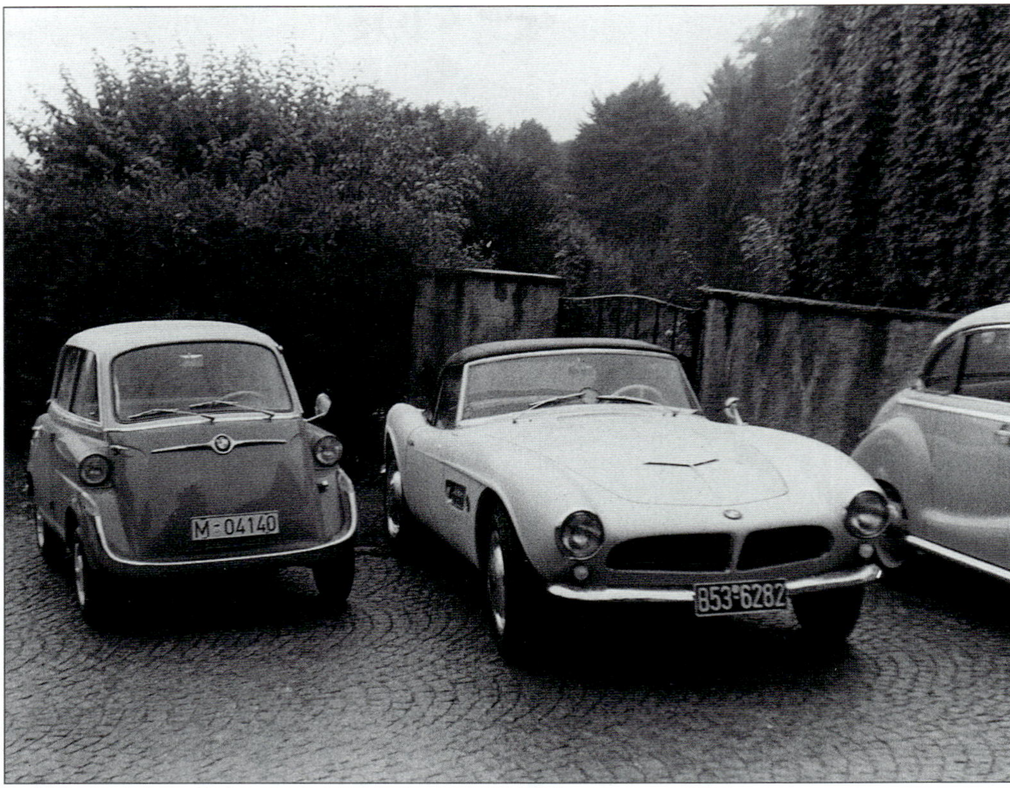

A BMW oddity, a 600, parked alongside a 507 roadster. Apart from the badge, these two have one thing in common – neither was a commercial success.

of the two-stroke units in the Italian original. It sold well – more than 161,000 were made, 1955–62 – and the BMW version was licensed to Brazilian and British manufacturers, the British types being three-wheelers. It also led to the 600, an unconventional small car that did not sell so well, but did lead to a proper small car, the 700. In volume terms, these vehicles made an important contribution in the lean years before the *'Neue Klasse'* models became established.

If there was a corporate aim in the 1950s, it was survival: nominally, BMW was a profitable company in only five years between 1946 and 1961. BMW did survive, shakily, then in the autumn of 1961 it was able to show its backers and dealers a sign of a promising future, as the first 1500 of a new generation appeared at the Frankfurt Motor Show.

Understandably, this was not an exciting period in BMW's motor sports history. Private 328s were still raced, and the big saloons were seen in rallies, and remarkably Isettas were raced and rallied. The 700 proved to be a useful small-class contender, especially in its coupé form. As well as sports cars, the straight six engine was used in Veritas and AFM single-seaters, and powered cars by other German constructors such as Grifu, Heck and Reif, and minor French efforts, Jicey and Monnier.

Then there were British cars, usually with Bristol engines that were a little removed from the BMW original by the 1950s. These were used in Formula 2 cars (notably Coopers, whose drivers scored World Championship points with them) and the Bristol Le Mans cars. These were placed 7–8–9 in the 24-hour race in 1954 and 1955, to set against two retirements in 1953 – as good a 24-hour race record as BMW had achieved.

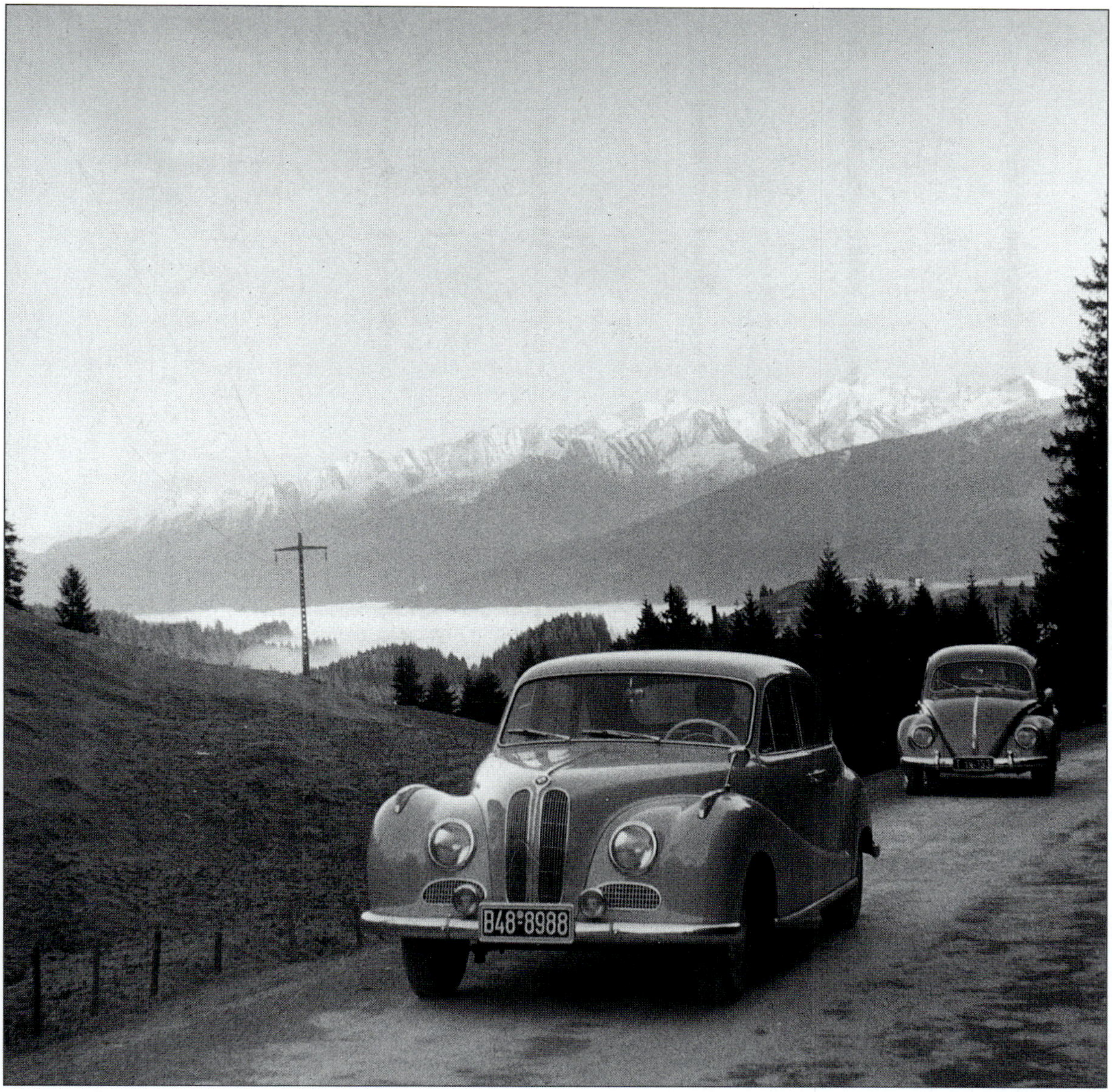

Unmistakably a BMW, this 501 is on a Bavarian road – appropriately, as it was the first production BMW car to be built in Munich. The in-house styling does not compare favourably with a Pininfarina design that was set aside. The straight six was still a 1971 cc unit, rated at 65 bhp – far from enough for a four-door saloon weighing in at 2,955 lb (1,340 kg). There was a sturdy box-section chassis, generously cross-braced, with double wishbone independent front suspension, a live rear axle, and torsion bars all round. It could not quite reach 85 mph (136 kmh).

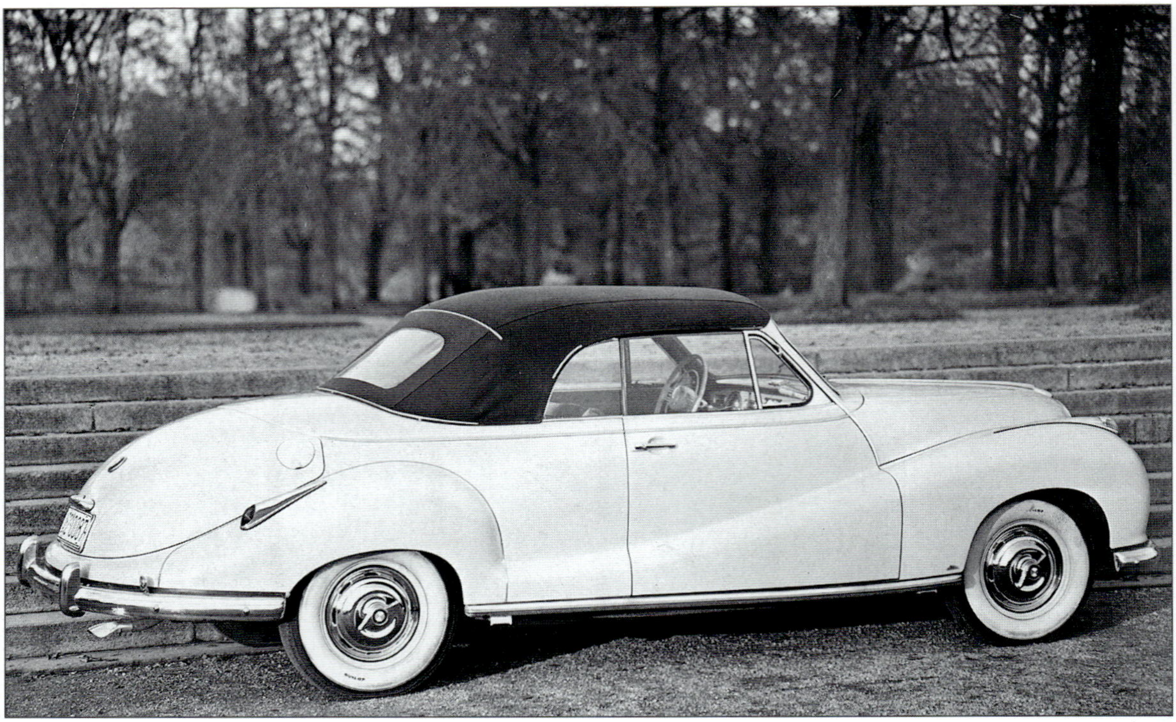

In 1954 the 501A was introduced with an engine giving 72 bhp and slight performance improvements. Only eight were completed with this bulky cabriolet body by Baur. A few cars had the straight six bored out to give a 2077 cc capacity; this version had a modified camshaft, and problems in this area led to some loss of reliability.

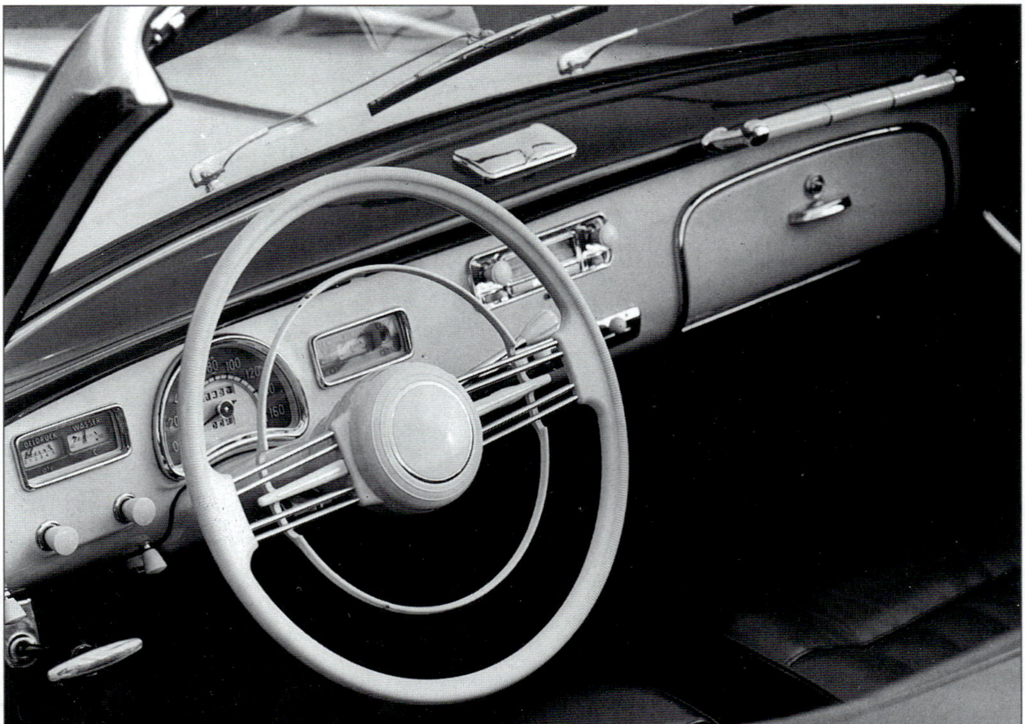

The fascia was simple, with a 160 kmh (100 mph) speedometer seen through the wheel, oil and water gauges to the left, fuel gauge to the right, and an elaborate grab handle for the passenger.

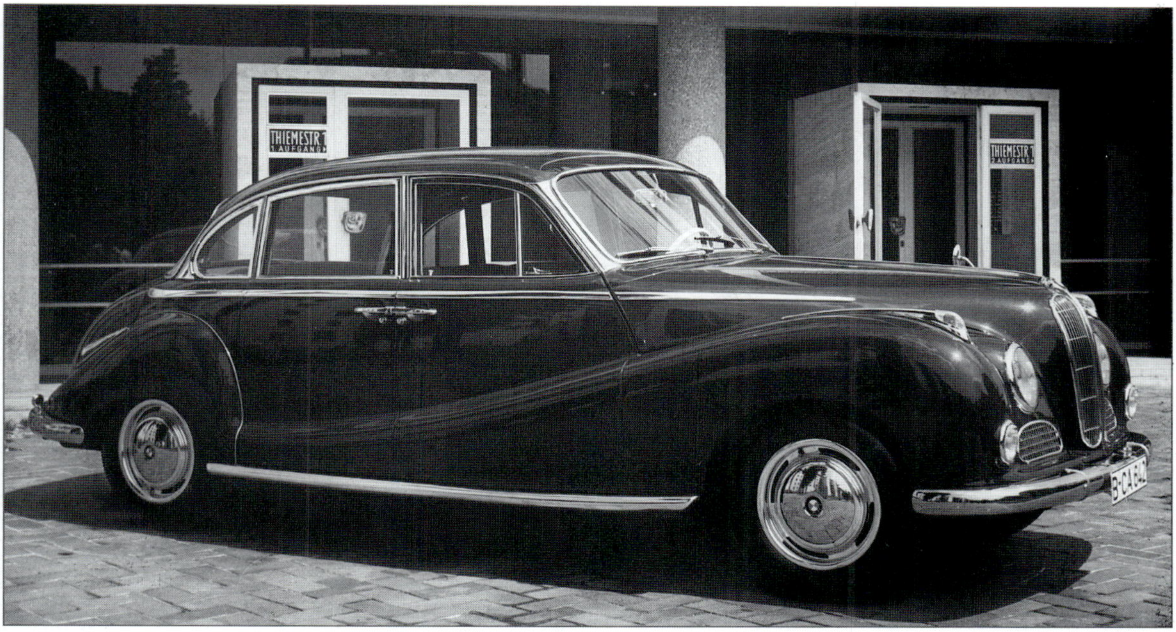

The first V8-engined BMW, and indeed the first post-war eight-cylinder German car, the 502 came in 1954. The external lines show that in most respects it was a 501 with the new engine; there was a little more trim embellishment and built-in fog lights to distinguish it. This is a 1955 model, with a large rear window.

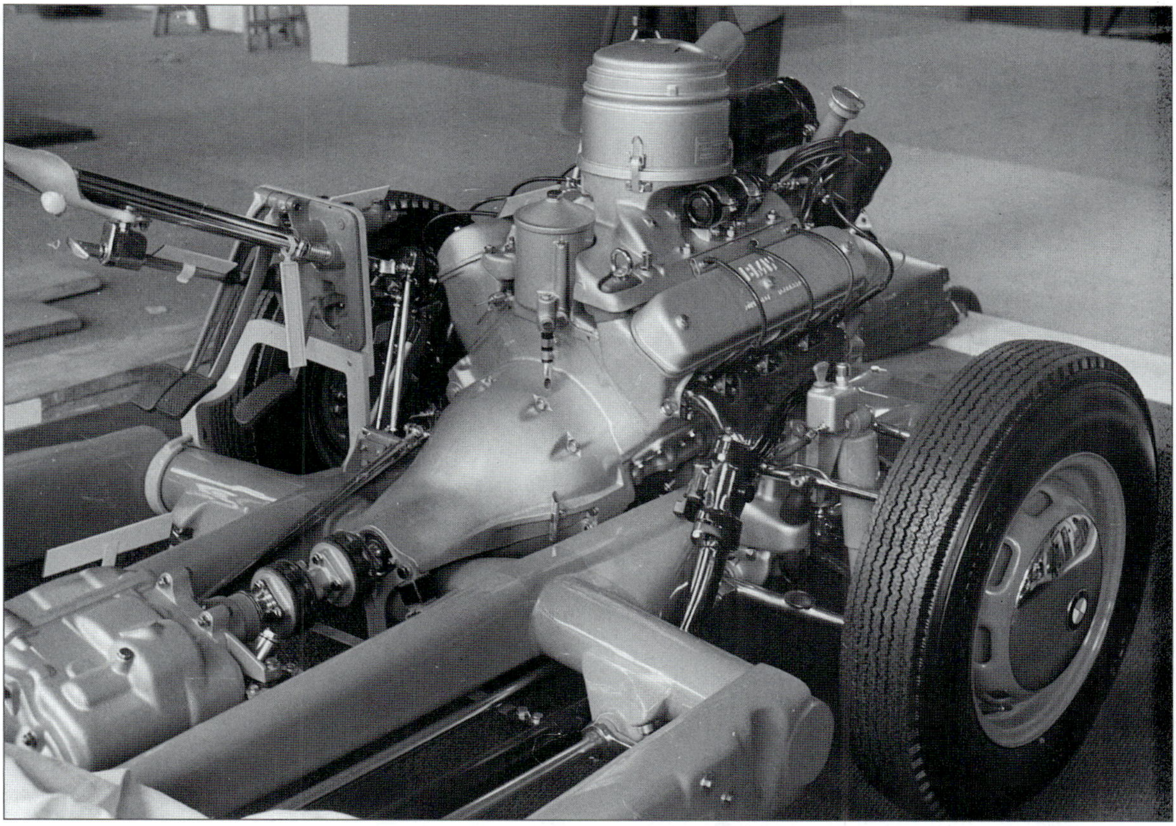

The light alloy V8 fitted snugly into the engine compartment and this motor show display chassis also shows the short shaft running to the four-speed gearbox under the front seats. The 2.6-litre engine in the 502 was rated at 100 bhp and gave a top speed nudging 100 mph (160 kmh).

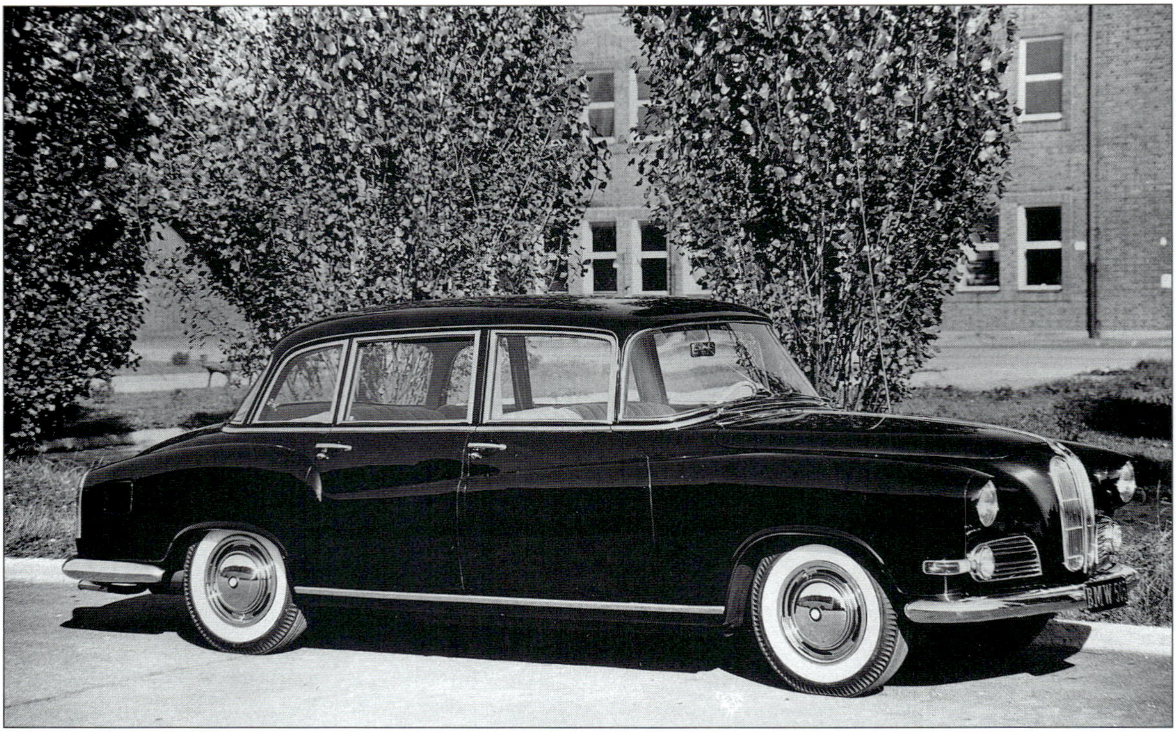

Only two 505 limousines were built, on lengthened and strengthened 502 chassis, with 3.2-litre V8s. The interiors were fitted out to high standards, but to BMW's disappointment the 505 was not adopted as a Presidential car. The 505s were built in 1955.

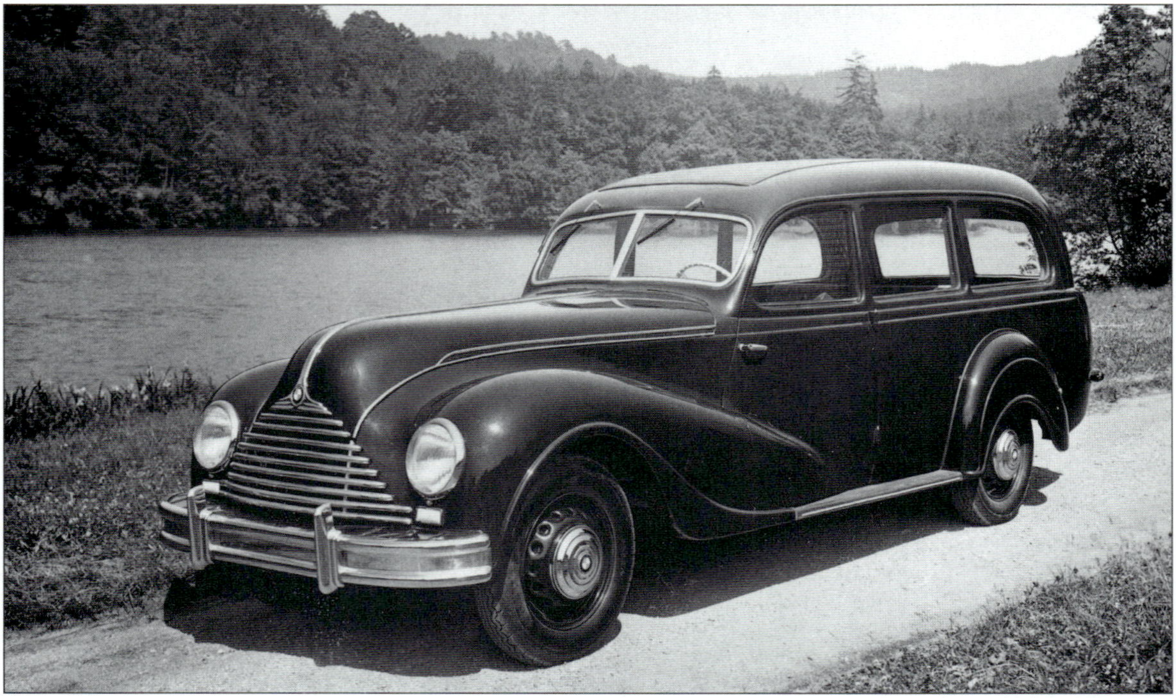

The 340 had been introduced as a BMW in 1949, but when this estate car version appeared two years later the East German company was becoming EMW. Its badge was similar, but with red and white quarters in place of BMW's blue and white. This oddly styled vehicle had the straight six in 1971 cc form, rated at 55 bhp, and it was built until 1955.

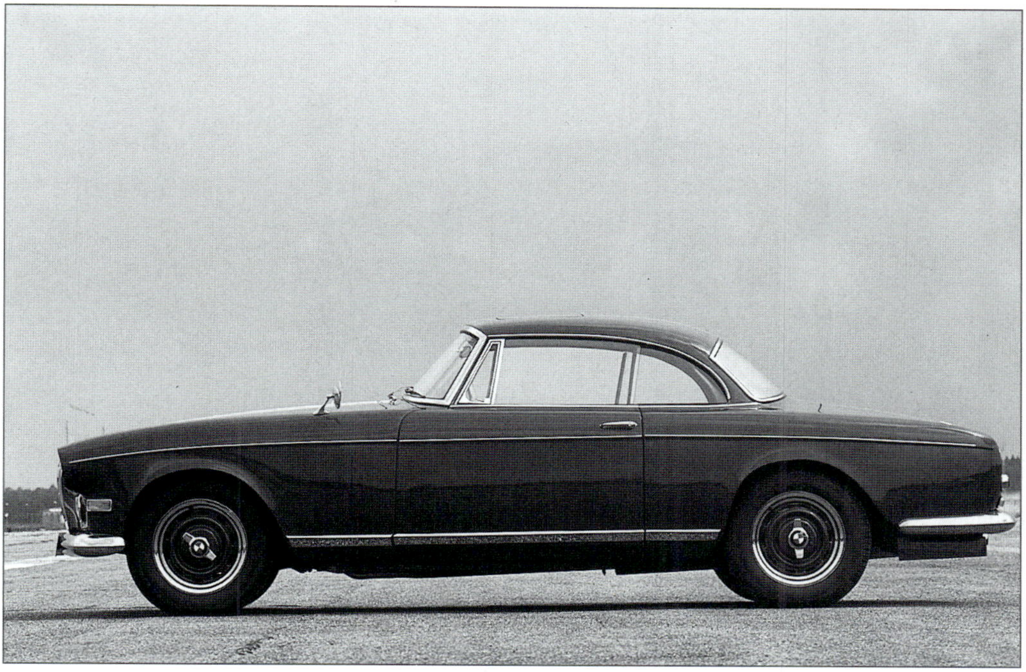

A 502 chassis, the alloy V8 in 3.2-litre 140 bhp form, and Count Albrecht Goertz's first design for BMW, made up the alluring 503 2+2 GT car. It was heavy, despite the use of some aluminium panels, and it looked big (the overall length was 187 in), but the long bonnet and boot made for balanced lines. Early cars had a column gearshift that was not highly regarded, but later examples had a floor-mounted gear lever. Just over 400 were built, 1956–59, in coupé and cabriolet forms.

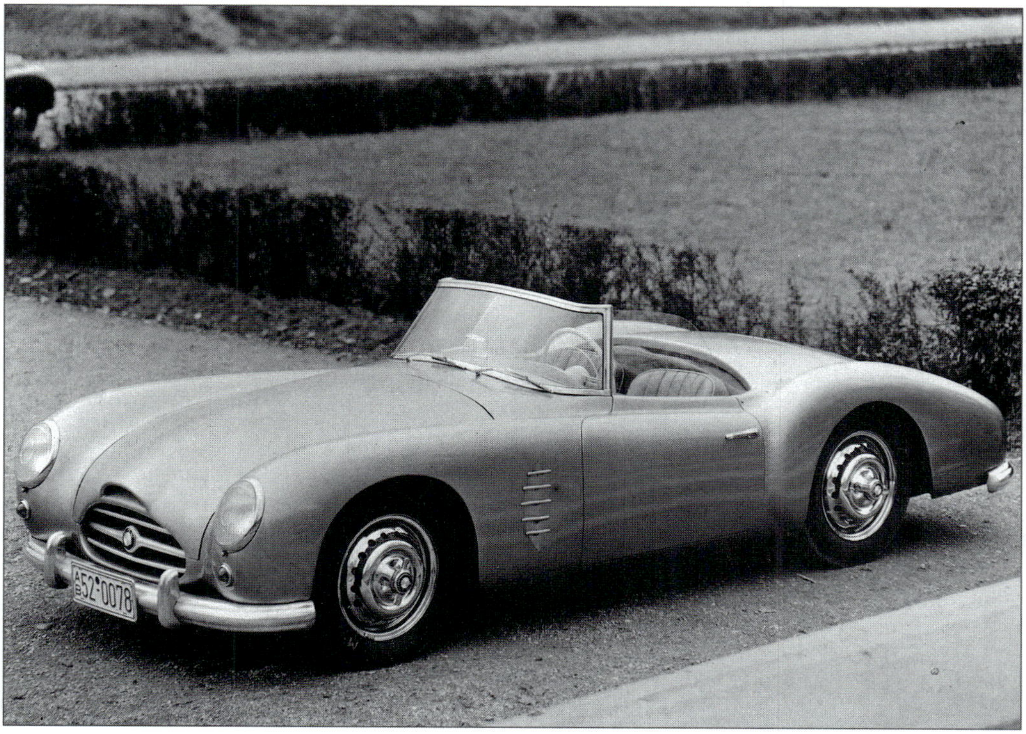

Ernst Loof was responsible for this 1954 roadster on a 502 chassis, which was fitted with a 2.6-litre V8. Loof was a BMW consultant when this prototype was built, but the Goertz design that became the 507 was preferred.

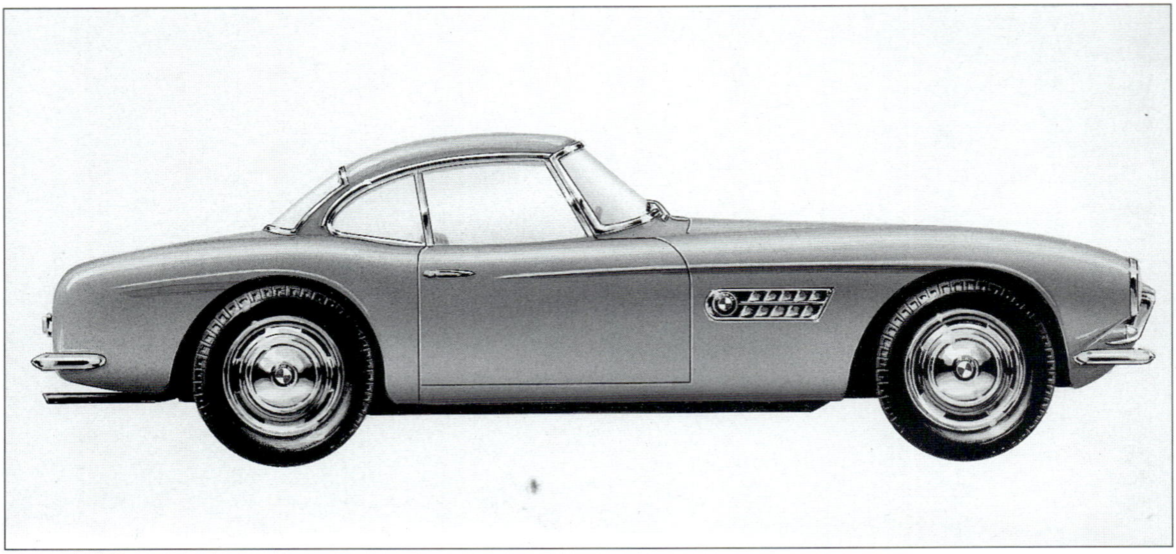

The 507 was the first post-war BMW to attain 'classic' status, but when it was built (1956–59) it was not particularly successful, or profitable, and only 252 were completed. Goertz's body was carried on a short version of the saloon chassis, with a 3168 cc 150 bhp V8, driving through the stock four-speed gearbox. However, three final drive ratios were listed, and depending on choice the top speed was of the order of 118/124/137 mph (190/198/220 kmh). A few late 507s were the first German production cars to be fitted with front disc brakes. The hard top on this car does not appear to be an afterthought, unlike so many. . . .

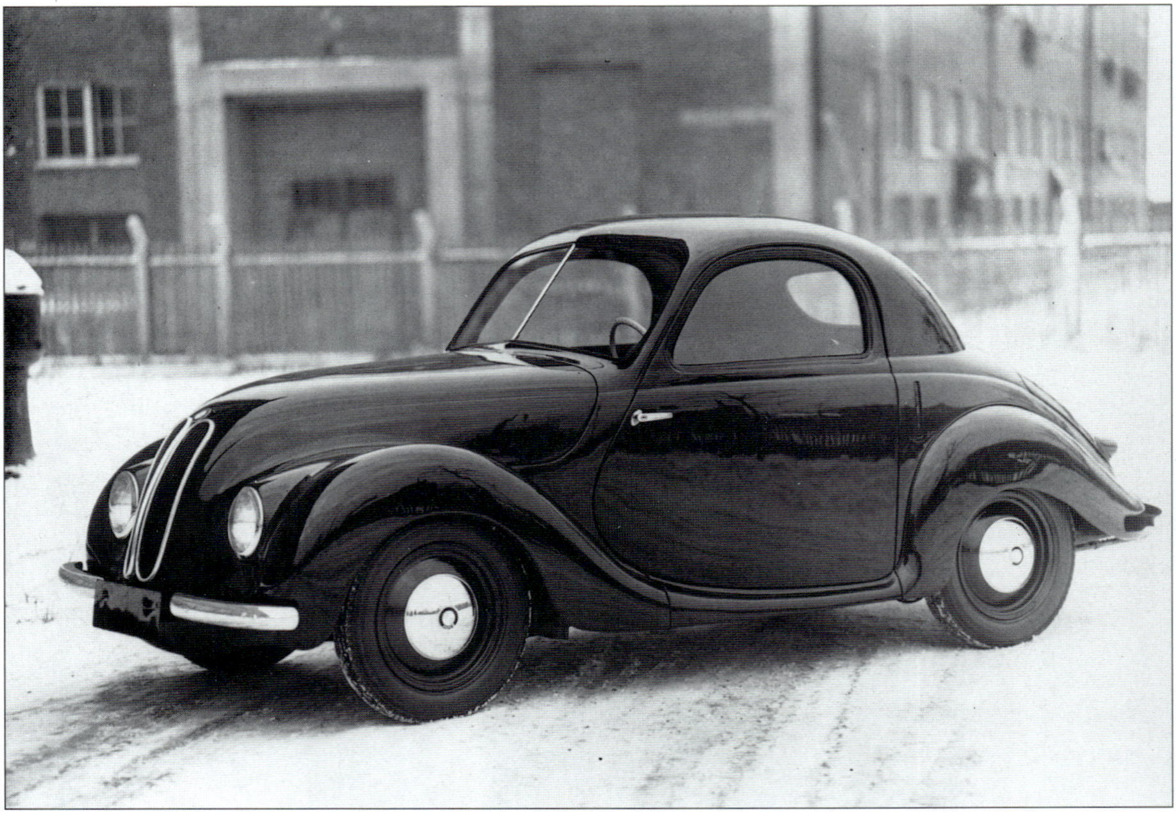

Small cars. The 531 prototype of 1950 was very similar to a contemporary Rover. A flat twin motorcycle-type engine drove the rear wheels. It was set aside as the BMW management chose not to enter the small-car market . . . and five years later BMW was building the Isetta 'motocoupé'.

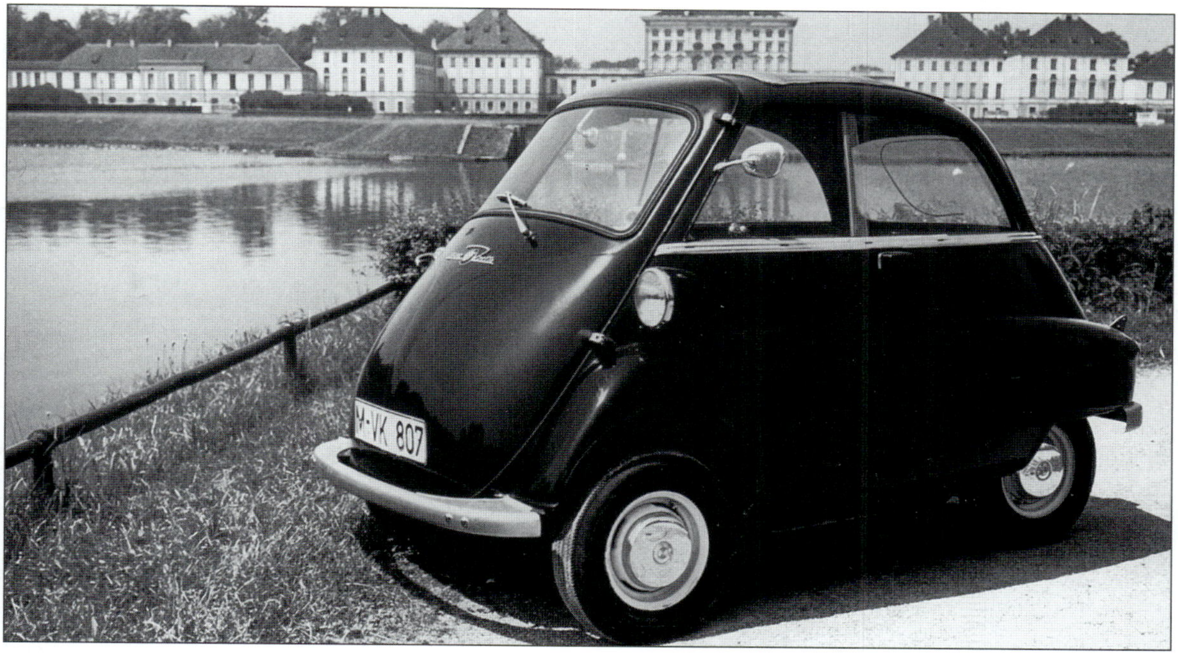

The Isetta was a valuable addition to the product range in 1955. BMW re-engineered it, and the German 245 cc engine gave a dizzy 53 mph (85 kmh), which seemed quite enough! From 1956 a 298 cc engine giving 13 bhp instead of 12 bhp was listed, less than the 1929 Dixi – but the bubble car weighed in at 794 lb (360 kg). This is one of the normal types, a four-wheeler with a very narrow rear track.

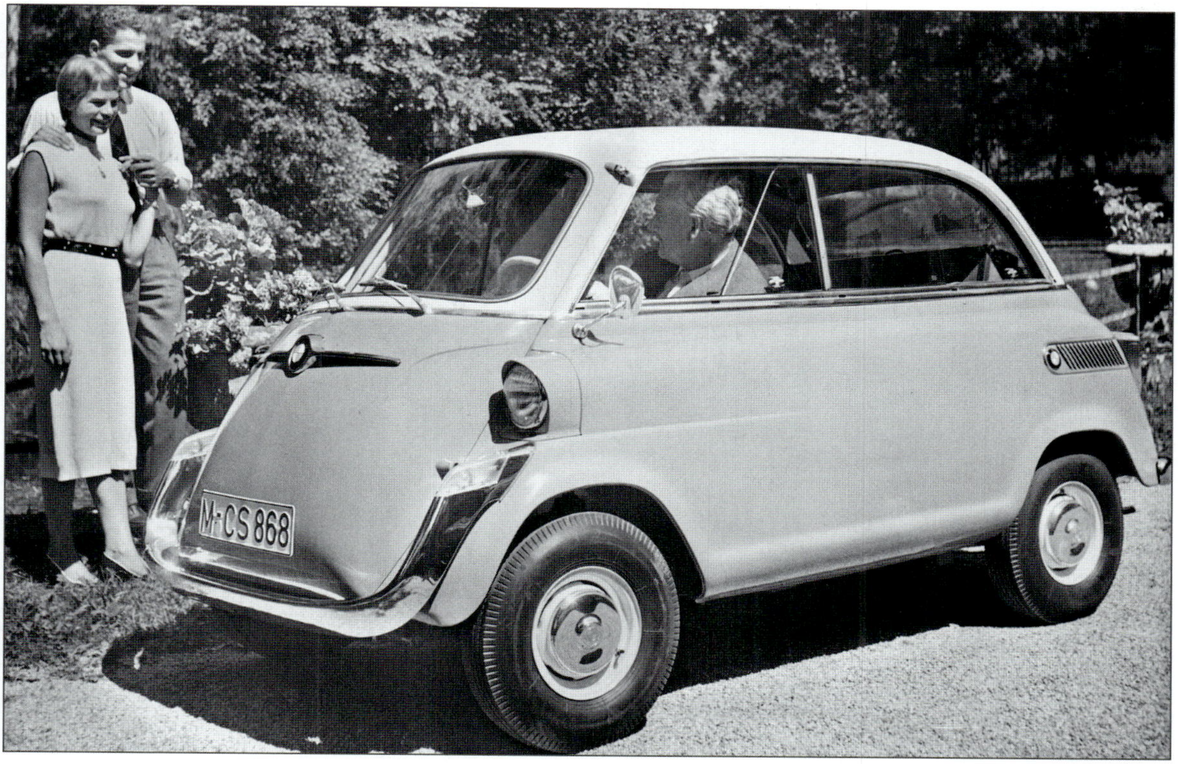

A charming publicity photograph from the late 1950s, with a family obviously delighted by a 600. There was an Isetta-like single front door and a single side door giving access to the rear bench seat. The 582 cc engine derived from the R60 motorcycle unit gave 19.5 bhp and a claimed 100 kmh (a nice round figure, which converts to just over 60 mph). Around 35,000 were sold in 1957–59.

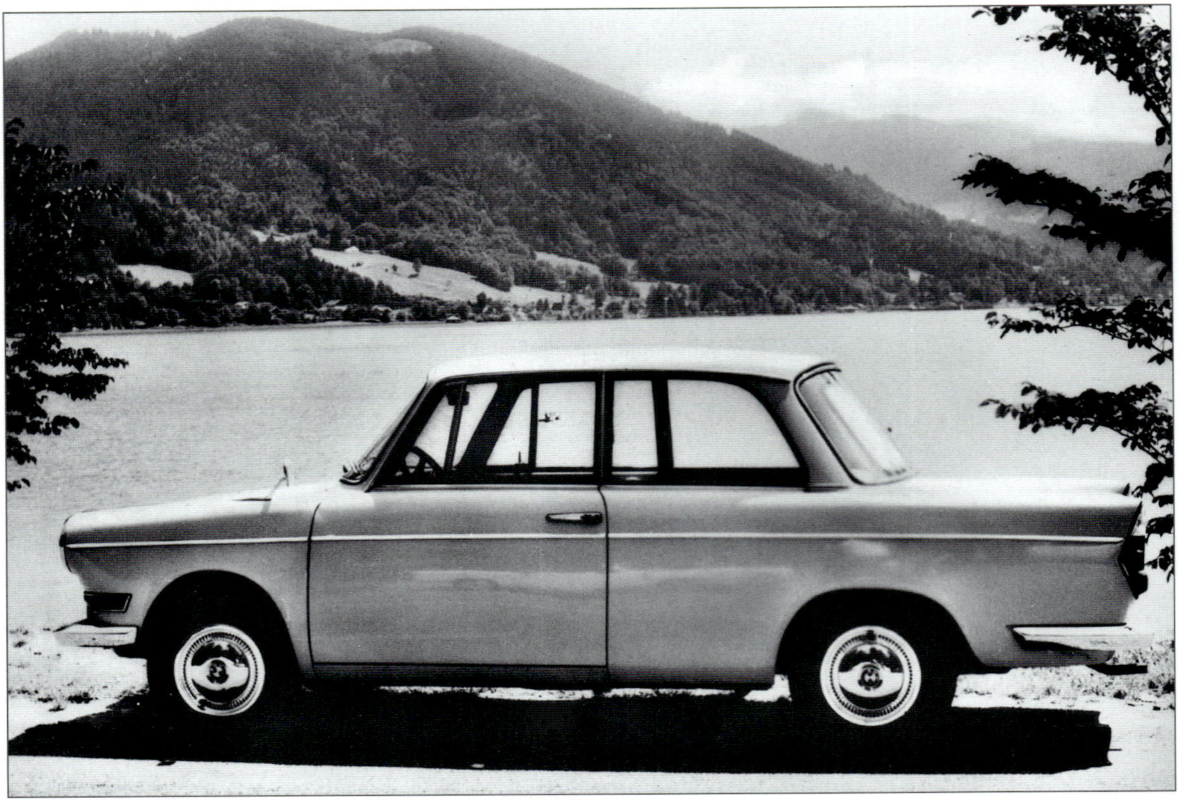

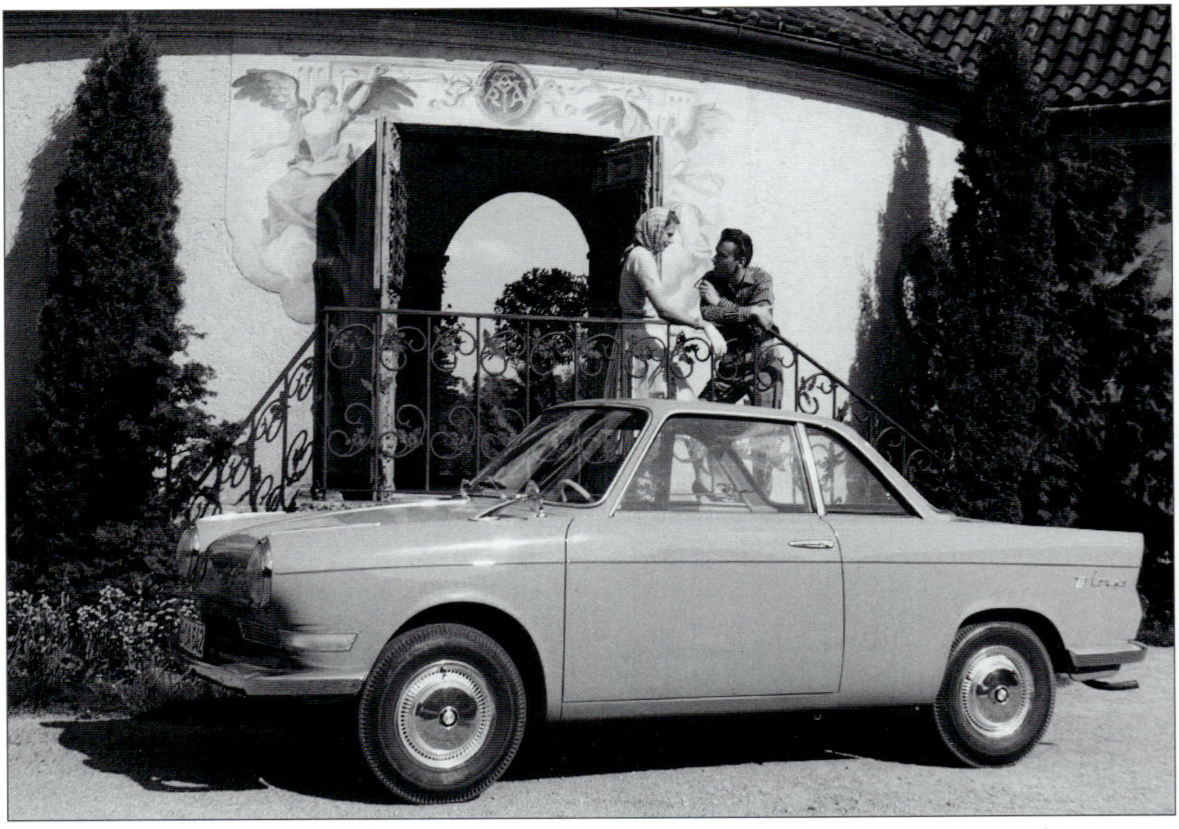

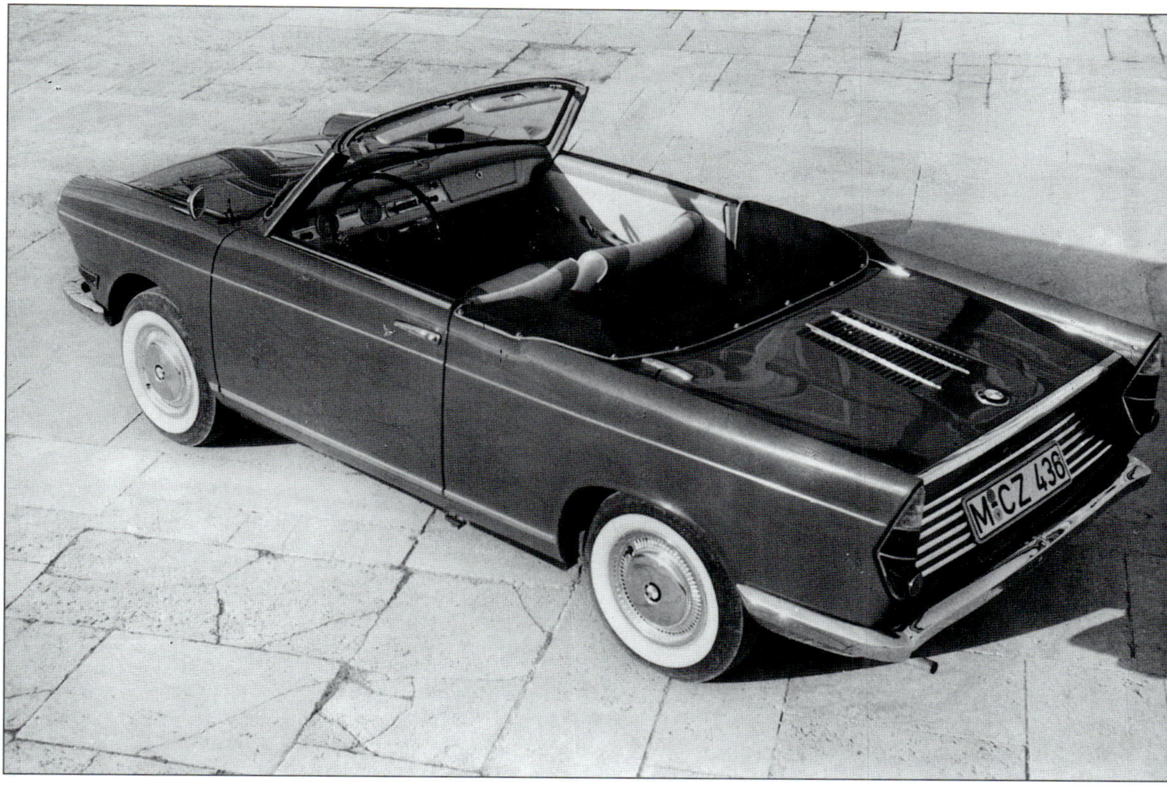

In that it derived from the 600, the 700 can be seen as a duckling to swan transformation. Austrian sports car constructor Wolfgang Denzel conceived it, around the mechanical elements of the 600, with a 697 cc engine. Giovanni Michelotti was responsible for the production bodies, in two-door saloon, coupé and convertible forms. The spare wheel was in the nose, with a little luggage space behind it, thin seat backs gained a few centimetres of cockpit space and the engine occupied the tail. Apart from odd Glas-originated cars in the 1960s, the 700s were the last BMWs for many years not to have a kidney grille. The saloon (top left) is one of the 1962–65 LS series, with an extended wheelbase also used for the coupé (left) and convertible (above). The saloons had 30 bhp engines, the coupés 40 bhp units which gave a claimed top speed of 84 mph (135 kmh).

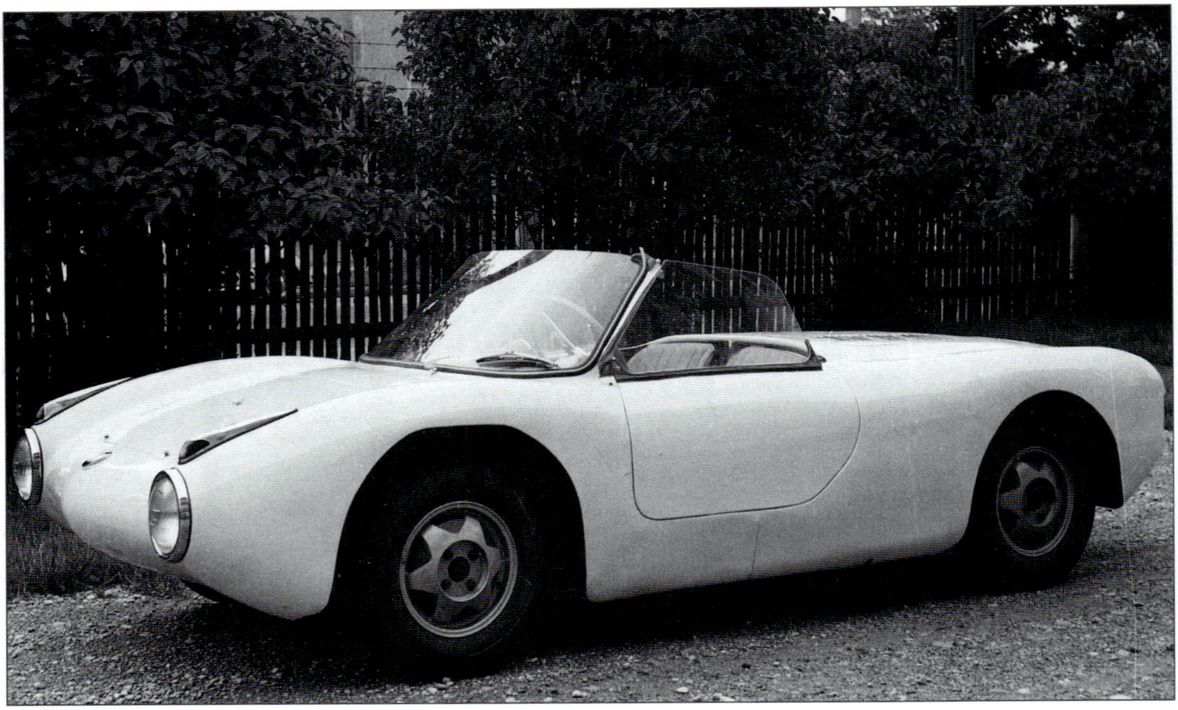

The 700RS was a pretty sports-racing derivative of the 700, successful in German and Austrian races and hill climbs in the early 1960s. An 800RS with a larger flat twin engine was also built.

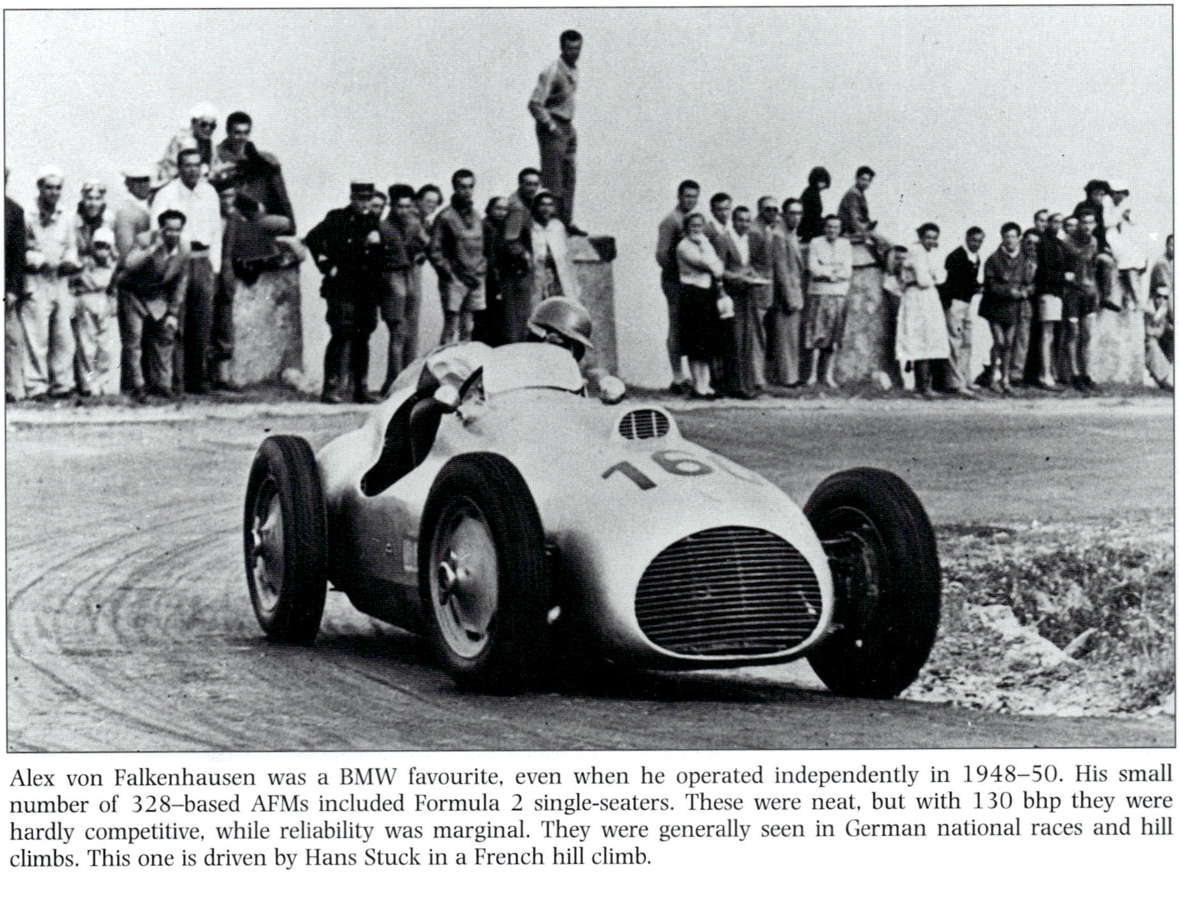

Alex von Falkenhausen was a BMW favourite, even when he operated independently in 1948–50. His small number of 328–based AFMs included Formula 2 single-seaters. These were neat, but with 130 bhp they were hardly competitive, while reliability was marginal. They were generally seen in German national races and hill climbs. This one is driven by Hans Stuck in a French hill climb.

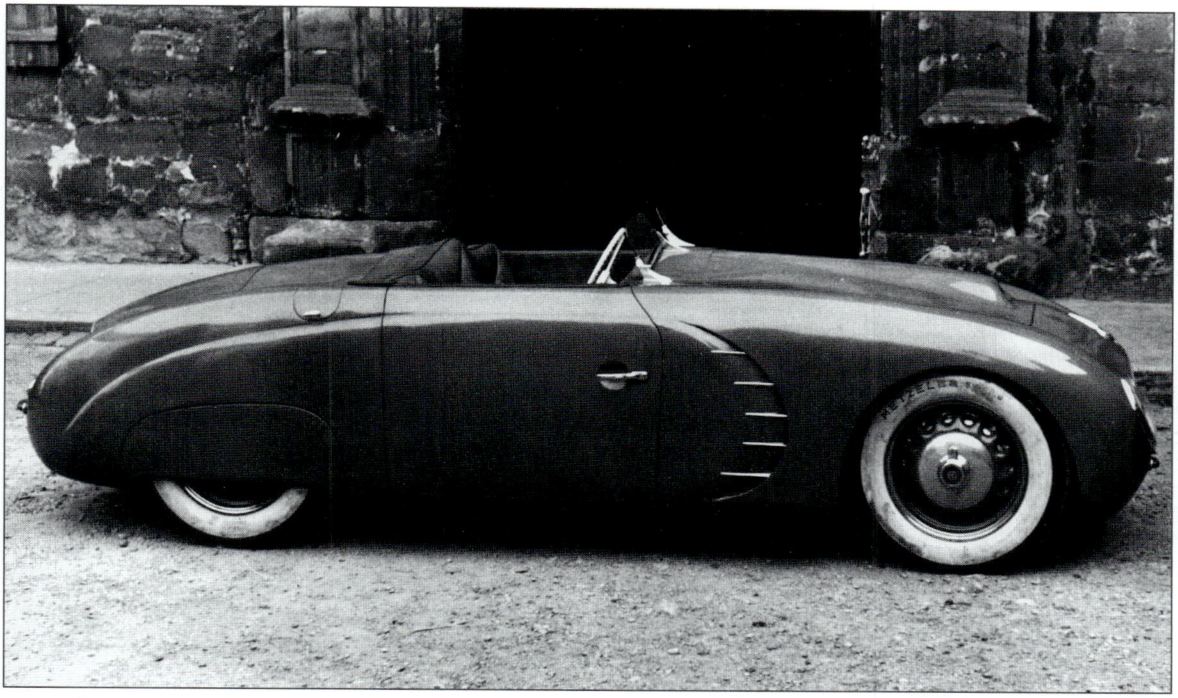

An early Veritas Komet (above) with a curious but quite possibly effective body – it came at a time when companies such as MG were building square-cut sports cars. Some of these Veritas had 328 engines. A coupé version was named Saturn. A more advanced Nürburgring coupé (below) came in the last months of the Veritas company, which just survived into 1953, building a few cars at the famous German racing circuit.

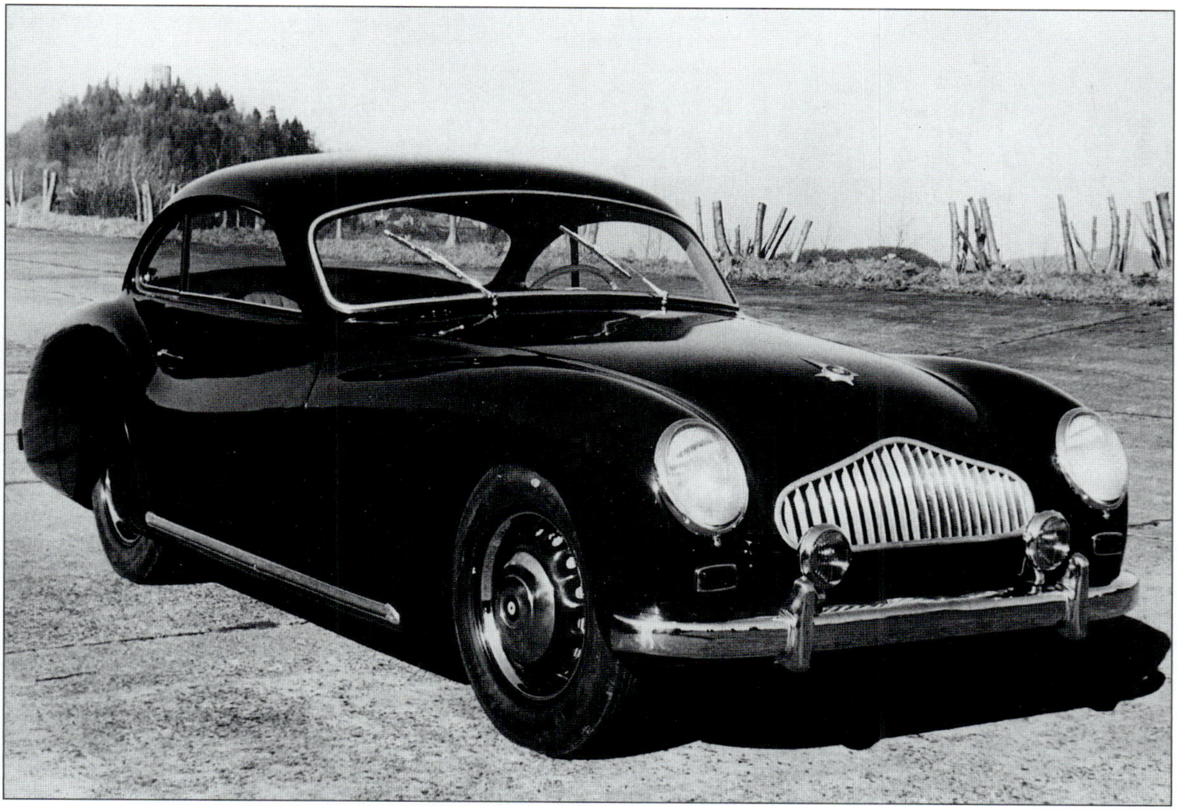

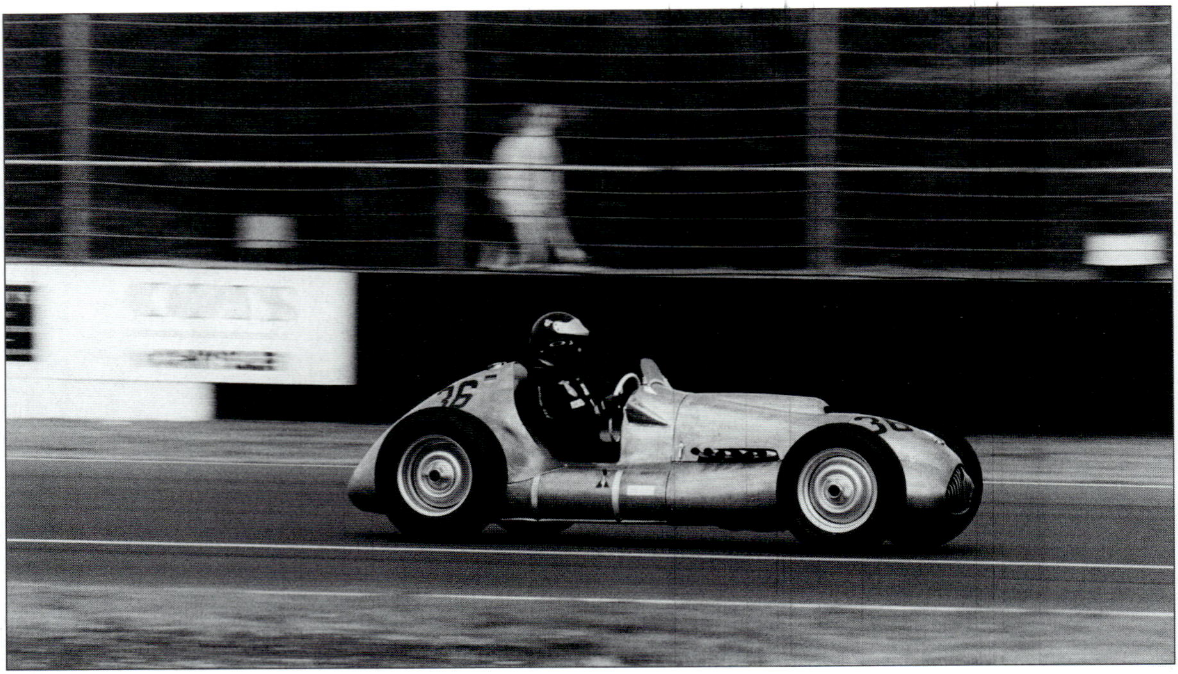

BMW components were vital to German constructors as motor racing recovered in the late 1940s, and as they entered International events early in the next decade, especially Formula 2. Veritas was set up by former BMW engineers Ernst Loof and Lorenz Dietrich in 1948. Among their single-seaters, the open-wheel cars such as Michael Rudwig's 1950 car (above, in an historic race at Silverstone) were most familiar, but after the original company collapsed Loof built this streamlined car (below) in 1952; it reputedly influenced the design of the mid-1950s Mercedes-Benz GP cars. Veritas cars were successful in German national events, but the marque's only international victory came when Paul Pietsch won the 1951 Eifelrennen. The best placing in a World Championship race was Riess's seventh in the 1952 German GP.

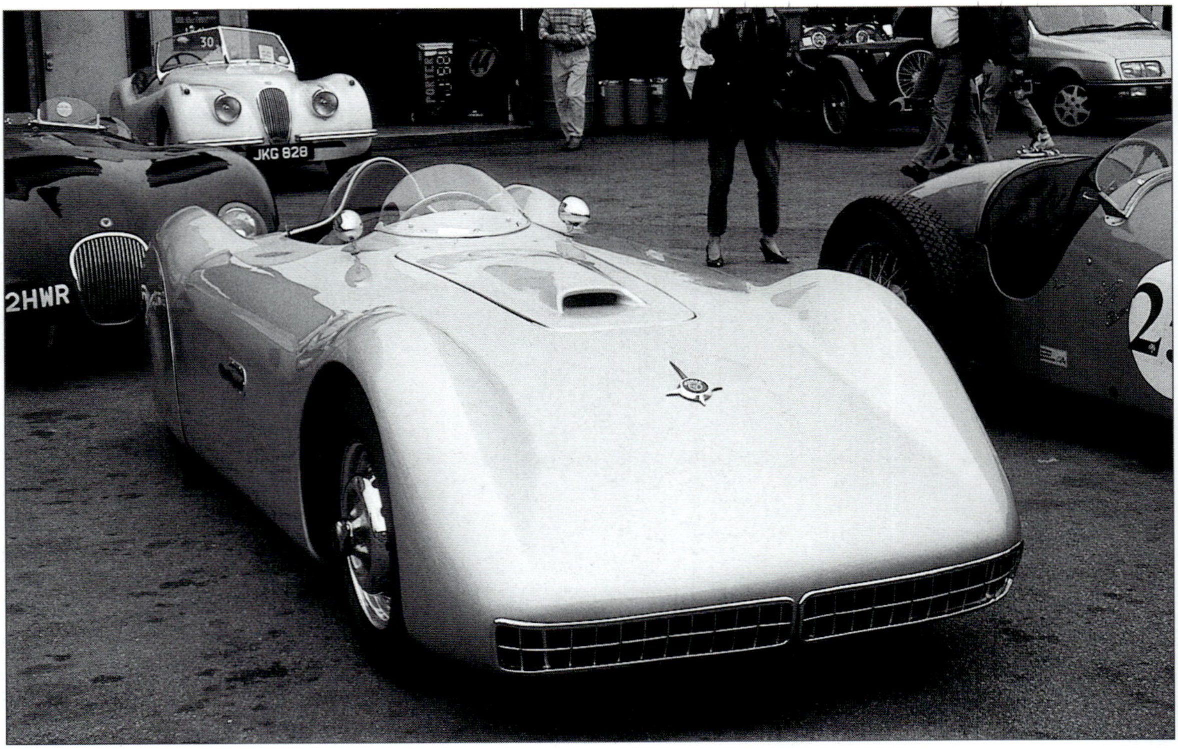

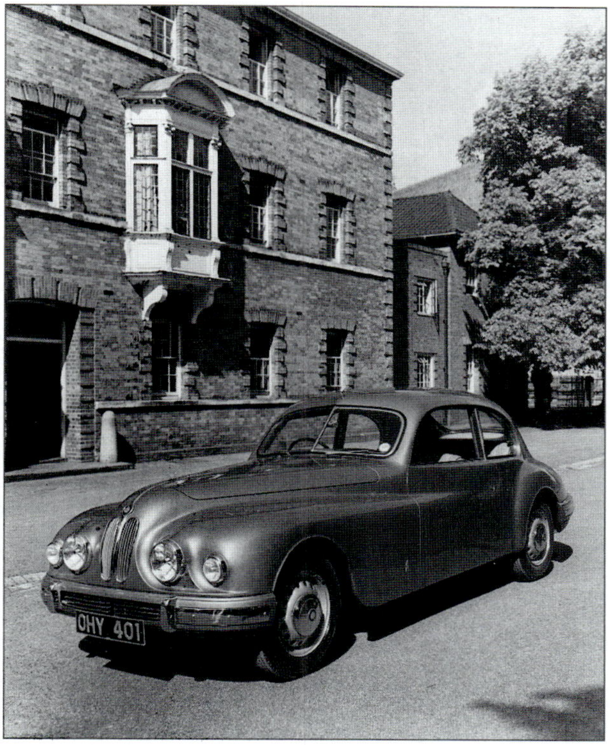

The Bristol connection. Bristol followed its own development path with the straight six, using it for ten years from 1947 and supplying it to other manufacturers and racing car constructors. The Bristol 401 (1948–53) was an elegant two-door saloon. Around the nose it looked as if it could have been a BMW, and under its Superleggera body it was to a degree an anglicized BMW.

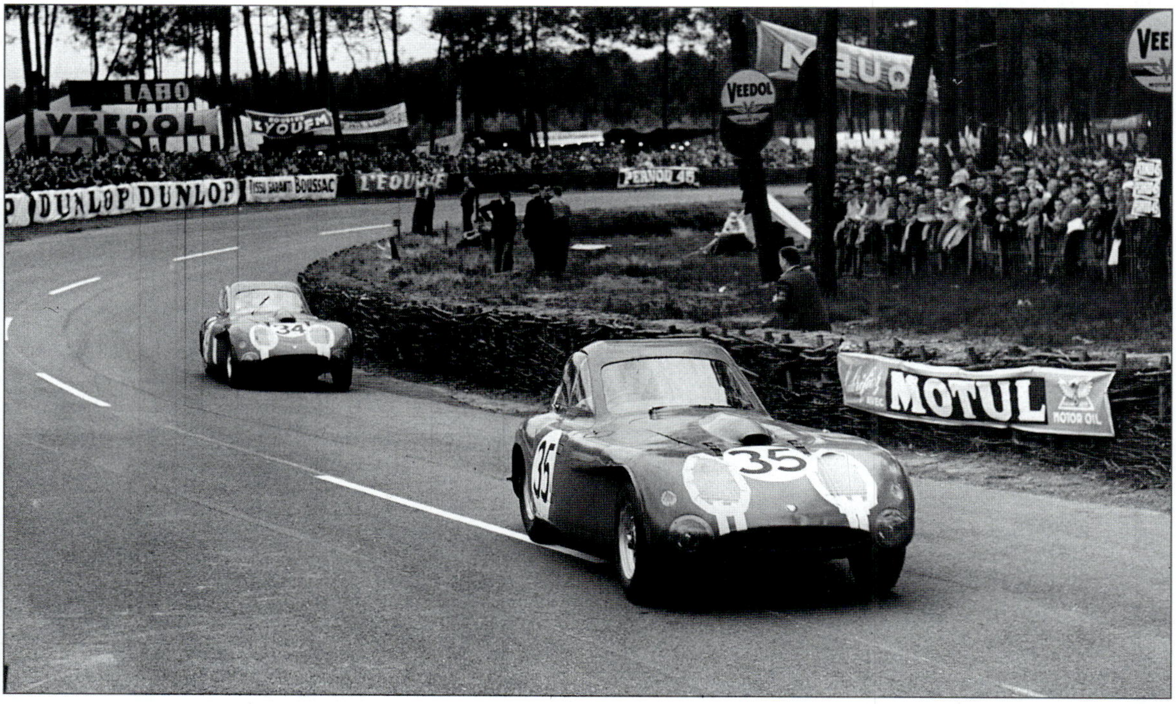

Most Bristols were luxurious GT cars, and the Type 450 came as a surprise. The ERA G-type chassis provided the basis, with a finely streamlined but odd-looking coupé body, a 1971 cc straight six giving some 140 bhp, and a BMW-originated gearbox. The two cars run at Le Mans in 1953 retired. In 1954 the engines were developed to give 155 bhp, and at Le Mans the Bristols placed 7–8–9. The car that finished seventh (number 35, above) covered 2,178 miles (3,505 km), 200 miles (322 km) more than the BMW 328 coupé in 1939. In 1955 the Bristols ran with open bodies, as 450Cs, and were again 7–8–9.

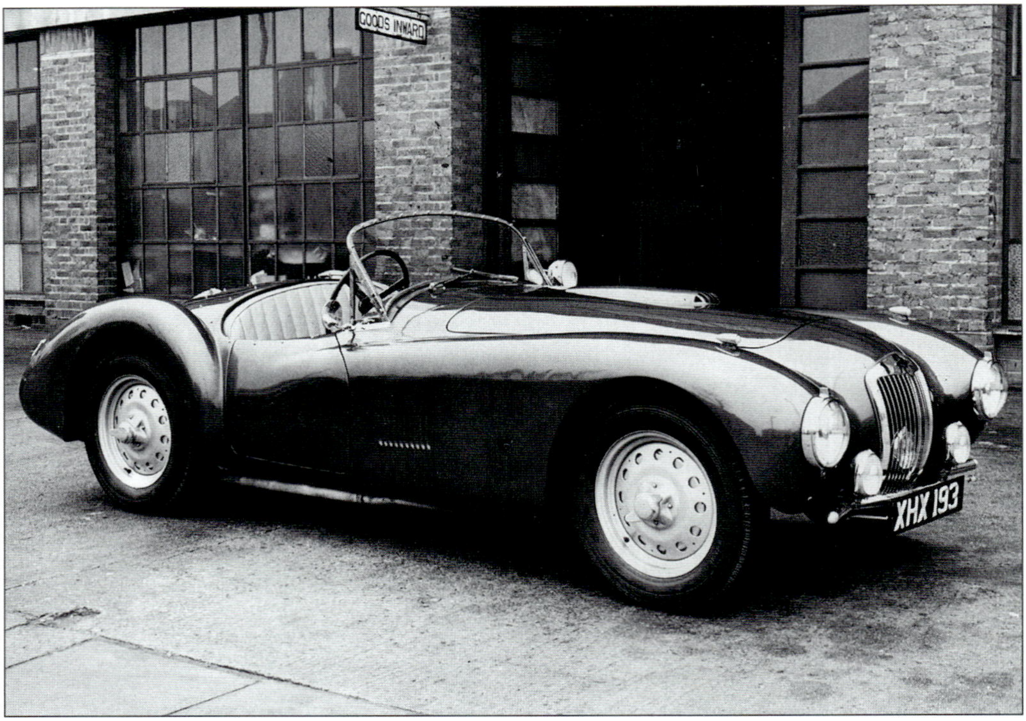

The Frazer Nash association continued after the Second World War, almost by proxy via Bristol. This Mille Miglia (1950–52) used the straight six in 90–100 bhp form and the one-time BMW four-speed gearbox with a freewheel, in a body as handsome as the racing 328's.

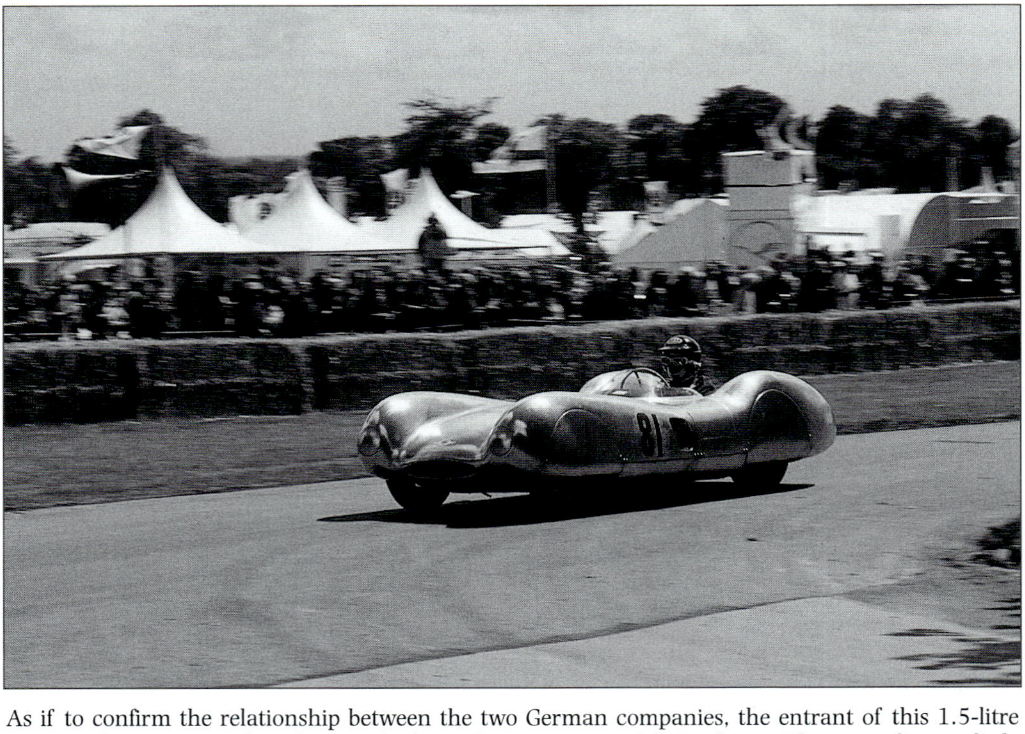

As if to confirm the relationship between the two German companies, the entrant of this 1.5-litre EMW at the 1997 Goodwood Festival of Speed was BMW Mobile Tradition. This very slippery little sports-racing car dates from 1955, and Paul Thiel, an EMW works driver in the 1950s, drove it at Goodwood.

CHAPTER THREE

FRESH START

A priority in reshaping BMW for the 1960s, and beyond, was a real mass-production model, a middle-market car. To this end, the 1500 was designed quickly and shown to potential customers well before production got under way. In the previous decade this might have suggested uncertainty, but now there was a sure touch to BMW marketing, and the approach paid off in an encouraging number of deposits on firm orders – cash in the bank.

Broadly, as the 1500 was first shown at the 1961 Frankfurt Motor Show, the 3200S with a Bertone body was also introduced, the 502 had been reworked to become the 2600, the 700s were still selling well, and the Isetta was in the last months of its life. The 1500 was seen as the wholly new BMW. That seemed much more important than the very limited production possible in 1962, when only 1,737 were completed.

The 1500 was an uncomplicated unitary construction car, with some new features to reaffirm the company's engineering image. BMW turned to MacPherson strut front suspension and a semi-trailing arm rear suspension that was to be used through the range, and for many years. A very small team headed by von Falkenhausen was responsible for the oversquare single overhead camshaft four-cylinder engine. Despite the limited development resources this was to prove exemplary, a very willing power unit and largely trouble-free in service. The three-box body had uncluttered lines, the interior was simple. Only 23,807 of these first new-generation saloons were to be built before the 1600 came, but the 1500 set a pattern for the years ahead.

It could be seen as the basis for a family of models. And it fitted neatly into a market sector where the competition was not intense, at least for a while: Ford and Opel were the German rivals, and their Taunus 12M and 17M or Rekords did not have the technical sophistication of the new BMW. There was no challenge to Mercedes here, save marginally to the 180, while Borgward had gone.

Derivatives with larger engines were to come, and appeared in several forms – a taster for the extensive ranges in the years ahead. Among them the 1964–66 TI or ti (Touring Internationale) may have forestalled the German tuning industry's interest in BMW, for a while.

The 1800 TI/SA was introduced for competition work, as the value of racing success was still recognized; a works effort was not feasible, so this lighter and more powerful variant was sold to holders of competition licences. Another variant, incidentally, was the last to have a Frazer Nash-BMW association, a modified and sporting 2000ti.

There were sporting coupés from the mid-1960s, using 2-litre engines and the running gear of the contemporary saloons, with bodies on the lines of the earlier Bertone cars. These were built in small numbers, but they foreshadowed the 3-litre coupés of the early 1970s which culminated in the famous 'Batmobile' 3.0CSL.

Two-door saloons were introduced in 1966, with the 1600–2, later the 1602 as its designation was brought into line with the 2002 and the 1502. Three of the 02 models were to be listed until 1975 (the 1502 only in 1975–77), and total production was 861,944 – a staggering number when earlier BMW figures are recalled. There were flaws, primarily a lack of rear-seat room, and there were to be many detail changes. There was

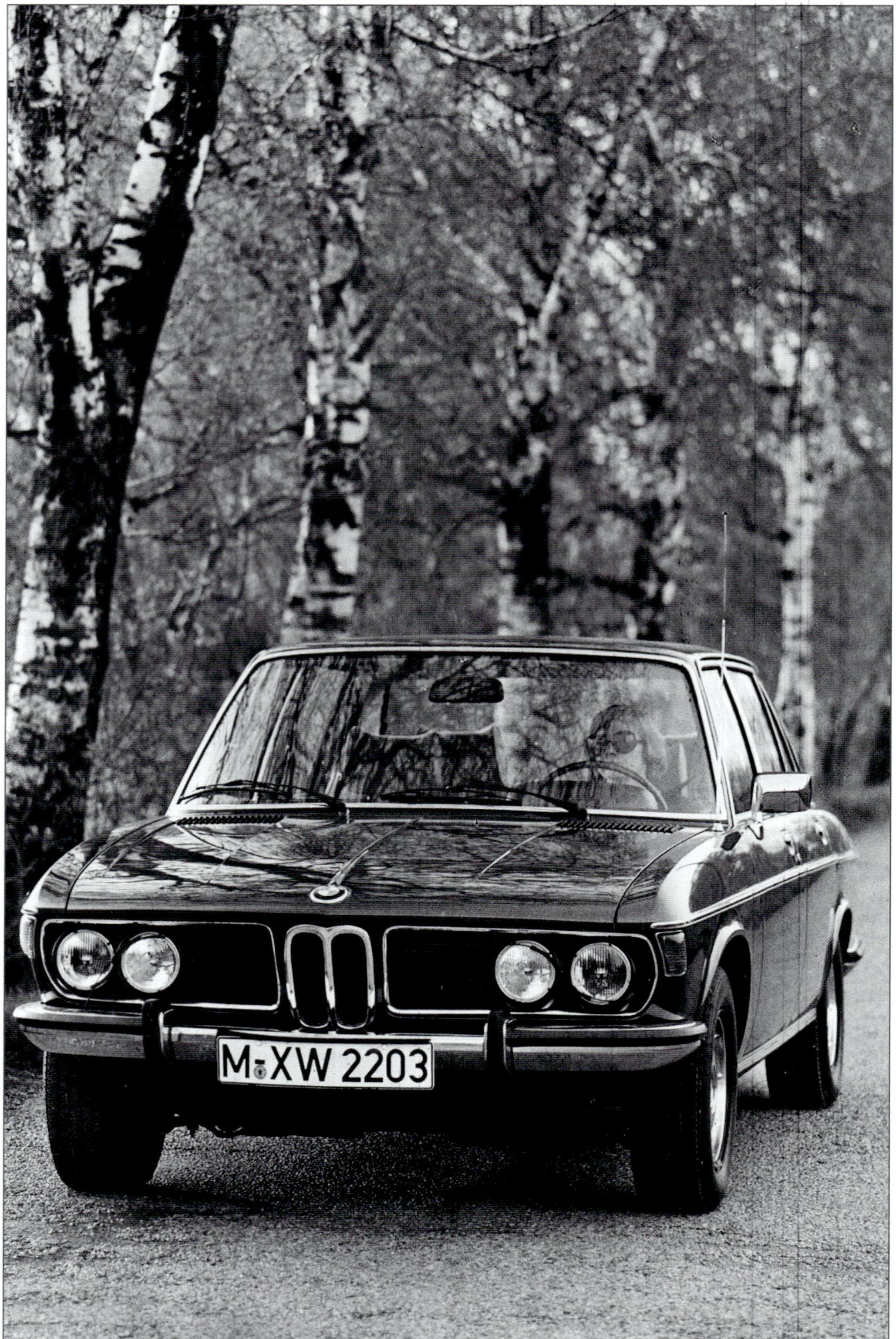

Early 1970s flagship saloon. The '*Neue Klasse*' period was notable for effective big saloons complementing the coupés in its closing years. The saloons had 2.8- or 3-litre engines in identical bodies and were 200 kmh – say 125 mph – executive cruisers.

an automatic transmission option, there was a hatchback 'Touring' version ('Touring' was to be used later for BMW's estate cars, which the company never referred to as estate cars), there was a clean-lined cabriolet by Baur, there was the 2002 Turbo. This was the first road-going turbocharged production car, a motoring landmark that was flawed and in any case came at the wrong time, during a fuel and economic crisis. . . .

Soon after the 3200CS was discontinued, another V8-engined coupé came into the BMW ranges. In the mid-1960s the Glas company's days were numbered, and BMW seized the opportunity to take it over, for its Dingolfing factory could provide much-needed extra production facilities. The handsome Frua-styled Glas 3000 was continued briefly, and the smaller Glas 1700 for a little longer, with a BMW engine as the 1600GT. The Glas middle-class saloon tooling was shipped to South Africa, where production of the 1800SA began in 1968; this car combined the Glas body and running gear with BMW engine and gearbox. The 2000SA followed, and both saloons were modified with 'BMW noses'. Production of these Glas-BMW hybrids continued until 1975.

Through this period many – most – BMW developments pointed towards high performance, with cars such as the 1800TI/SA (SA – *Sport-Ausfuhrung*, effectively sports specification, or according to some meaning *Sonder-Ausfuhrung*, or special specification), although high-performance should not be seen as synonymous with racing.

A six-cylinder engine, based on the existing four and intended for similar canted installation, was introduced for larger BMWs. The 2500 was followed by saloons with larger engines, the 2800 and the 3.0, which in America had a rare BMW model name, Bavaria, to complement the splendid coupés of the early 1970s. But in 1972 there were signs of a move on from the '*Neue Klasse*', as the first 5 Series cars were introduced, to be followed by the 3 Series. With these, BMW could build on the very sound foundations laid by the new generation.

Meanwhile, there had been a works effort in motor sport again in the second half of the 1960s. The rather special 1800s were effective then, and they have been just as competitive in historic races three decades later. BMW's Formula 2 efforts in 1967–70 brought modest rewards, then from 1973 the M12 engine based on the four-cylinder production unit was to power champions' cars in this category.

Rallies have generally had low priority with BMW since the 1930s, on the realistic grounds that returns seldom justify the costs of competing at the top level; there was just one World Championship event victory in the 1970s (in Austria in 1973), but there were successes in secondary European events.

Real competition fireworks exploded as a BMW versus Ford battle for European Touring Car Championship honours was fought out in 1970–4, between CS and CSL coupés and Capris. This all-out saloon car racing faded during fuel crises. BMW Motorsport GmbH was set up when competition was intense, and was to play important roles in the next decades.

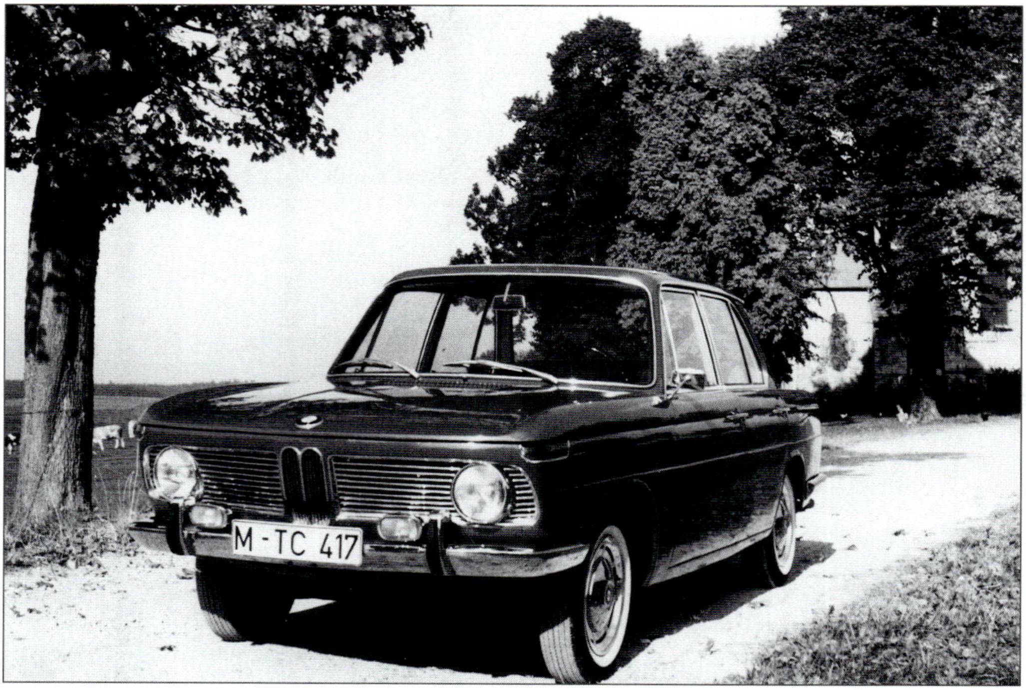

The new shape that saw BMW on the way to real success. This early 1500 shows the clean simplicity of the overall lines, and the careful avoidance of fussy detailing. In 1962 its sohc engine was rated at 80 bhp, to give a 92 mph (148 kmh) top speed. BMW lore has it that when the final design was presented to the management for approval, Herbert Quandt required the re-introduction of the traditional grille style.

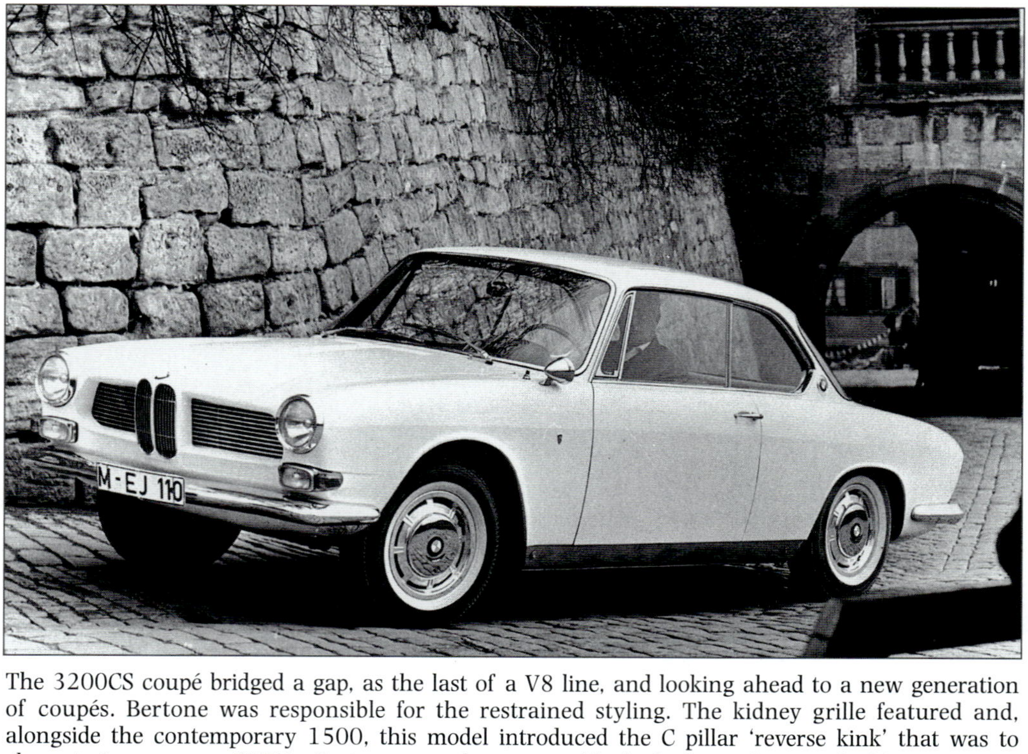

The 3200CS coupé bridged a gap, as the last of a V8 line, and looking ahead to a new generation of coupés. Bertone was responsible for the restrained styling. The kidney grille featured and, alongside the contemporary 1500, this model introduced the C pillar 'reverse kink' that was to characterize so many BMWs. Bertone was also responsible for body production, building just over 500 between 1962 and 1965. His convertible was a one-off variant.

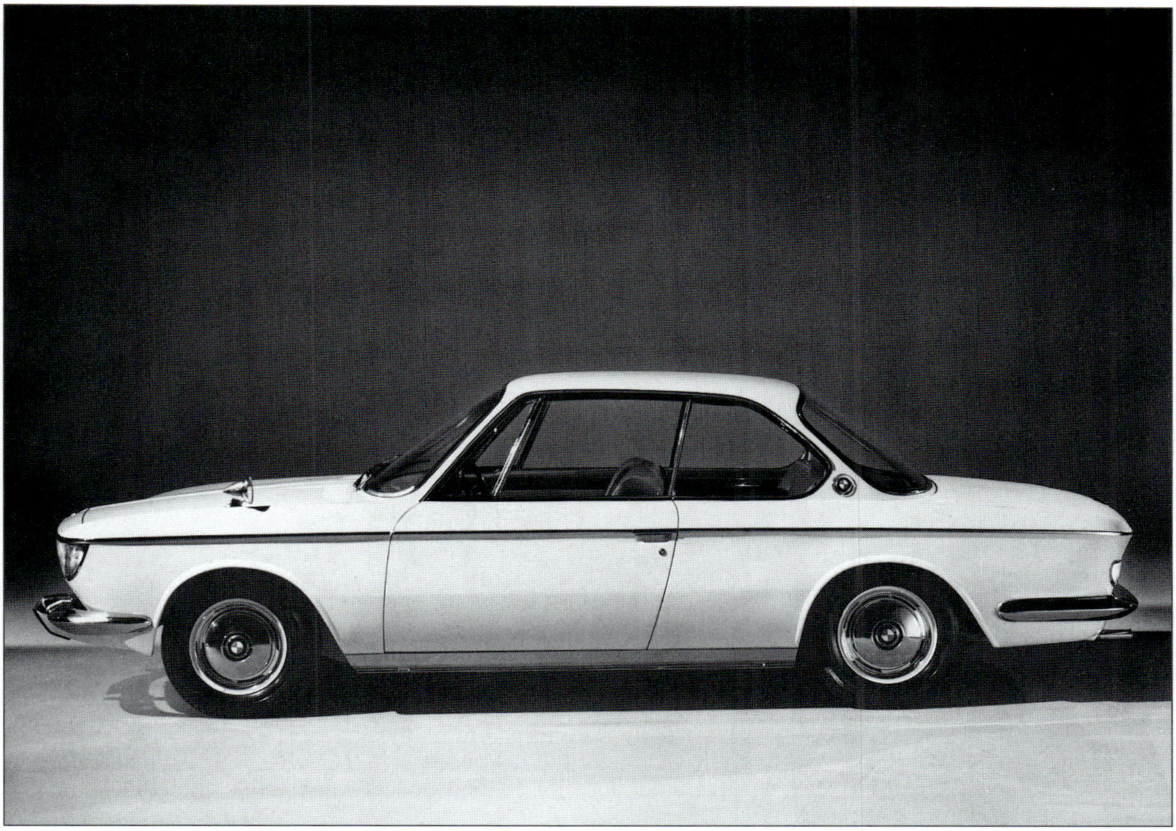

The 2000C and 2000CS coupés of 1965–69 looked larger than the 2000 saloons, but shared wheelbase and track dimensions, and they foreshadowed the better-known coupés of the 1970s. The photograph shows the large glass area achieved and the positive belt line holding the design together; it does not show the tail, perhaps the only weak feature and one which some critics suggest resembled the rear end of the Chevrolet Corvair. The 2-litre engine was a tight fit, but there were four proper seats and the boot was generous. Four-speed gearboxes were normal, with a ZF three-speed automatic optional on the 2000C. That was a 104 mph (166 kmh) car; with twin carburettors and a higher compression ratio, the engine of the 2000CS gave 120 bhp and 115 mph (185 kmh). Coachbuilder Kamm produced a cabriolet version with lines aft of the A pillar that did not quite work.

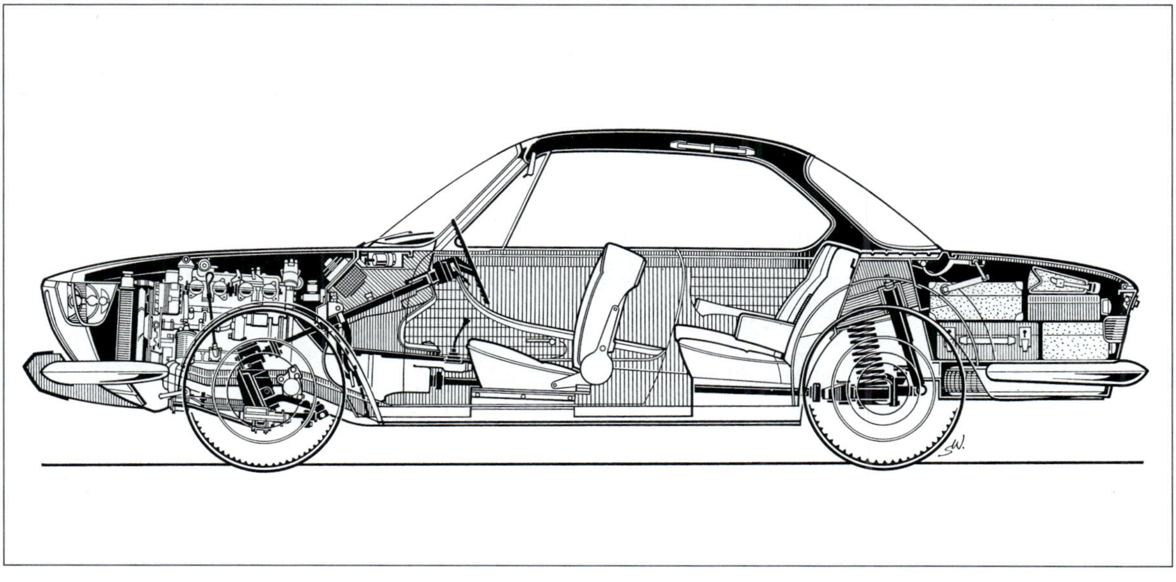

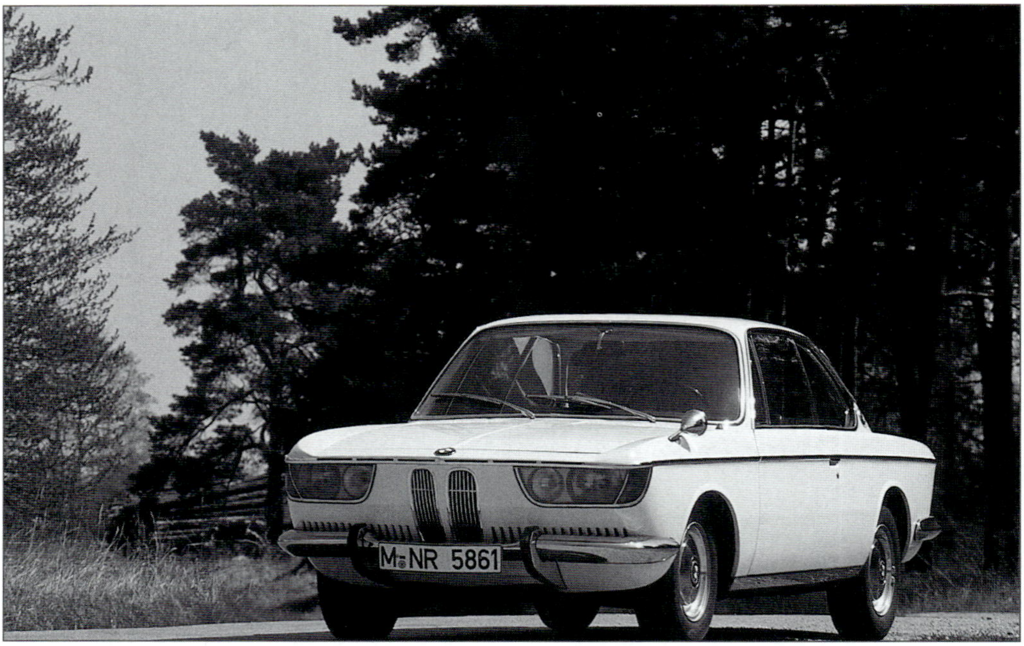

The 2000C/CS had an overhanging nose, which seemed curiously unfinished as the kidney grille was not flanked by horizontally barred intakes. This was one BMW that really did justify the 'aggressive appearance' often attributed to the company's cars.

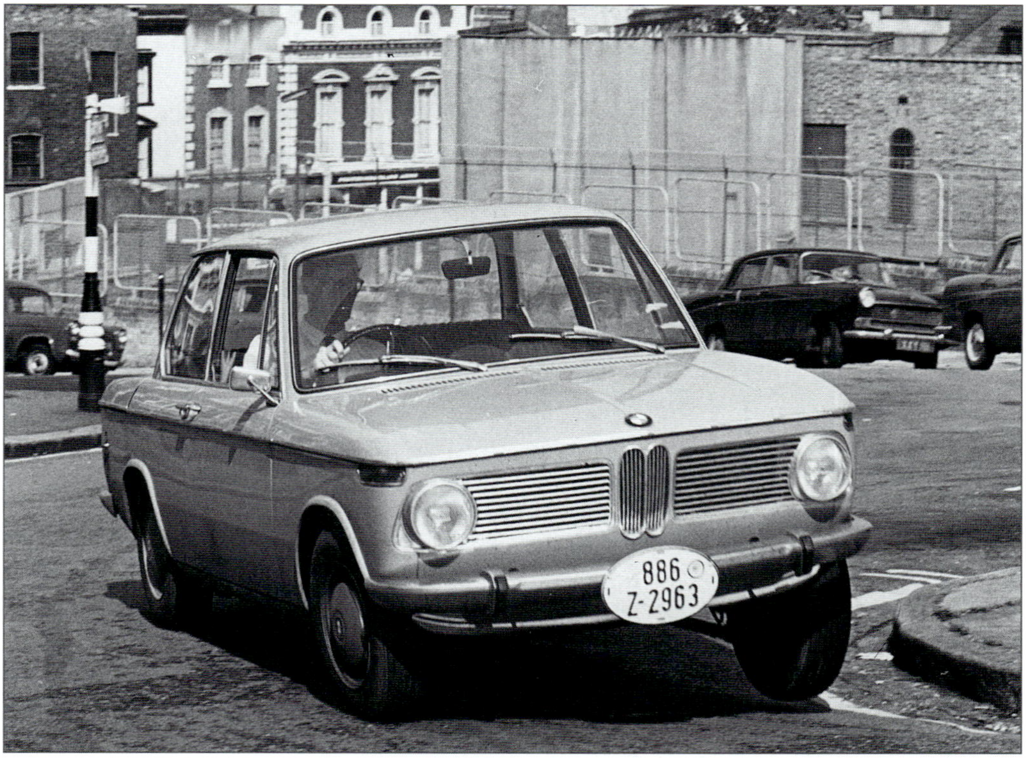

The 1600–2 introduced a two-door small model to the '*Neue Klasse*' in 1966. On a shortened floor pan, it was also lighter than the four-door saloons, and more agile – at a price in rear-seat leg room, so that realistically it could be considered a coupé for two adults and two children. This early right-hand-drive car had a single-carburettor 86 bhp engine, giving a top speed a whisker shy of 100 mph. A twin-carburettor 1600TI in 1967 had an extra 20 bhp, and reached 109 mph (175 kmh).

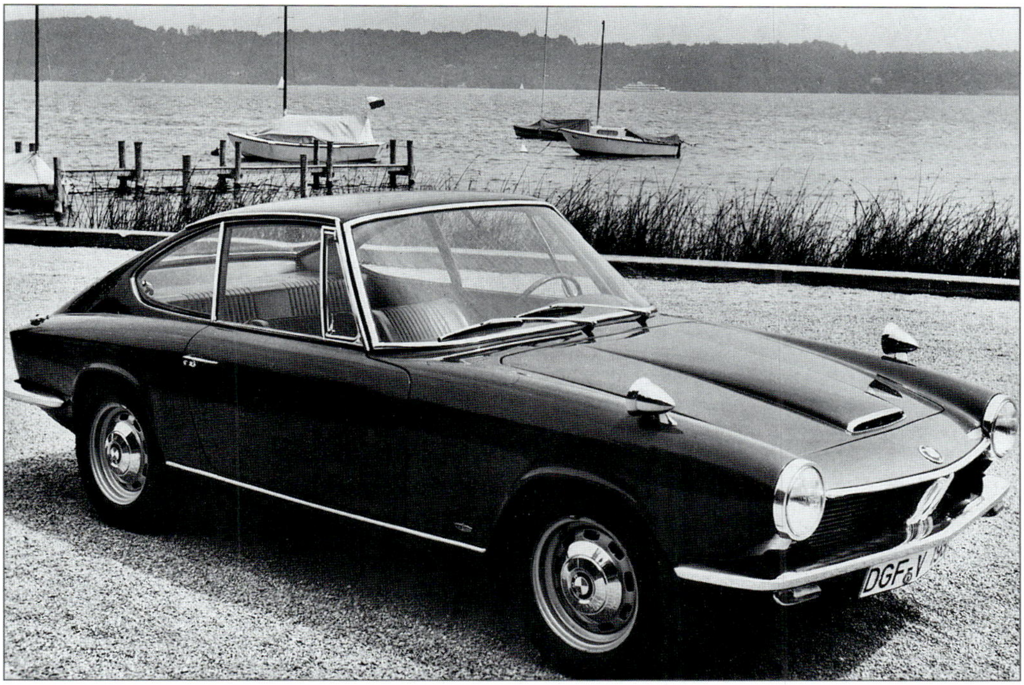

The BMW production version of the small Glas coupé used 1600TI twin-carburettor engine, transmission and rear suspension, while its Frua-designed and built body was given a BMW nose and badges as it became the 1600GT. Seemingly this pretty little car was not accepted as a BMW; poor sales meant that it was built for less than two years, and the stock was jobbed off at 'remainder prices' in 1968.

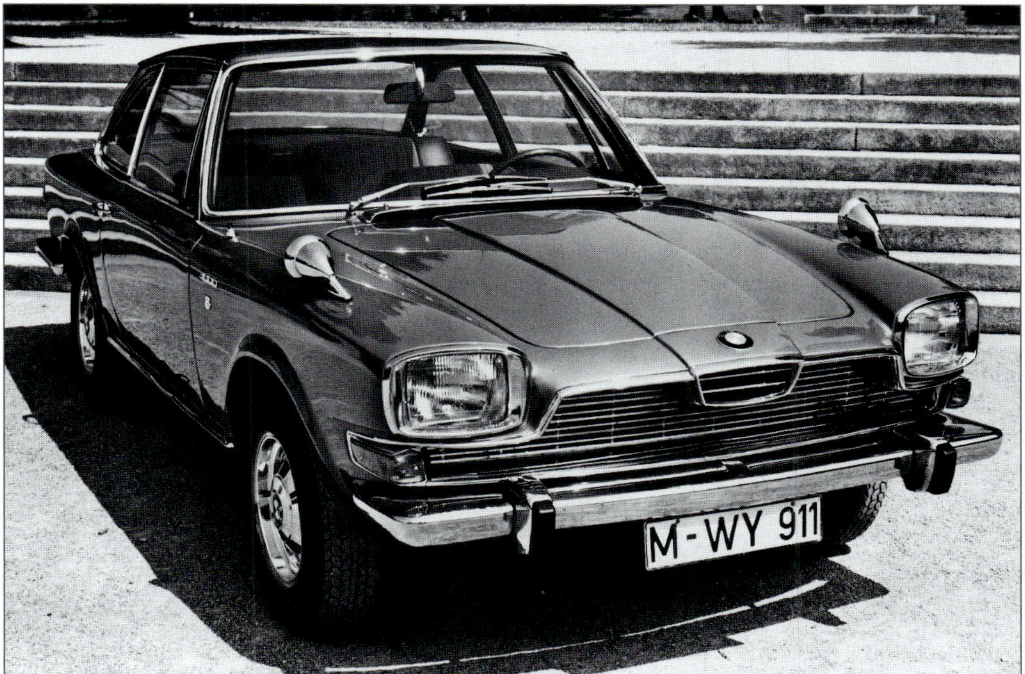

The larger Glas V8 was a potentially quick GT car, with a 3-litre V8 giving a top speed of over 120 mph (192 kmh). But its handling was suspect and from this angle it was not handsome. This is one of the 389 BMW-badged cars completed, 1967–68.

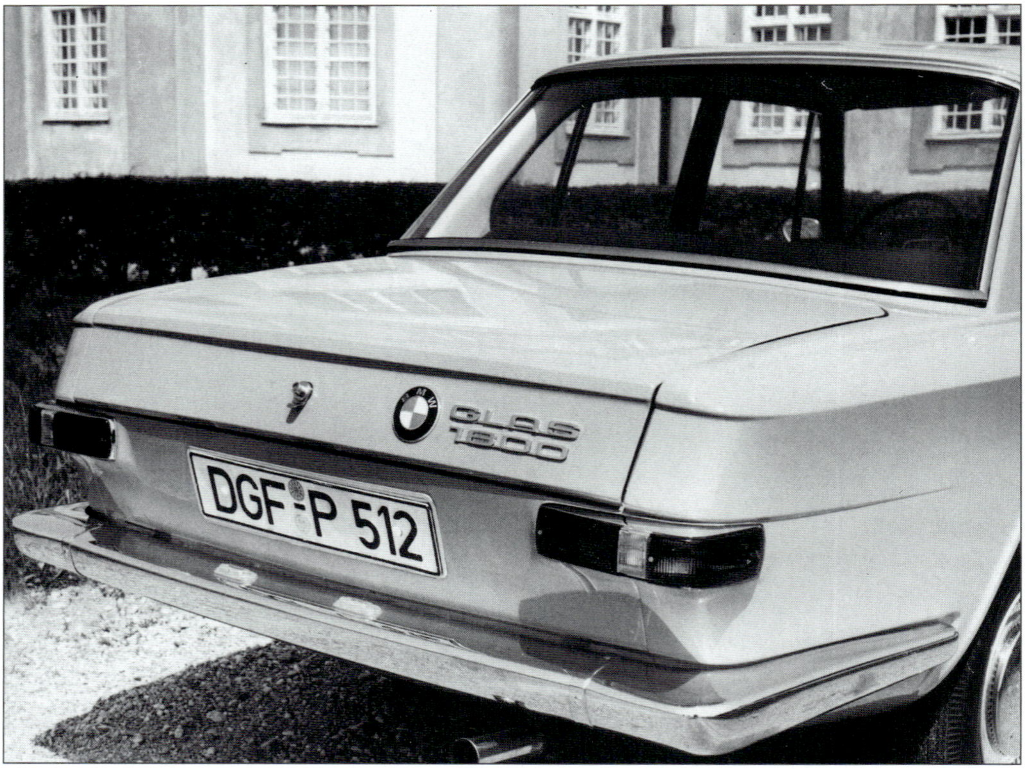

The rather complex rear of an erstwhile Glas 1800 with an ill-fitting boot lid shows two marque loyalties. . . .

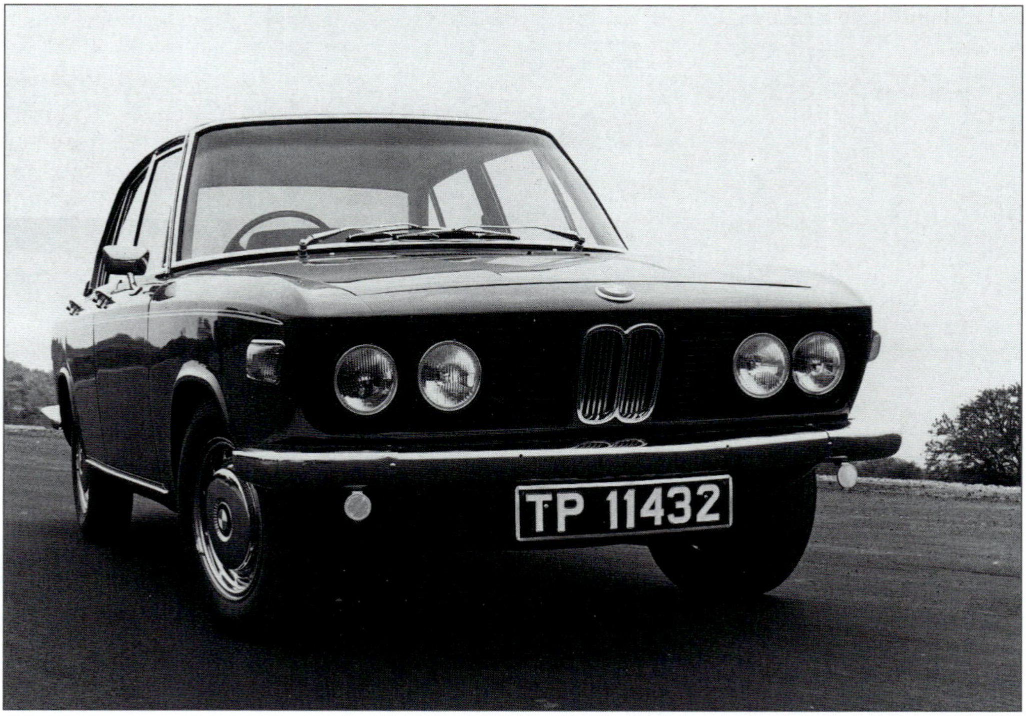

Some of the Glas tooling was shipped to South Africa, where production of the 1800SA saloon began in 1968. In the mid-1970s the 1804 and 2004 were given a face lift, with BMW grilles and inset quad lights (and rather large turn indicators). These cars were built until 1976.

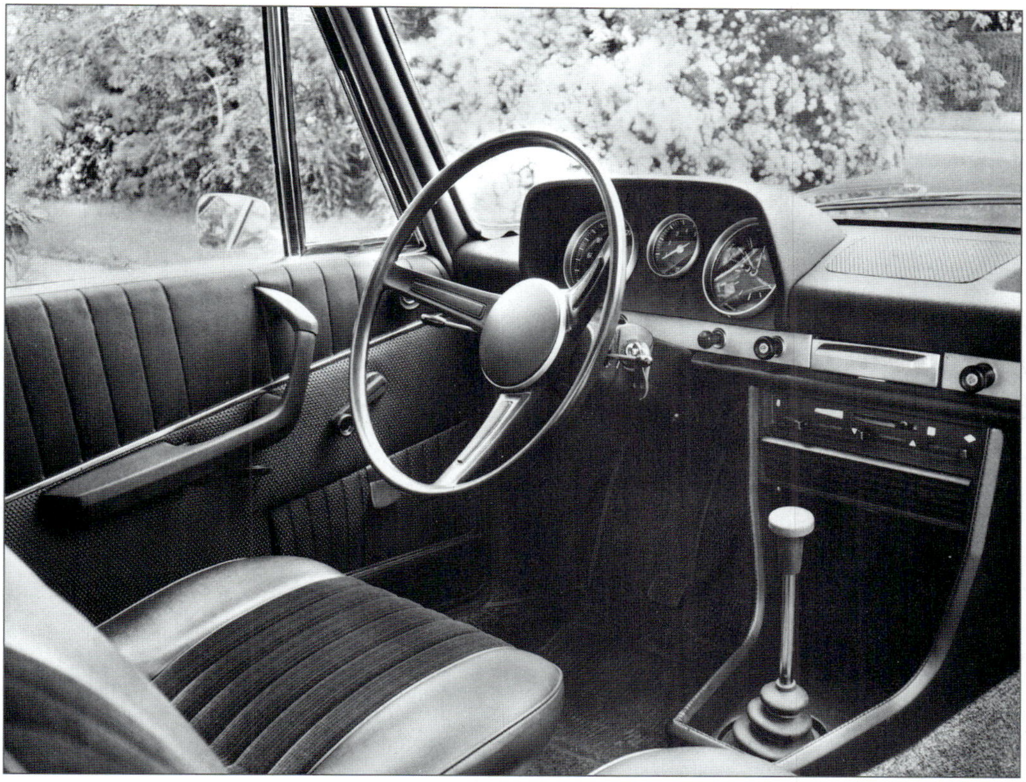

The cockpits of BMWs in the 1960s were essentially straightforward. This is a 1969 2000ti lux, with speedometer, rev counter and clock (its face incorporating a fuel gauge). Switches were mounted on the centre strip, and the heater controls were simple. The dished steering wheel with its padded boss was new in 1969, and shared with other models such as the 1800.

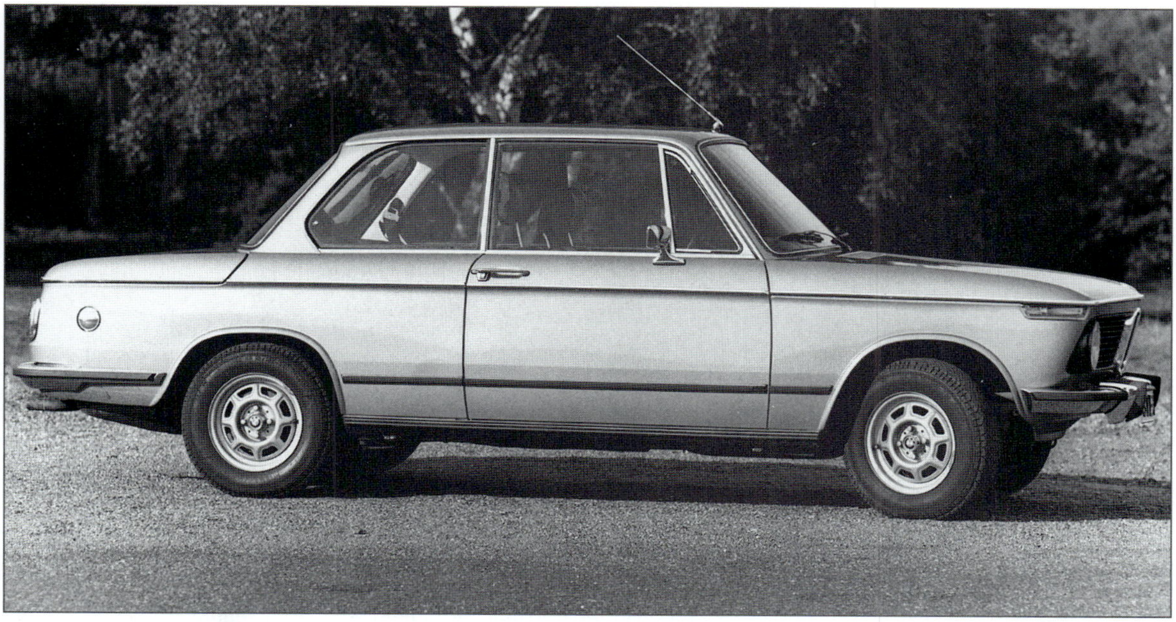

An early-1970s development in the two-door sports saloon line was the 2002ti, the first fuel-injection BMW, with a Kugelfischer system feeding the 2-litre engine, replacing the twin-carburettor unit run 1968–71. With 130 bhp, this was a 116 mph (185 kmh) car. The simple lines of the 1960s cars are retained, in this case with unattractive wheels. A Bertone coupé on a 2002 basis actually carried an *Entwicklungs* number (E19).

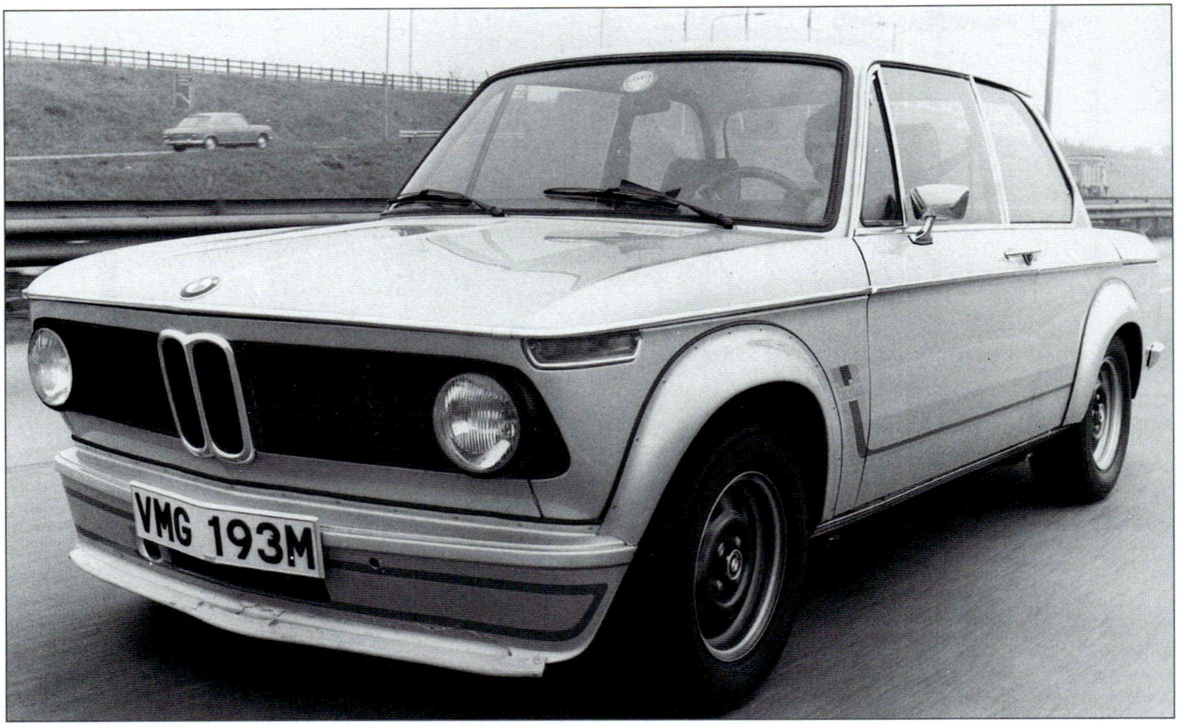

The world's first road-going turbocharged production car, the 2002 Turbo, could not have come at a worse time. In 1973 there was a major fuel crisis and the German public mood had turned against 'performance cars', so only 1,672 were built before production was cut short late in 1974. A KKK turbocharger was used, raising the output of the 2-litre fuel-injection engine to 170 bhp. That gave 130 mph (210 kmh), with potentially excellent acceleration but pronounced turbo lag. The car was also criticized for some aspects of handling and braking.

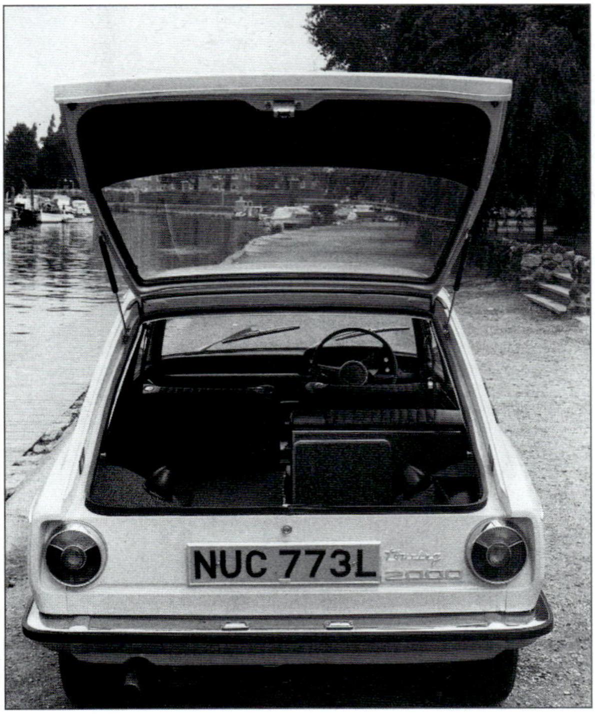

Touring bodies – three-door hatchbacks – came on the 1600 (briefly), 1800/1802 and 2000/2002 types in 1971, and BMW chose to perpetuate the term for its later models on estate car lines. There was no loss of measurable performance, but some drivers felt that handling suffered. Sales reached just over 30,000 before these first Touring types were discontinued in 1974.

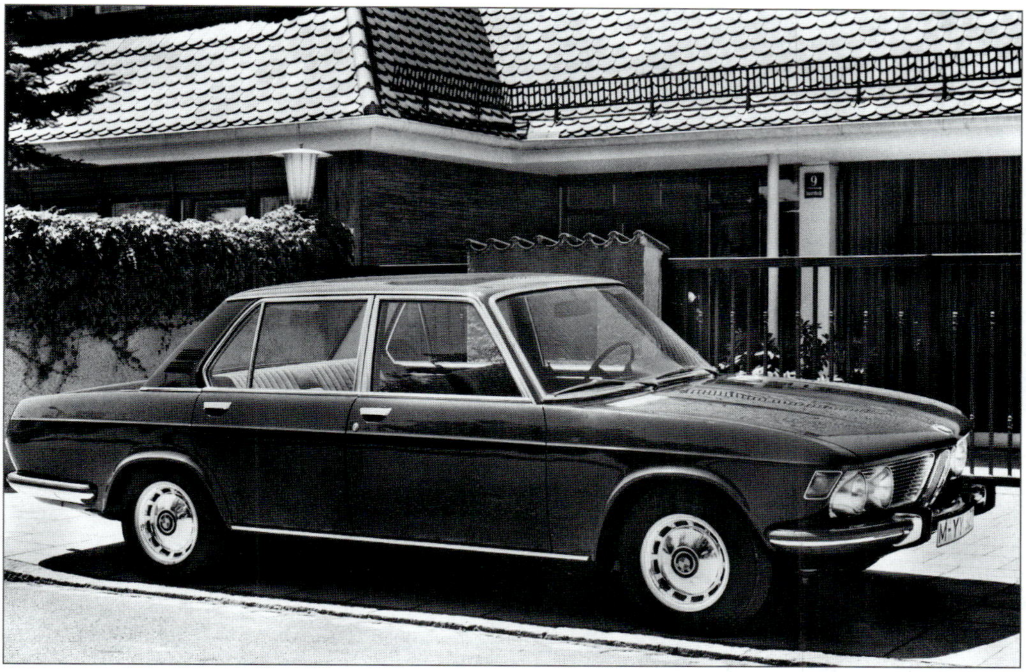

A move back towards large cars was signalled by the 2500 and 2800 in 1968. As the designations implied, these had a 2.5- or 2.8-litre straight six, and combined sports saloon handling qualities with 'executive' interiors to match Mercedes's mid-sector models. Outwardly there was perhaps an excess of chrome, and certainly there is more glitter to this 2500 rear end than other BMWs.

The engine was effectively the sohc 'four' with two more cylinders, and it gained a reputation for smooth performance. With automatic transmission the 2500 reached 115 mph (185 kmh) while the 2800 was a 124 mph (201 kmh) car. The 2500 lasted until 1977, while the 2800 was listed until 1974, when it gave way to the 2.8L.

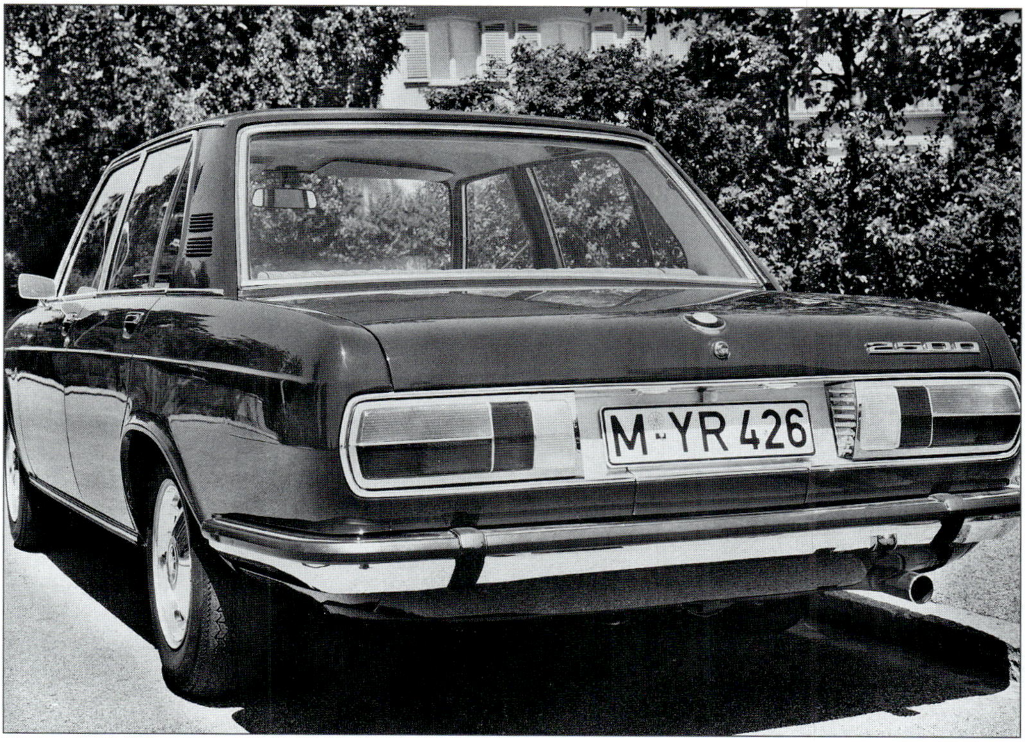

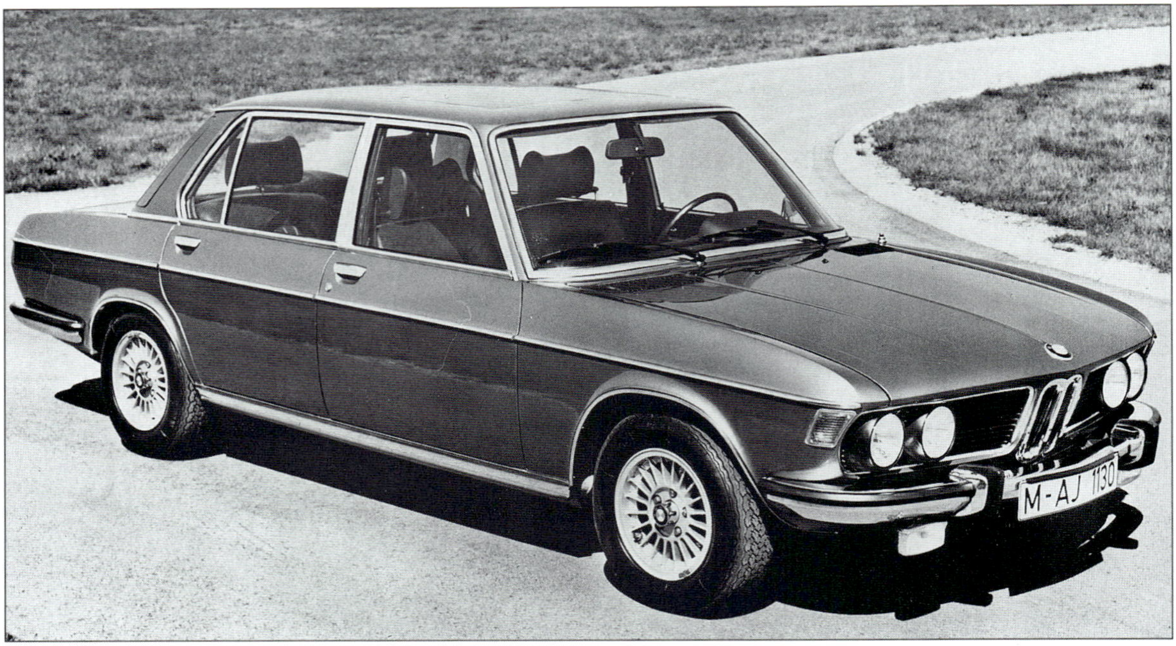

Largest of the mid-1970s BMWs was the 3.0L and 3.3L (L for *lang* – all saloons were limousines in Germany!). The 3.3Lia had automatic transmission. These cars had a 4-inch stretch in wheelbase and overall length, and the additional space was given to rear-seat passengers. Top speeds were around 126 mph (202 kmh). Those alloy wheels would have improved the looks of some of the big coupés of the time.

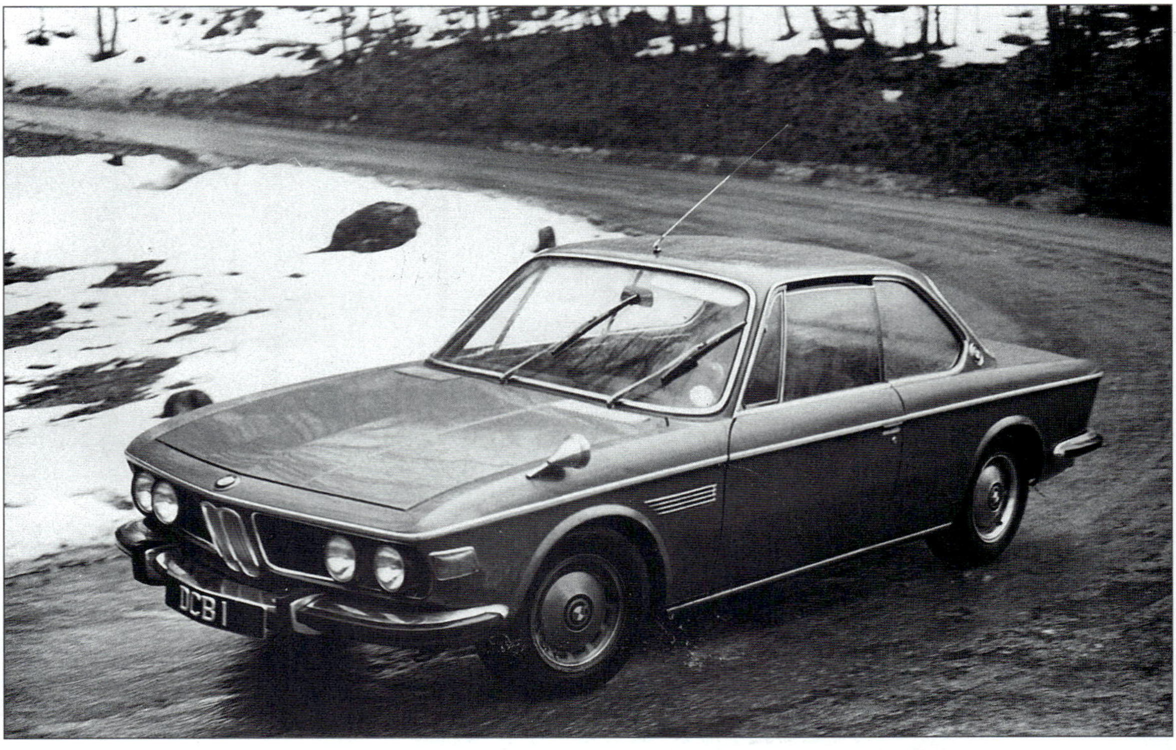

The definitive coupé lines of the famous 1970s CS, CSi and CSL models were first seen in the 2800CS in 1968. This had a 2.8-litre inclined straight six rated at 170 bhp and was a 128 mph (201 kmh) car with a four-speed manual gearbox (a three-speed automatic was an option). The 3-litre cars appeared from 1971, with an 'economy' 2.5-litre coupé in 1974–5 as a response to critics during a fuel crisis.

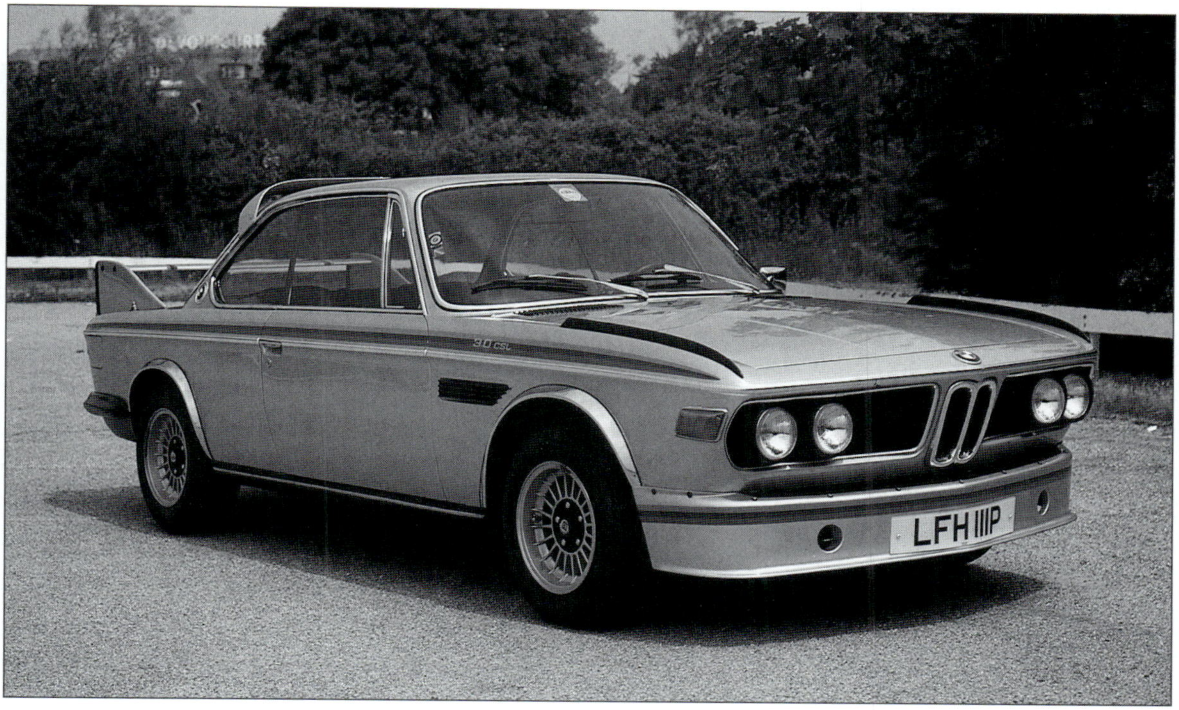

The 3.0CSL built in 1973–5 was the ultimate version in this series. Its fuel-injected 3153 cc engine was normally rated at 206 bhp, there were lightweight body panels and cockpit fittings, and the claimed top speed was 140 mph (225 kmh). This rare 'Batmobile' is road-registered, but cars with full aerodynamic kits were normally seen on circuits. (They were not allowed on German roads in this condition, and were supplied with the rear wing in the boot as a kit!)

Altogether, 44,254 coupés were built, 1968–75, and some were 'uprated' with cosmetic modifications to resemble the most desirable 3.0CSL.

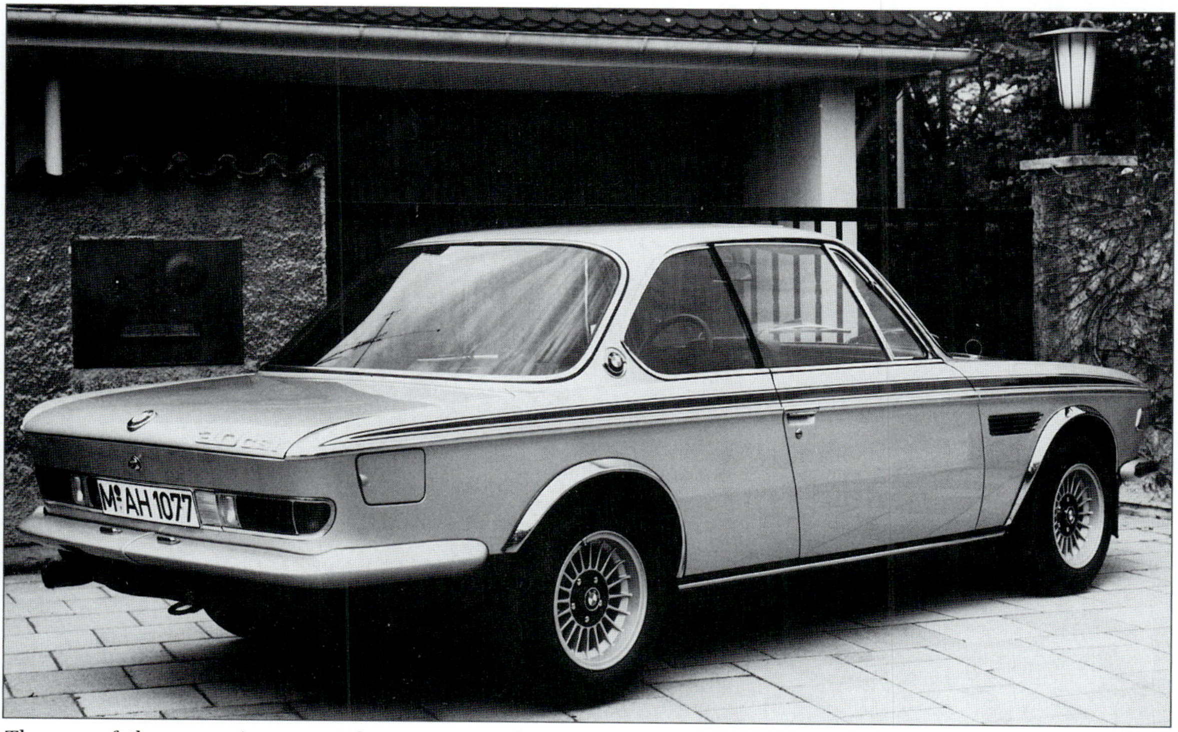

The rear of these coupés was not the strongest styling point. This is a 3.0CSL.

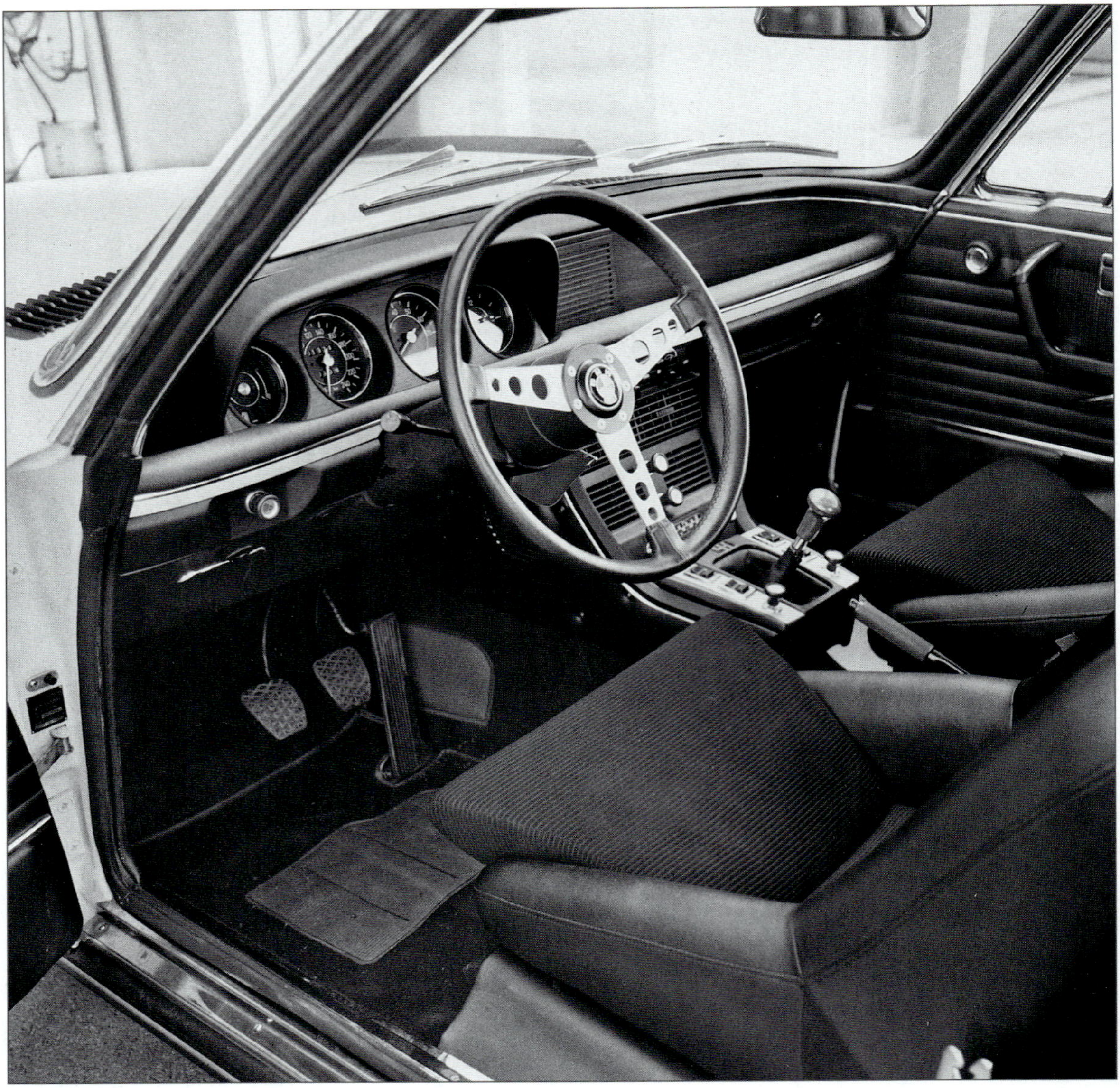

The 3.0CSL cockpit was distinctly sporting, but by no means spartan. The supportive seats, stubby gear lever and the wheel combine to state the car's character.

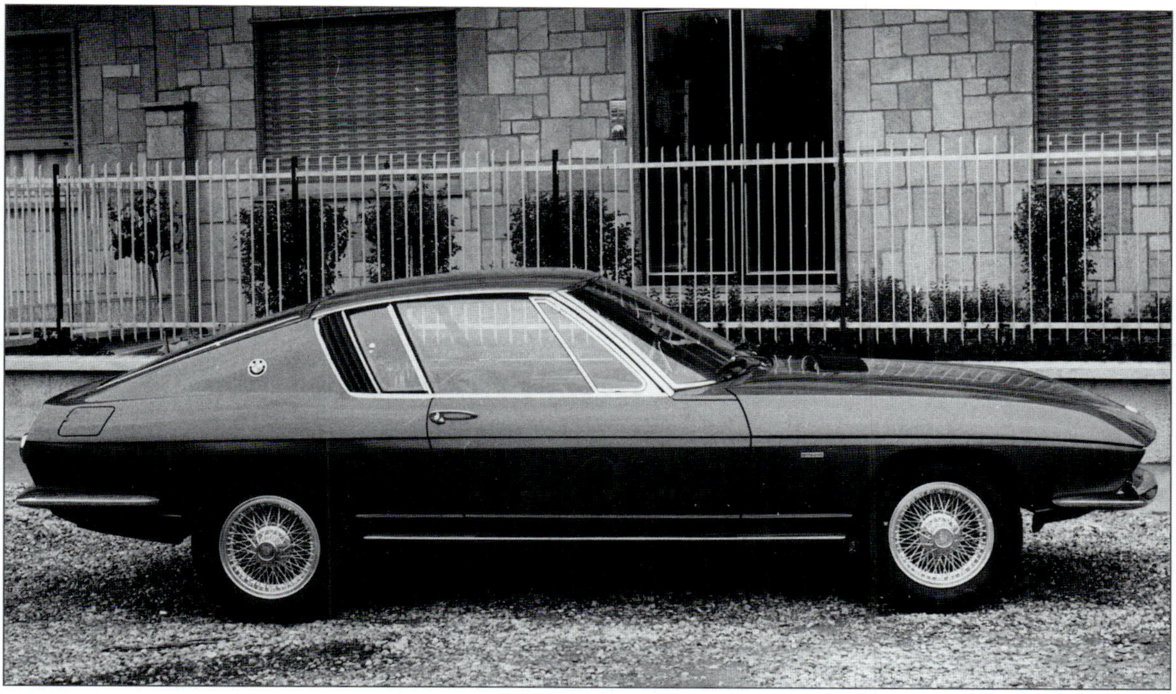

Italian coachbuilder Frua showed this svelte coupé on a 2800 chassis in 1969. Wire wheels were a most unusual feature on a BMW.

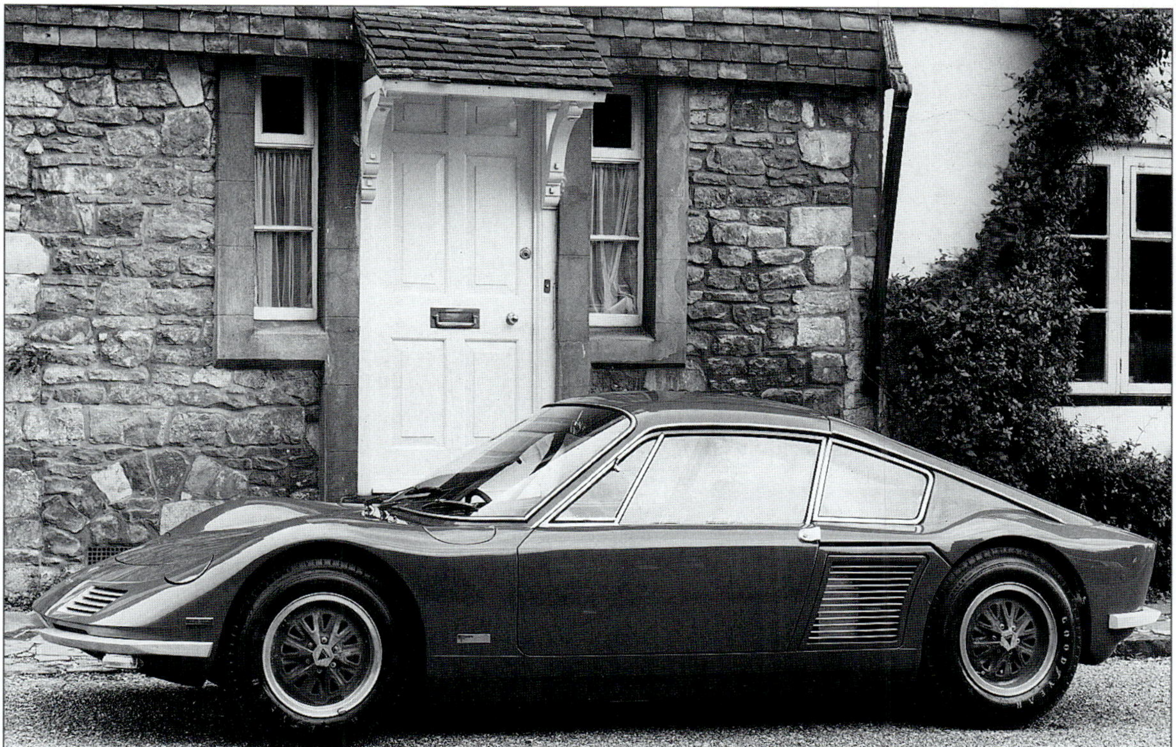

BMW engines have powered many high-performance cars built by specialist companies, and few of those have been more attractive than this mid-1960s Elva GT 160. The sleek 40 in high body was designed by 'Trevor Fiore' (Trevor Frost) and built in Turin by Fissore. The 2-litre engine gave a projected speed of 160 mph (255 kmh). Sadly, this Elva-BMW did not reach the production stage.

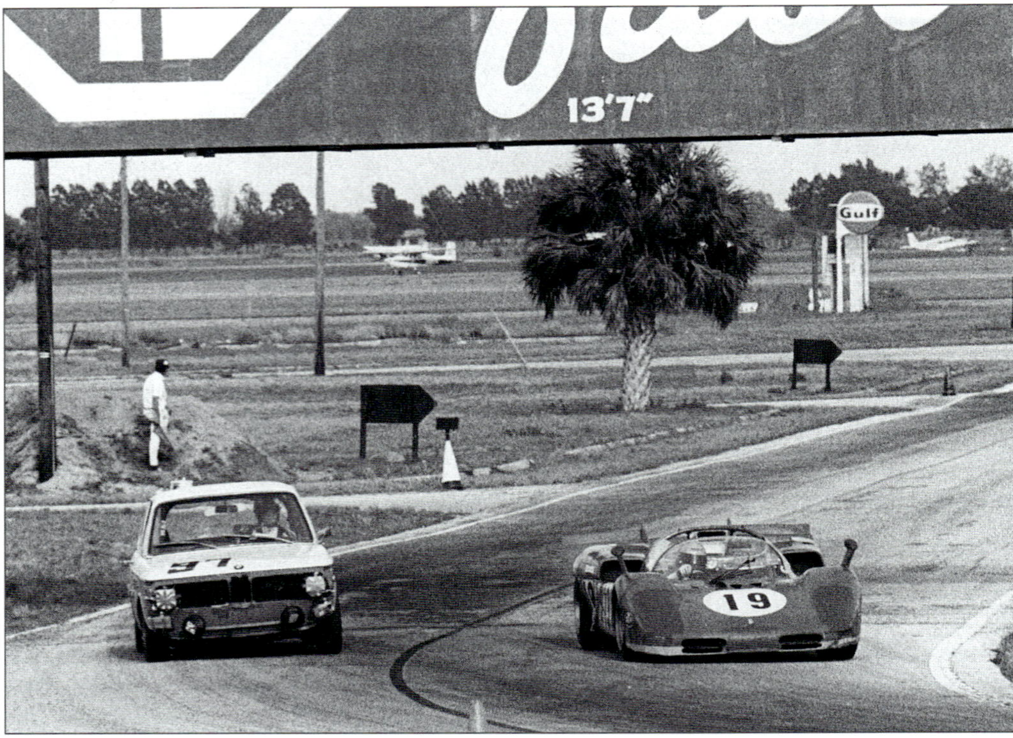

Sports car racing attracted mixed entries as the 1970s opened. Here a 2002 driver keeps well out of the way of a Ferrari 512S in the 1970 Sebring 12-hour race.

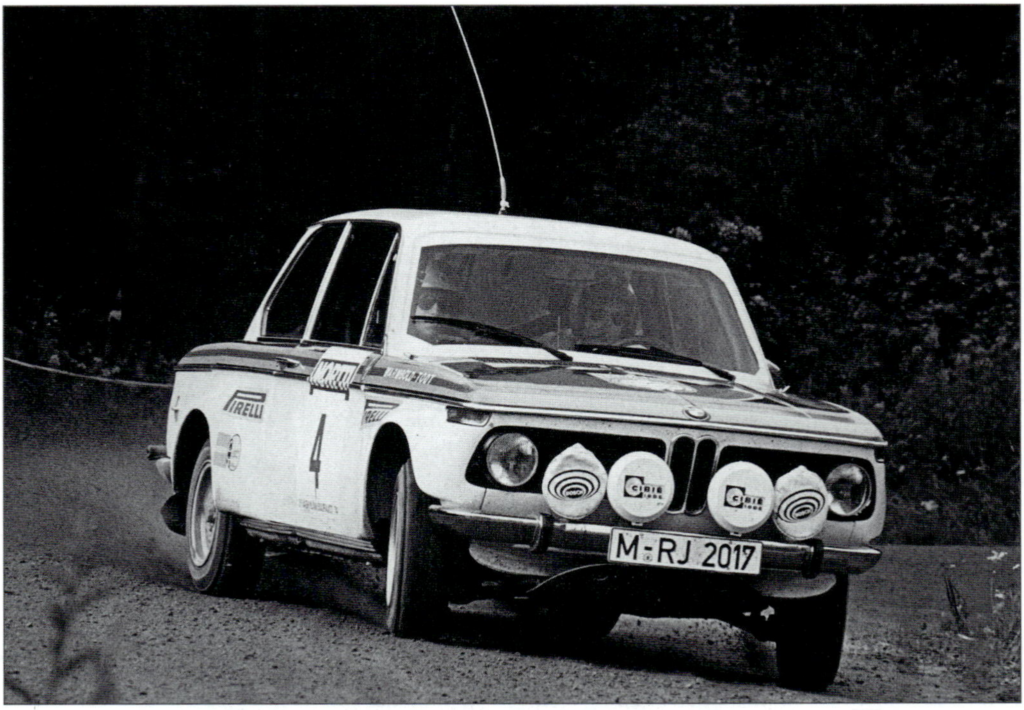

The 2002ti, with a Schnitzer-developed twin-cam version of the 2-litre engine giving some 220 bhp, showed some promise in the early 1970s rallies. (A major victory in the period fell to Achim Warmbold in the 1972 TAP Rally in Portugal.) Two years on, Warmbold is driving this car in Finland's 1000 Lakes Rally; his co-driver was Jean Todt, Ferrari GP team manager in the 1990s.

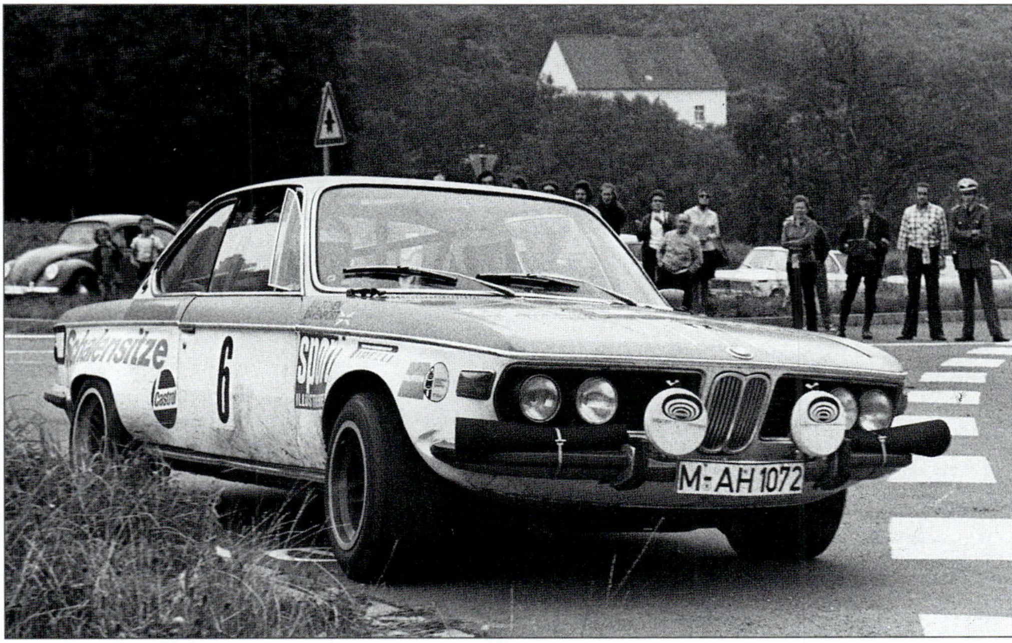

The big CSL coupés seem less plausible rally cars, but were used by German entrants in German rallies. This car in the 1972 Olympia Rally was crewed by Rauno Aaltonen and John Davenport.

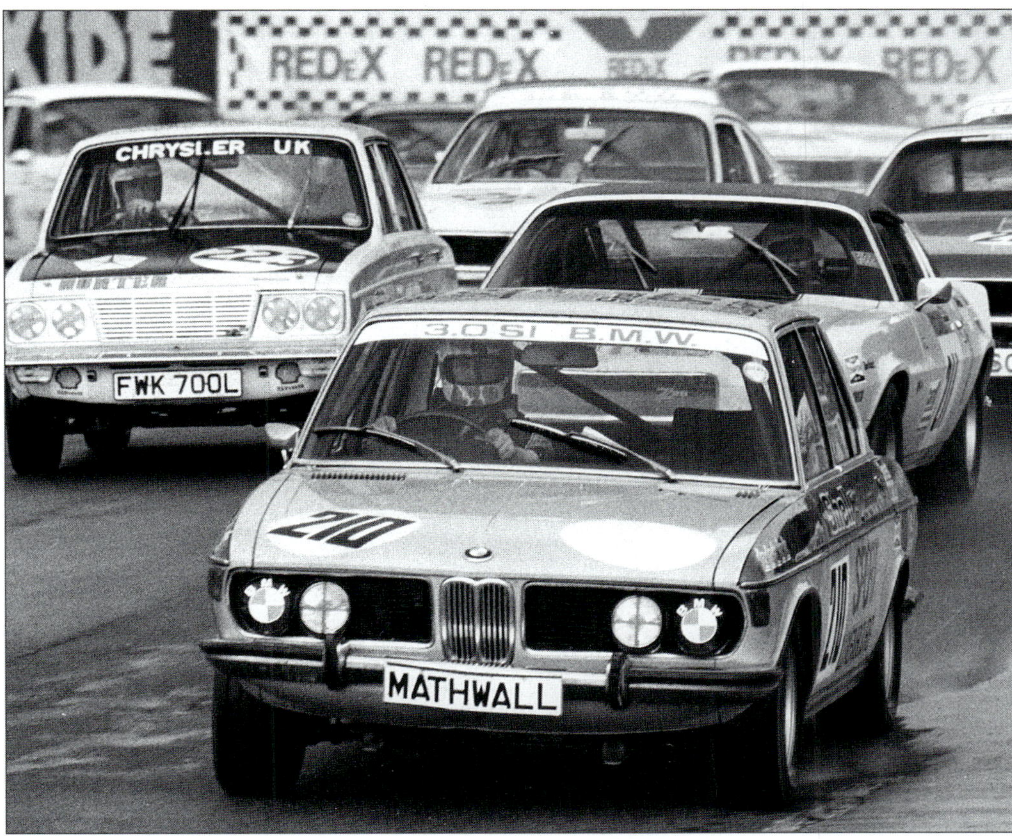

Group 1 racing, for more-or-less showroom cars, attracted BMW's British concessionaires to enter the national championship in the early 1970s. This 3.0Si prepared by Mathwall Engineering was raced by Tony Lanfranchi (driving in this shot) and Roger Bell in 1973, when it was a class winner.

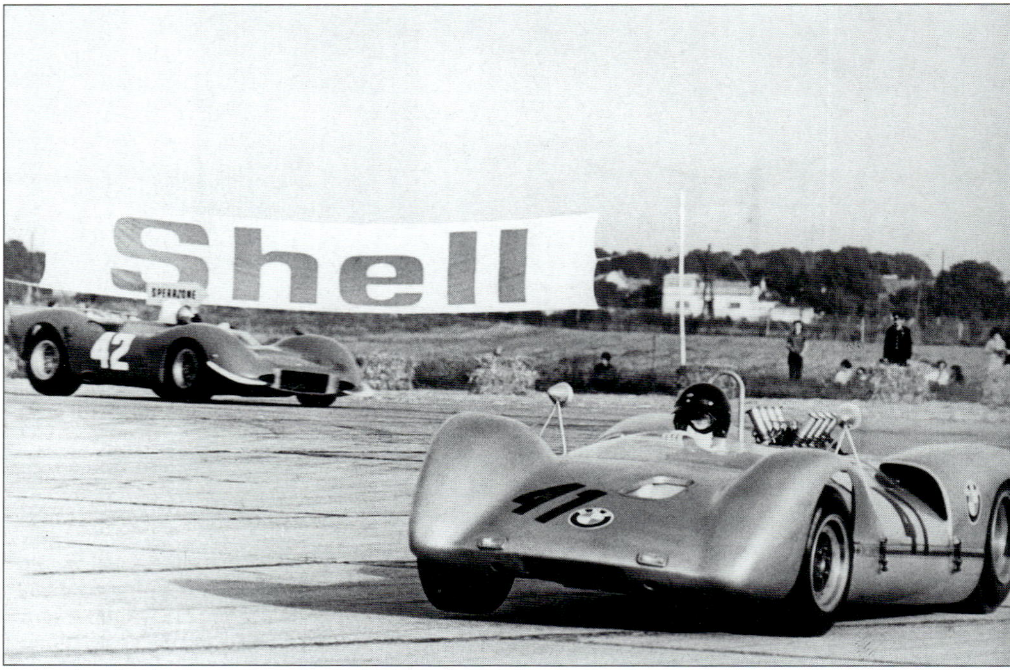

BMW tried hard to challenge Porsche in the 2-litre sports-racing class in 1967–68, using Lola chassis and self-handicapped with the complex Apfelbeck engine favoured by development director von Rücker. The valves in each pair for each cylinder were at 180 degrees to each other, thus calling for two sets of exhausts. The inlet trumpets project behind Dieter Quester's head in this late-1967 version in an airfield race at Aspern.

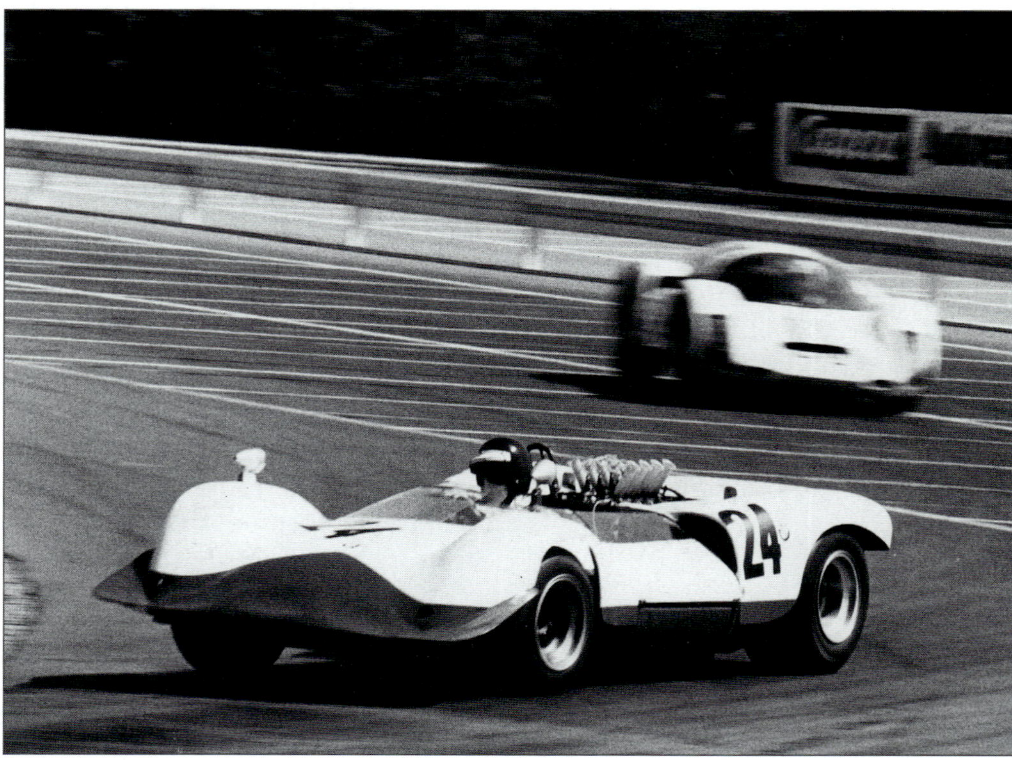

The Monti Bergspider version (Quester again, at the Norisring) in 1968 has more body variations, although not to the extent seen when it was run with a vast wing in hill climbs.

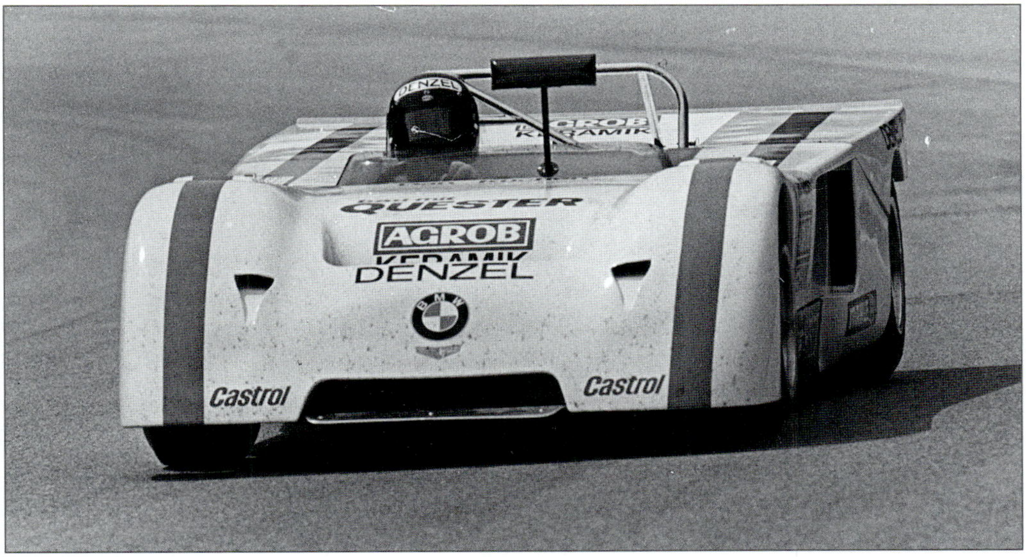

The BMW badge overwhelms Chevron's on this B21, devised and raced by Quester in 1972, when the European Trophy for 2-litre sports-racing cars was popular. At a time when BMW had officially withdrawn from competition, this car had an engine based on one of the M12 Formula 2 units, giving 275 bhp. Quester drove it to a victory at the Salzburgring and a second place at Dijon; a crash in a Sicilian race cut his season short.

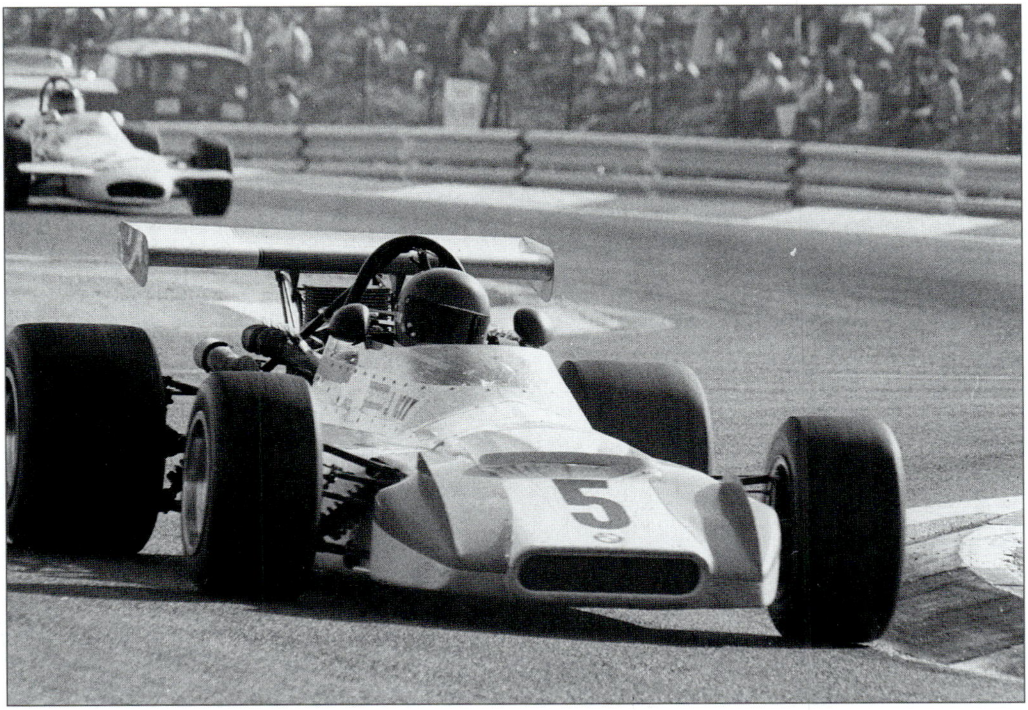

BMW entered single-seater racing during the 1.6-litre Formula 2 period, supplying engines used in Team Surtees Lola T100s in 1967. For the following year the T102 was commissioned, and often run as the BMW T102; the engines were the notoriously unreliable and cumbersome Apfelbeck radial valve types.

For 1969–70 BMW commissioned journeyman designer Len Terry to lay out its own car, and the chassis were built by Dornier. The Apfelbeck engine was set aside in favour of the M12/1-4. Four cars were built and run in several aerodynamic configurations – in this shot, Jacky Ickx is racing one with a large rear wing – and in 1970 five races fell to them.

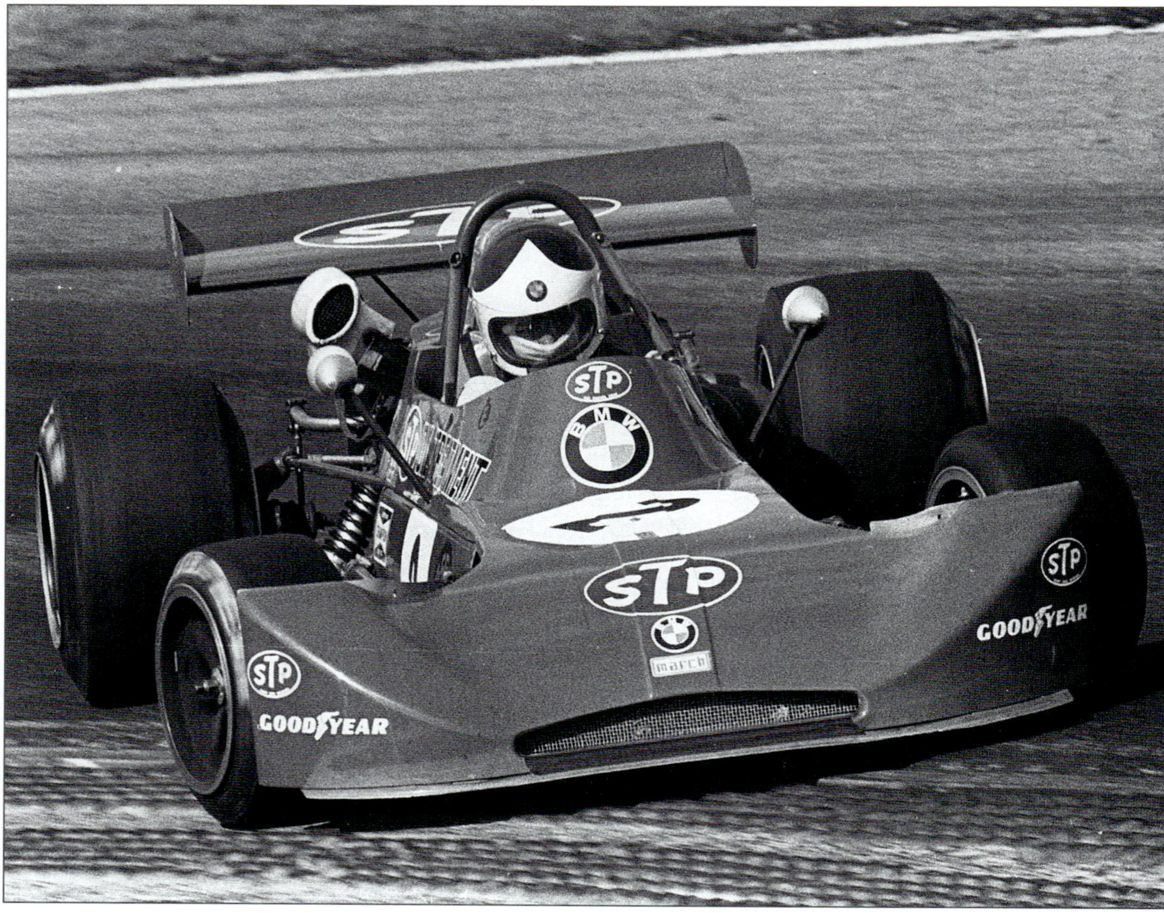

At the end of 1970 BMW withdrew from racing, but the interlude was brief and as far as single-seaters were concerned the company returned as a partner – a policy it has followed successfully since 1972. Jochen Neerspach and Max Mosley formed an alliance that was to make March and BMW almost synonymous in Formula 2. In 1972 Jean-Pierre Jarier dominated the European Formula championship with the March 732 powered by BMW M12/6 2-litre engines giving around 275 bhp. Jarier's 732 in an extravagant cornering pose carries prominent BMW badges. Incidentally, Quester was the first driver to win with an F2 March-BMW, in his independent 712 at Monza in 1971.

In 1973 BMW Motorsport GmbH built full-race lightweight CSLs under the direction of Martin Braungart, with 3.3-litre Kugelfischer-injected 360 bhp engines and by mid-season with full aerodynamic equipment – nose dam and bonnet strakes, and an aerofoil to smooth the airflow to that tail wing. The works cars were run in distinctive colours – white, with blue, purple and red stripes. The racing department and these cars were developed quickly, as the team smarted from defeats by Fords. Backed up by placings scored by the Alpina and Schnitzer teams, BMW won the then-important European Touring Car Championship in 1973. *Opposite top*: This is a team car at the Paul Ricard circuit in France. That year, Quester and Hezemans finished eleventh overall in the Le Mans 24-hour Race and won their class (*opposite bottom*).

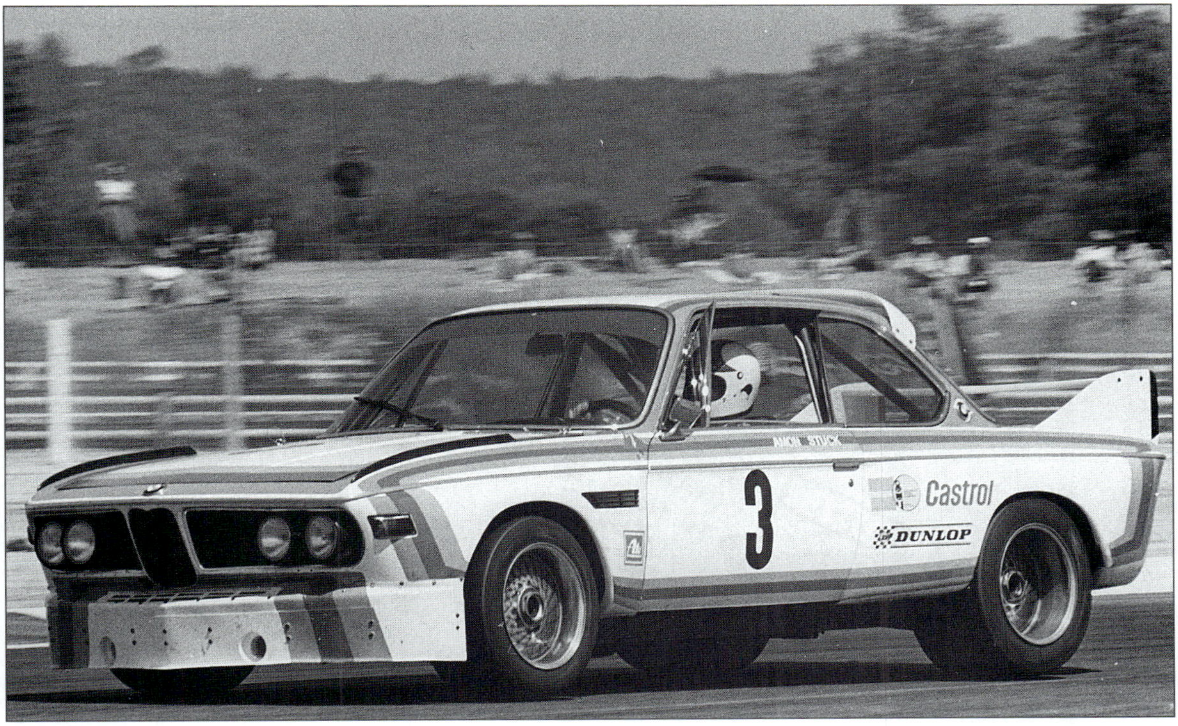
AMON STUCK
3
Castrol
DUNLOP

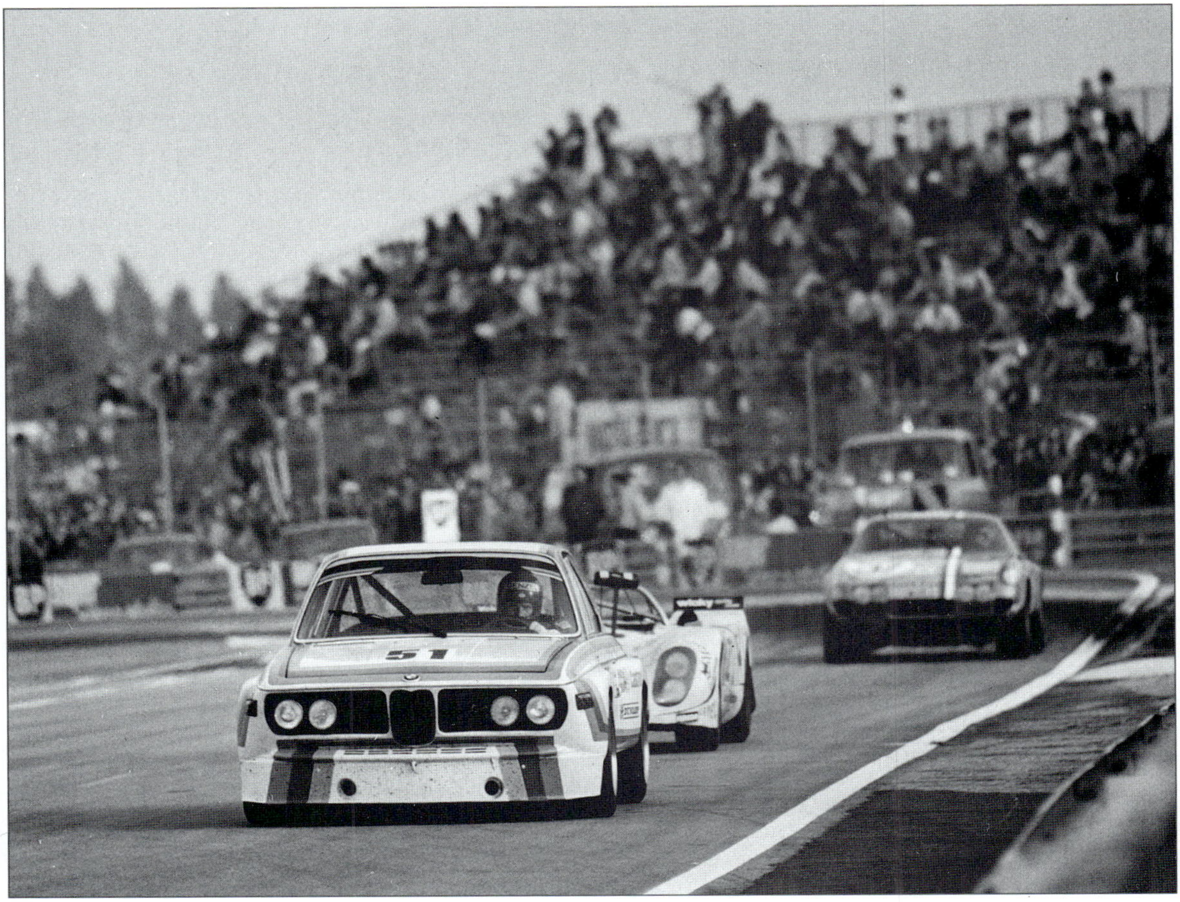
51

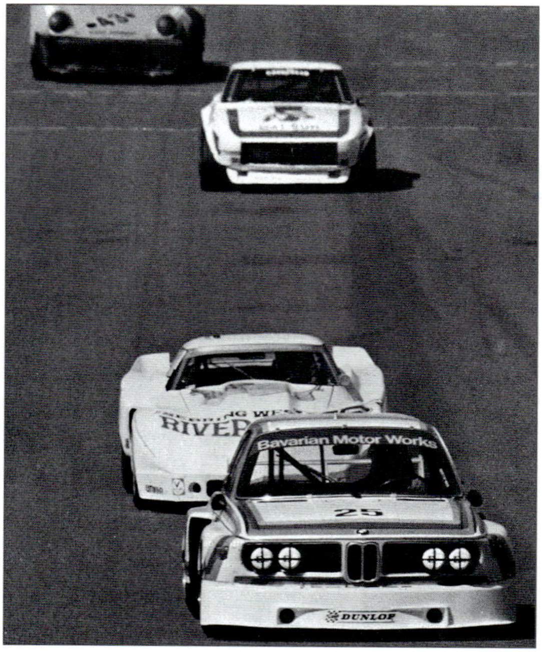

In 1975 BMW concentrated on American racing, to promote the marque as its US company became established, and as sales were sluggish. The IMSA series admitted substantially modified cars, and CSLs were run with engines giving up to 480 bhp. Hans Stuck won two of the races, and here is hounded by a Corvette Stingray with some 800 bhp, but much less agility in corners. 'Bavarian Motor Works' on the screen was presumably intended to educate the American public!

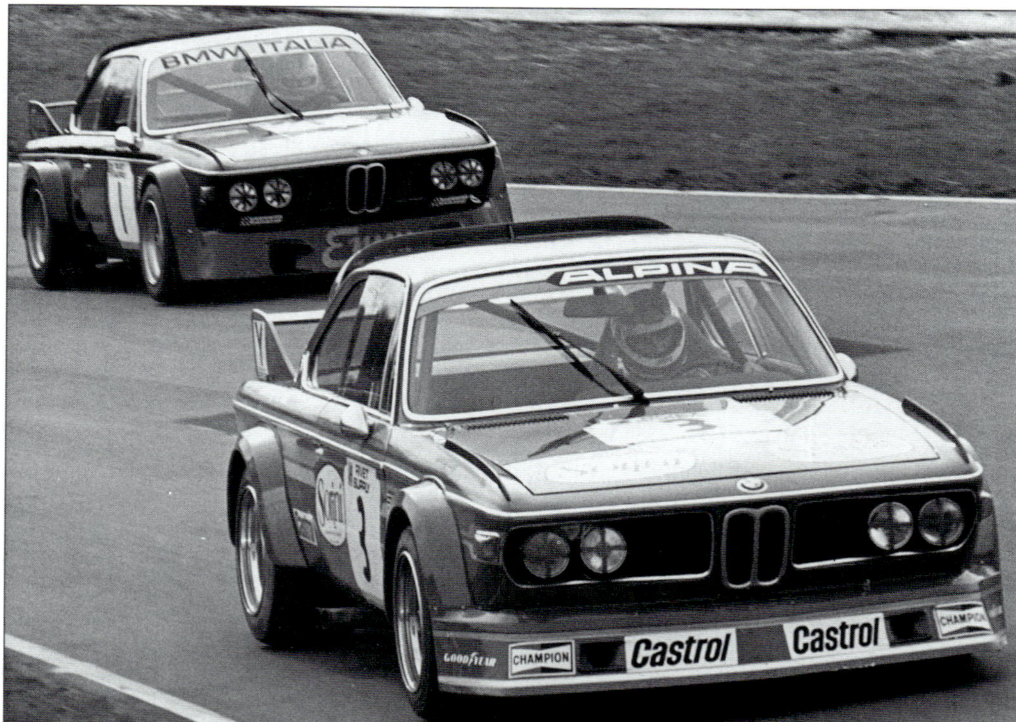

The works did not contest the European series in the second half of the 1970s, although a 24-valve 3.3-litre M49 engine giving over 400 bhp was ready in 1974. There was still development, notably by Alpina, and independents continued to race the big coupés, keeping them in the front line until the end of the decade. Two of the leading 1978 cars at Brands Hatch show aerodynamic refinements and extended wheel arches. Leading is an ex-Alpina 1977 car, followed by the Luigi car that won, driven by Umberto Grano and Tom Walkinshaw. Italian efforts with a Luigi CSL gained a sixth European title for these BMW coupés in 1979, although by that time the series had lost some of its importance.

Chapter Four

3 Series

The second of the principal model lines launched in the 1970s, the 3 Series arrived in 1975. These cars were the smallest in the BMW range, and numerically far and away the most important, with production running to millions and many permutations of bodies and engines – for 1997, BMW listed more than 50 variations, and that rules out a comprehensive survey of the types sold through a quarter of a century. . . .

As production started in May 1975 the E21 line was simple, with two-door saloons that were direct successors to the 02s. The '*Neue Klasse*' pattern was broadly followed, with inclined single overhead camshaft four-cylinder engines driving the rear wheels, MacPherson strut front suspension and trailing arm rear suspension. However, worm-and-roller steering was abandoned in favour of a rack-and-pinion system and, importantly, the design allowed for a six-cylinder engine. First-generation 3 Series cars had four-speed gearboxes. The cars were a little longer and wider than their predecessors, and heavier, which to a degree told against a sporting performance, until the M60 oversquare straight six was introduced. (Then the four-cylinder unit's reputation for being easy to maintain and cheaper to service was enhanced.) Body styling was crisp, with a hint of a wedge shape, but hardly avant-garde. Inevitably, in the mid-1970s cars destined for the USA carried ugly '5 mph bumpers'. A four-door type did not come until the first major 3 Series revision, for 1982, but Baur cabriolets were available earlier.

The six-cylinder engine was in production in 1977, and used in the 1978 model year 320–6 and 323i. It was based on the then-current four-cylinder unit, with belt-driven overhead camshaft and a single carburettor or Bosch fuel injection. Importantly, it met US consumption requirements. The fuel-injected unit was rated at 143 bhp, and the performance of the 323i did not quite match that of the 2002ti.

Some 1.36 million 3 Series cars had been sold when the second-generation model (E30) was launched late in 1982. The body was a little smoother and gave small aerodynamic gains, the suspension was revised and ABS braking was offered. (All models except the all-disc 323i had disc/drum combinations.) Interiors were still sombre, but the instruments now included a service interval indicator (SII). The engines were a 1.8-litre four and 2- and 2.3-litre straight sixes, with a 2.7-litre unit in the 325e (e for eta, thermodynamic efficiency) for the USA, and a diesel announced in 1983. The South African subsidiary created a 333i in 1985, using the 3.2-litre six in an E30 body, while another 'special' for a local market was the 2-litre 320is, falling into an important Italian taxation class.

The main-line cabriolet was factory-built, and so was a five-door Touring model, also announced in 1987, for 1988. Meanwhile, the four-wheel drive 324iX had come in 1985, but only in left-hand-drive form. BMW had experimented with four-wheel drive in the 02 days, but turned away from it until a mid-1980s need to match Audi and Ford arose; 325iX production passed 30,000, then four-wheel drive was set aside in 1991.

In 1987 any doubts about the sporting potential of the 3 Series were laid to rest as the M3 was introduced. Highly developed competition versions were to be immensely successful in racing.

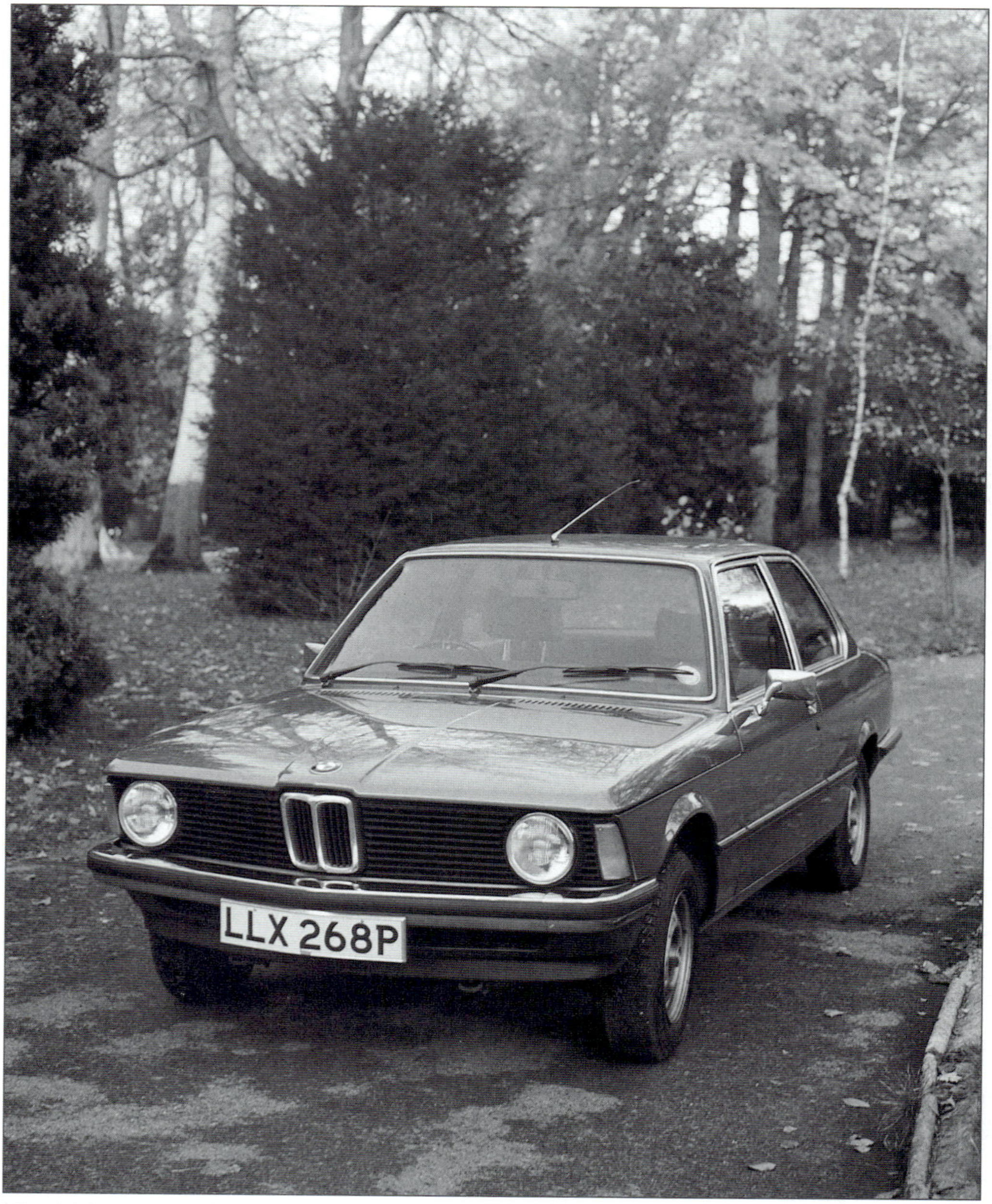

The 316 came as the base model of the new 3 Series range (E21) in the summer of 1975, outwardly with simple three-box lines and large glass areas, obviously related to the 5 Series. Interiors were equally simple, and sensible. The 1.6-litre sohc engine was rated at 90 bhp, and the top speed was just over 100 mph (160 kmh).

Total production of the E30 3 Series cars was more than 2.22 million. The first third-generation cars (E36) came in 1991, initially in four-door saloon form. The lines were sleeker, and the body a little larger (inside and out), while attention was paid to aerodynamic details, for example in the faired headlights. The 'Z axle' rear suspension with multi-link central control that had first been used on the Z1 sports car made for improved handling.

The entry-level saloons tended to look faster than they were, but the range was to be extended to include outstanding sports saloons, and exemplary diesels. A four-seat coupé came after a year, with completely new body panels; until it received the more flexible M44 1.9-litre engine in 1996, the 318i coupé was not quite competitive in performance terms, and it was costly. Naturally, the convertible was also a two-door car, stylish and with a totally effective and simple electrically operated soft top. Stowage meant that this cut into rear seat leg room and luggage space. A weight penalty was inevitable, but the reinforcement did mean that this car was almost free of the scuttle shake that can still dog cars in this class. A Touring version came next, cleanly restyled and again rated a 'driver's car'.

The three-door hatchback Compact gave BMW a new entry model that could be competitively priced against cars such as the Ford Escort and VW Golf. It was shorter than other 3 Series types, and less roomy, and the trailing arm rear suspension of the earlier 3 Series was used. The Compact came in 316i and 318i forms, with a 318ti late in 1994, while the other models in the range could have the four-cylinder engine or the straight six in 2- and 2.3-litre forms (3 litres in M3 variants).

A new 1.7-litre four-cylinder diesel engine came with the 318tds for the 1995 model year, complementing the 2.5-litre six-cylinder diesels in the 325td and 325tds. It was to be followed by the new 1.9-litre DI engine in the saloons for the 1999 model year. (In the increasingly complex range, Touring models retained the earlier diesels, while the four-cylinder type became available in the Compact.)

A version of the Compact with the six-cylinder engine (323ti) was a 143 mph car; there was a Sport Compact with the M sports suspension, and a variant with a landaulet fabric soft top that opened over the entire length of the roof while the rigid sides were retained.

By no means finally in the ever-growing 3 Series, the 316g was listed from the end of 1995. An 80-litre natural gas tank took the place of the rear seats (but luggage space was increased by some 50 per cent). With a range of 125 to 150 miles (200–250 km), the 316g was essentially a city car, and the response from the German car-buying public was not enthusiastic.

Evolution continued. The E46 'new saloons' were launched in 1998, with reworked engines. BMW liked to trace the lineage back to the mid-1960s, but as far as the defined 3 Series was concerned these were fourth-generation cars. At first there were four types, with 1.9-, 2.5- or 2.8-litre engines; traction control and cornering brake control were standard, and cabin equipment was uprated, with subtle improvements – and six air bags. Cornering brake control adjusted the pressure to individual front wheel brakes if understeer while braking was detected. The size of the saloons was increased again, and so was weight (justified as the body was stronger, and hence safer).

These were still patently 3 Series models, with the qualities that the BMW faithful expected – above-average balance and handling, excellent steering and stability, responsive and smooth. But they no longer stood alone as their 1970s predecessors had, and faced serious competition from cars such as the Alfa Romeo 156, Audi A4, Lexus IS200 and Mercedes-Benz C180. The sophisticated 328Ci in 1999 had to face even tougher competition, and this coupé seemed further removed from the 3 Series cars of the mid-1970s.

By 1998 BMW was building 3 Series cars in three factories, and they continued to account for the bulk of its sales. That year a little ground appeared to be lost, in home and overseas markets, but it was a period of model changes. Increased Compact and Touring sales gave better indications of the lasting popularity of BMW's core line.

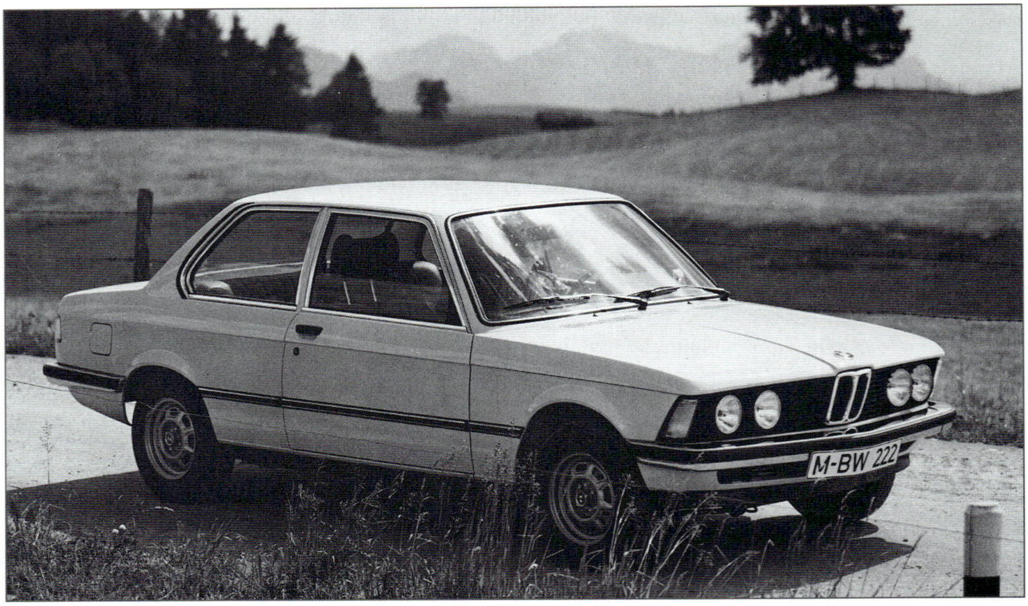

Four headlights outwardly distinguished the 320, which fell between the 316/318 and 320i in the first E21 quartet. All-round performance more than justified the extra 417 cc in its engine, not so much in top speed (106 mph/171 kmh) as in acceleration and fuel consumption. The 109 bhp engine was smooth, but noisier than the smaller units. In 1977 this 320 was discontinued, in favour of a six-cylinder 320.

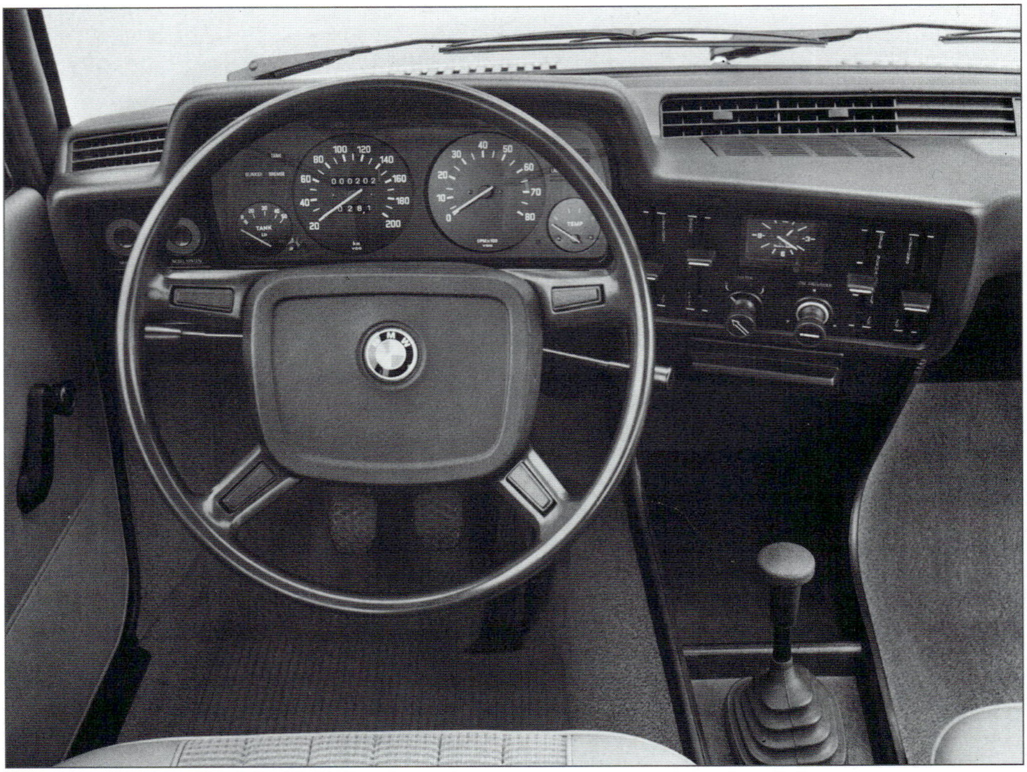

First-generation 3 Series cars had severely functional fascias and controls layouts. There were small fuel and temperature gauges, a large and clear speedometer, and a matching rev counter on all save the base models (where a large clock took the tacho's place). The centre section was angled towards the driver, with vertical slide controls for heating and ventilation.

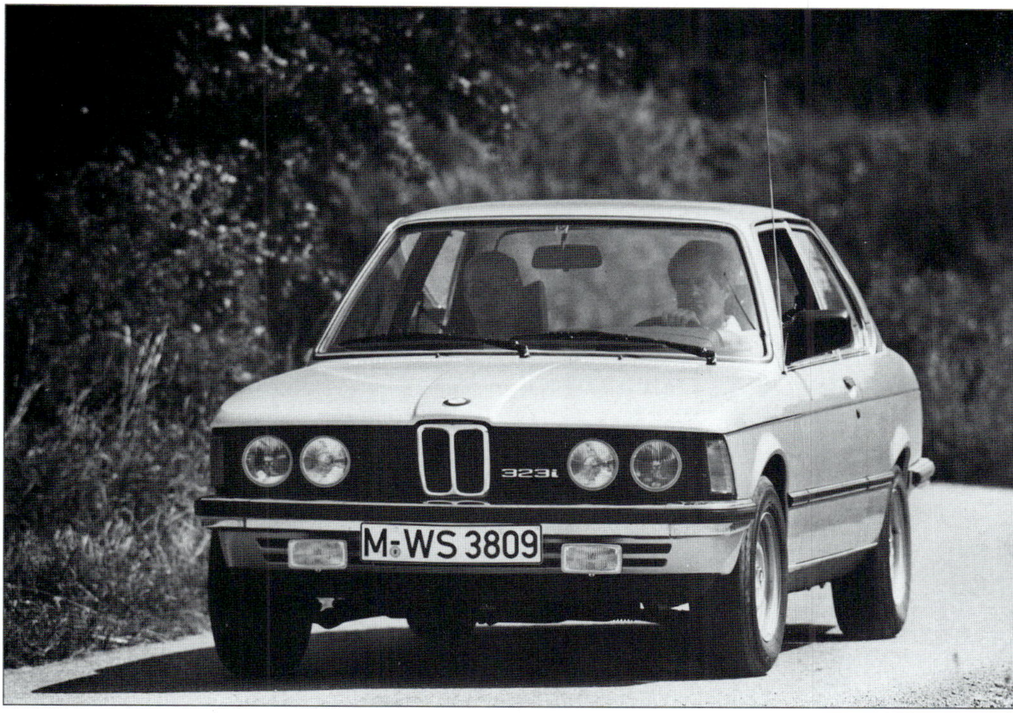

The 323i was the top model among the first six-cylinder 3 Series cars, in 1977. Its 2.3-litre Bosch fuel-injected 143 bhp version of the M60 engine was a straightforward and willing unit. A four-speed manual gearbox was normal, with the option of a ZF automatic transmission. This is a 1980 car, which had a 118 mph (190 kmh) maximum speed.

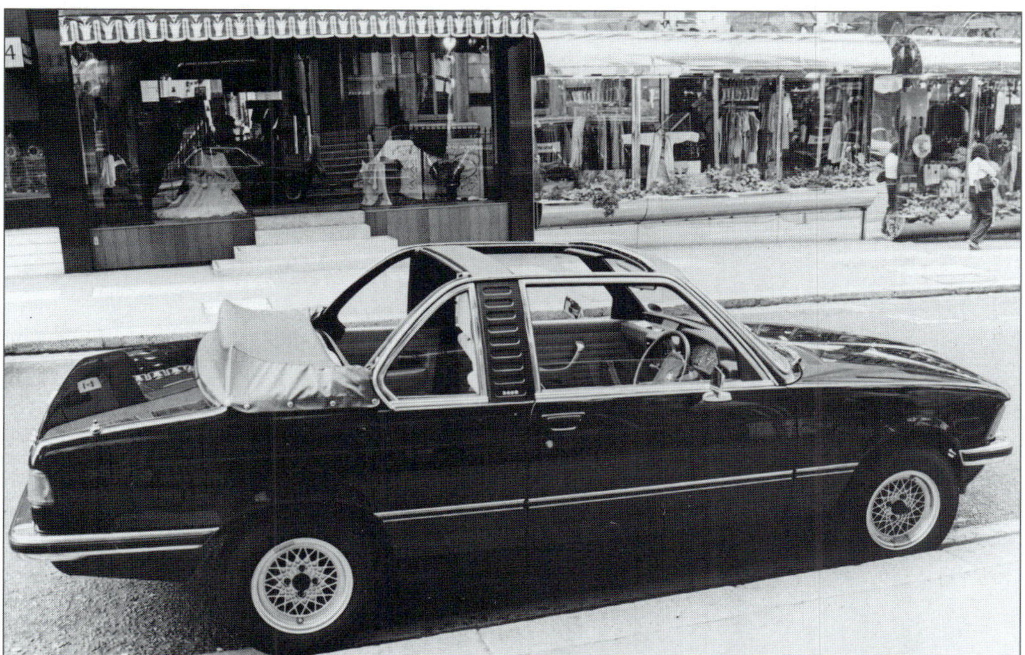

Stuttgart coachbuilder Baur had a long association with BMW, and built this neat 3 Series cabriolet conversion from the late 1970s. (This rare right-hand-drive car dates from 1979.) There was a substantial roll bar, and as the door frames were retained little rigidity was sacrificed. A removable centre section matched the rear folding section.

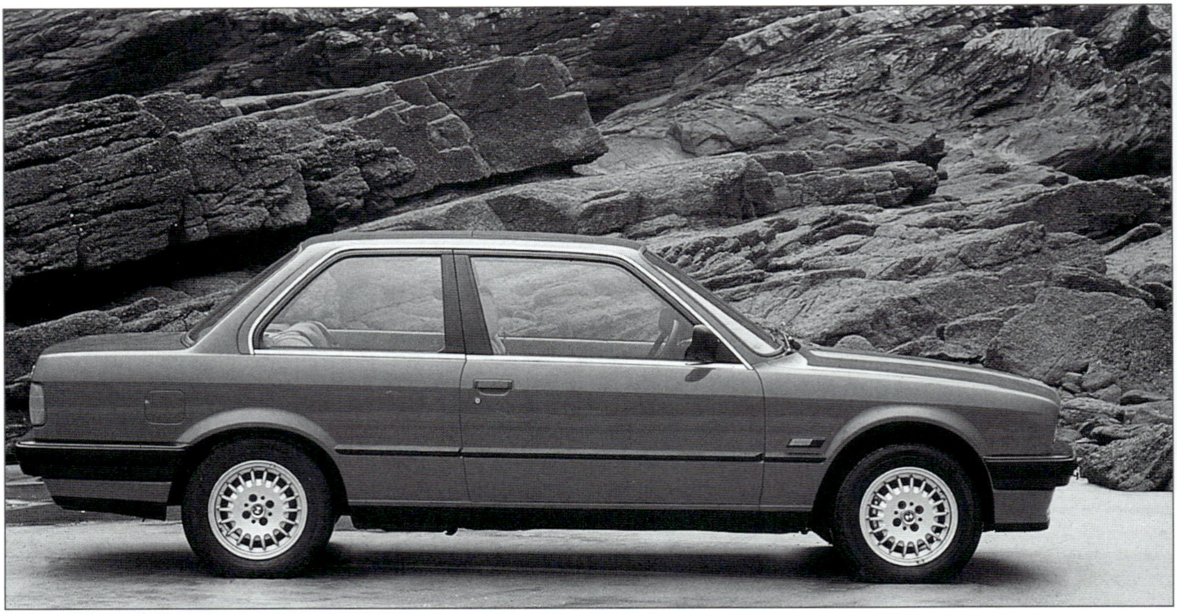

The E30 cars were a little sleeker than the E21s, and had a stronger image. The early 316 had a 1766 cc engine, which seemed to contradict the designation. (This came into line when the 1596 cc unit was introduced in 1988.) The equipment level was not generous, but as ever there was compensation in the build quality. The performance of the first four-cylinder cars was not outstanding, but with the M40 engine a 1988 316i was capable of 112 mph (180 kmh).

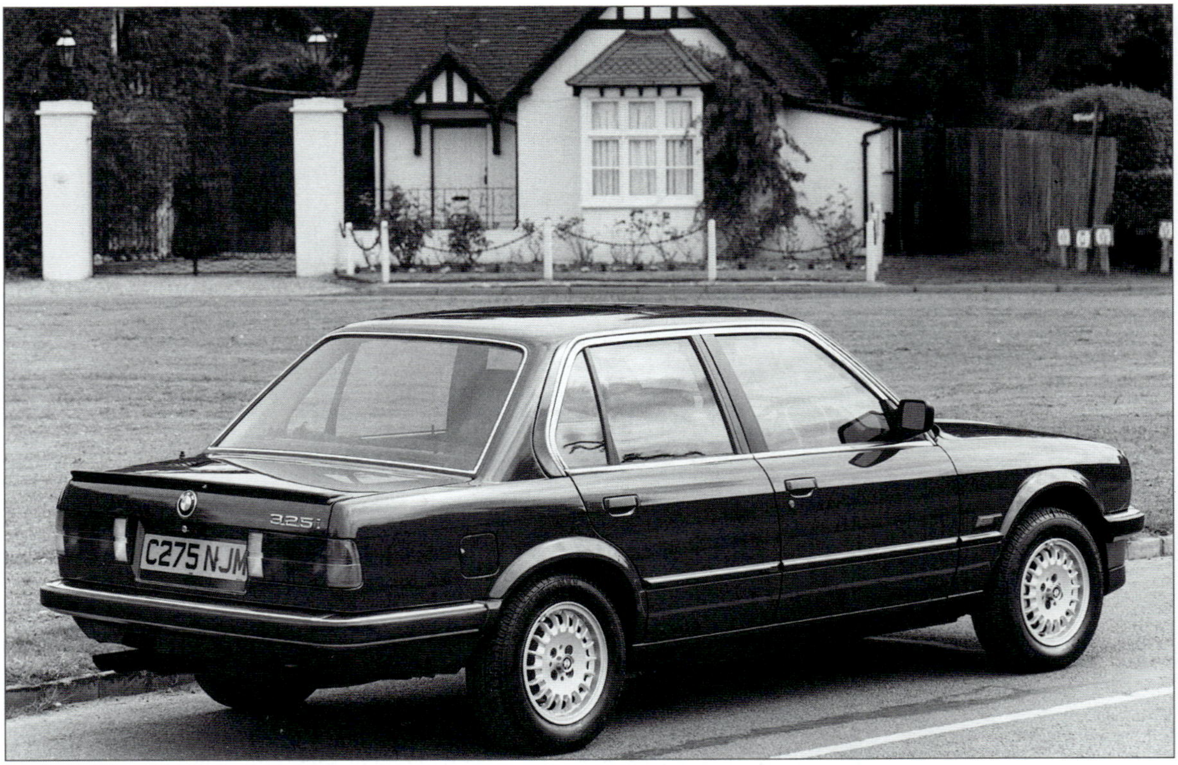

Four-door bodies were available in the E30 range. The top model was the 325i, which replaced the 323i in 1987, and could be identified by its standard fog lights and small boot-top spoiler. The drag factor (0.38) did not compare well with rival German cars, but the 2492 cc 171 bhp straight six gave it a 130 mph (208 kmh) top speed and good acceleration.

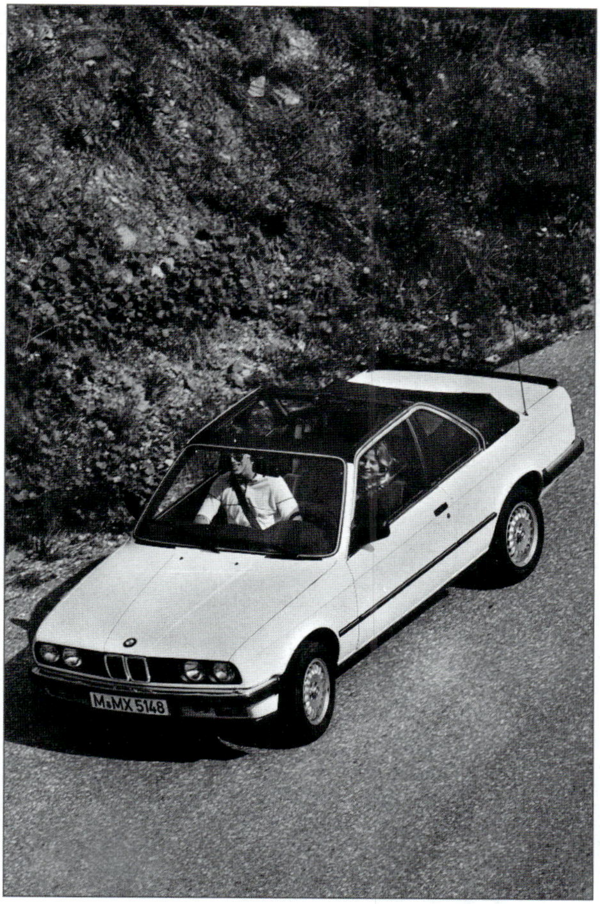

Baur continued its cabriolet bodies on E30 3 Series cars, retaining the existing door frames in a sub-type known in Germany as a '*topcabriolet*'. Effectively, it was a 2+2. This car also shows the air dam incorporating fog lights.

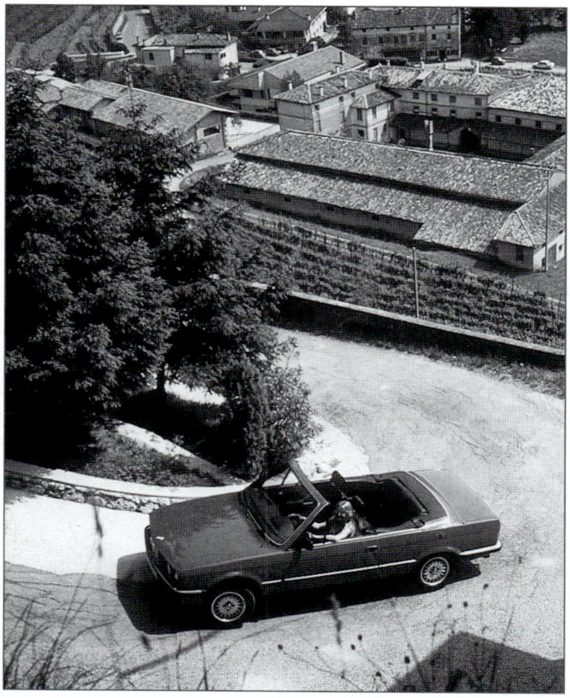

BMW produced its own sophisticated (and expensive) convertible in 1986. Despite reinforcement to maintain rigidity there was some body shake on rough surfaces – Baur scored in that respect. The hood stowage in a recess under a cover was very neat.

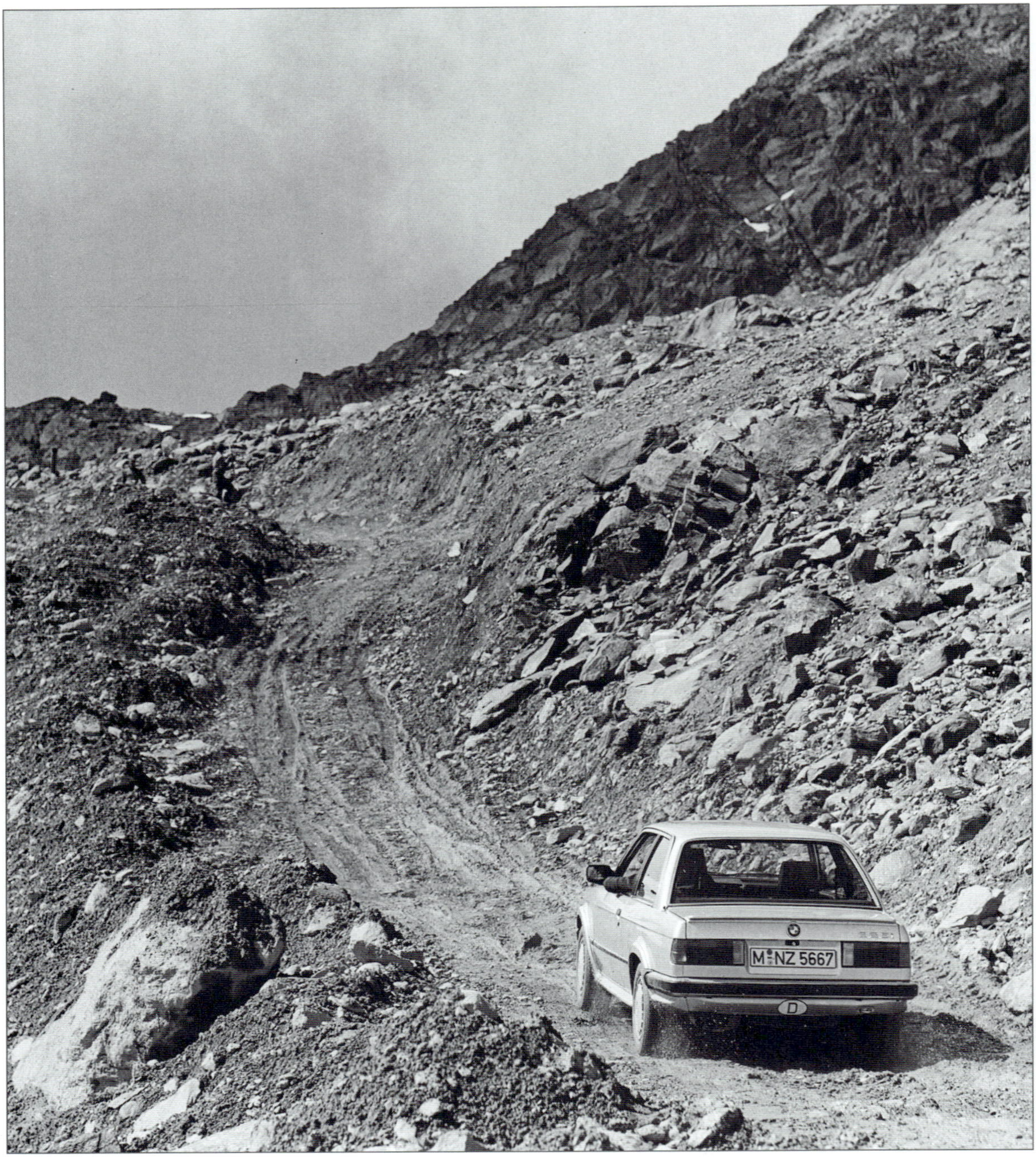

The four-wheel-drive 325iX was launched in 1985, and built in left-hand-drive form. Unlike some of its contemporaries, it owed nothing to rally requirements. The transmission split was 37/63 front/rear and viscous differentials were fitted, as was ABS. Some suspension modifications were called for, and these entailed minor unseen body changes.

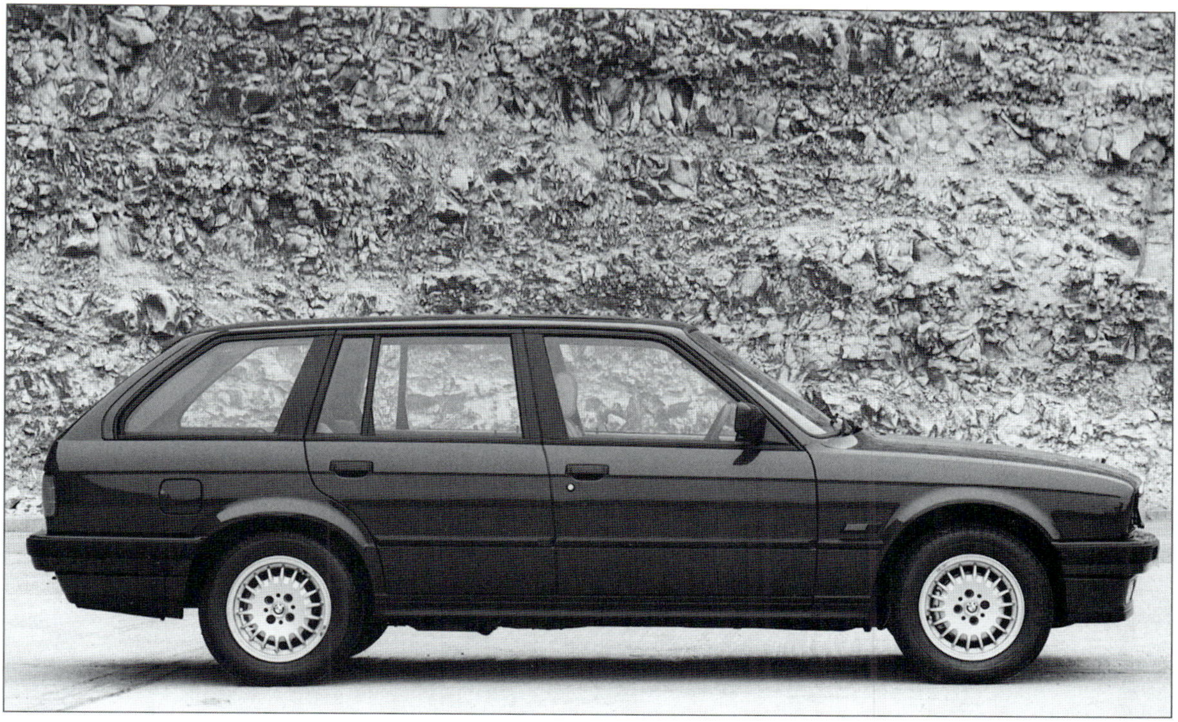

'Touring' was revived for 3 Series models on estate car lines: the company never used the term, preferring these models to be accepted as superior touring cars rather than load-luggers. The 325i Touring shown here illustrates the very clean lines. Luggage space was not generous or easily loaded with bulky items, although the floor-level space between light clusters helped to a degree, and 50/50 split rear seats were normal. Performance was on a par with that of other 325i models. Production reached 104,000, 1988–94.

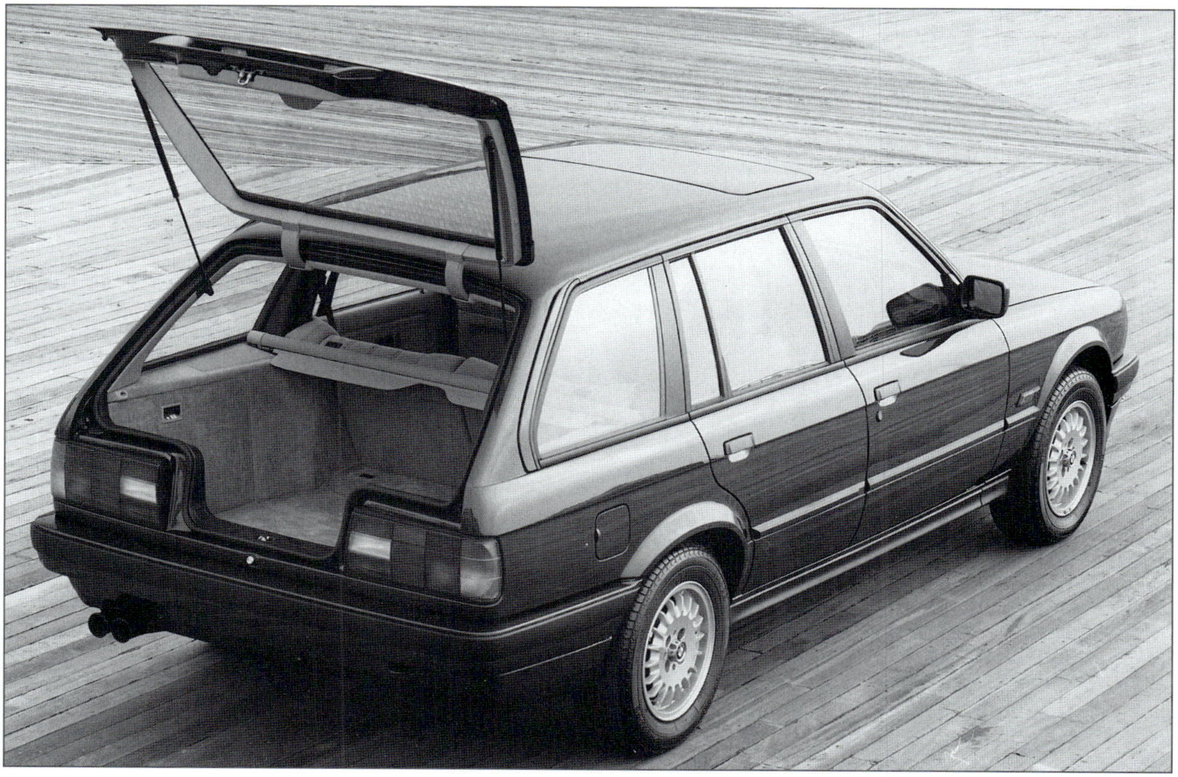

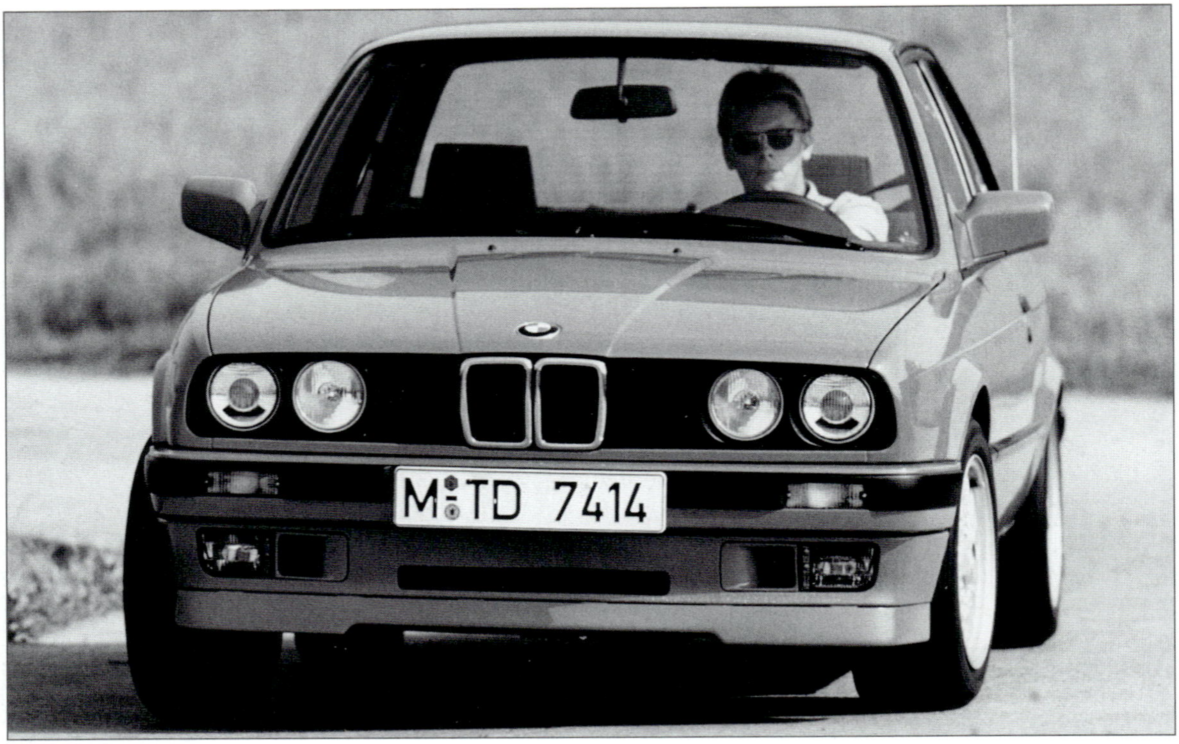

The 318iS was introduced in 1989 as a sporty model to compete with hot hatch types, with a new four-valve 1795 cc 136 bhp derivative of the M4 engine, M-Technic suspension, and disc brakes all round. It could be picked out by its lower ride height. Top speed was of the order of 125 mph (200 kmh). This proved to be something of an interim model, listed until spring 1991. Baur built a few cabriolet conversions.

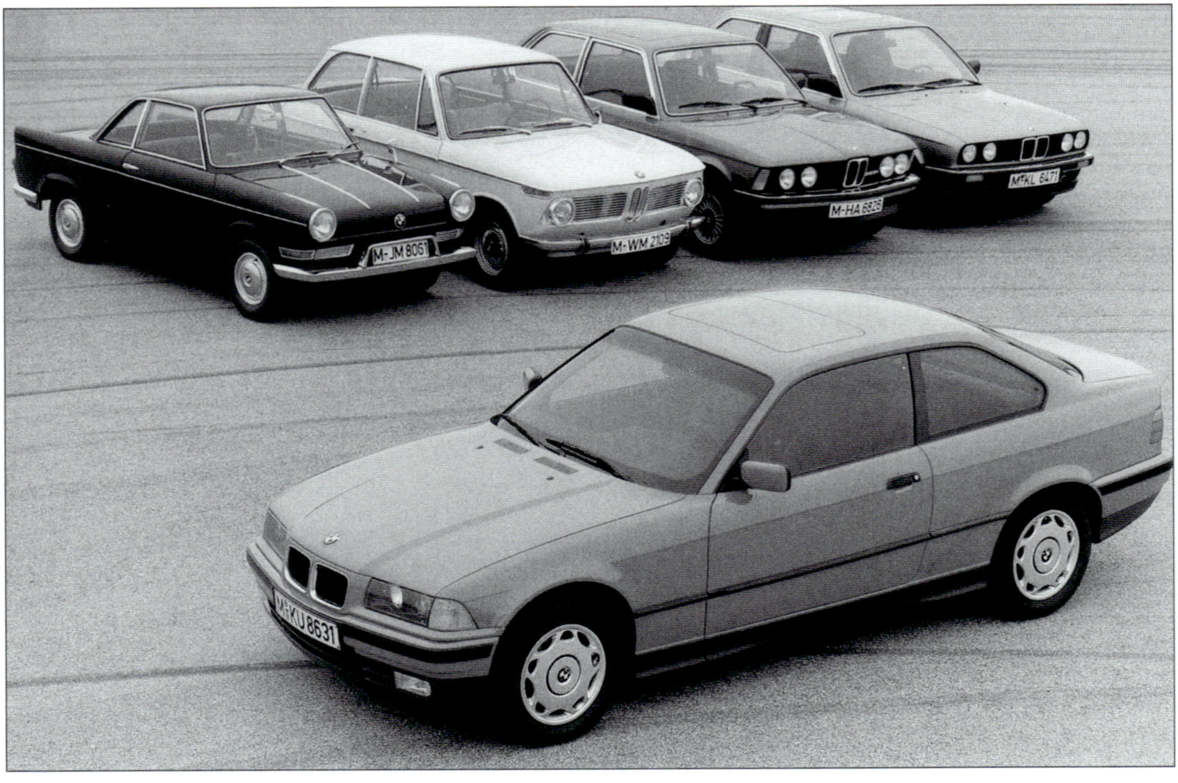

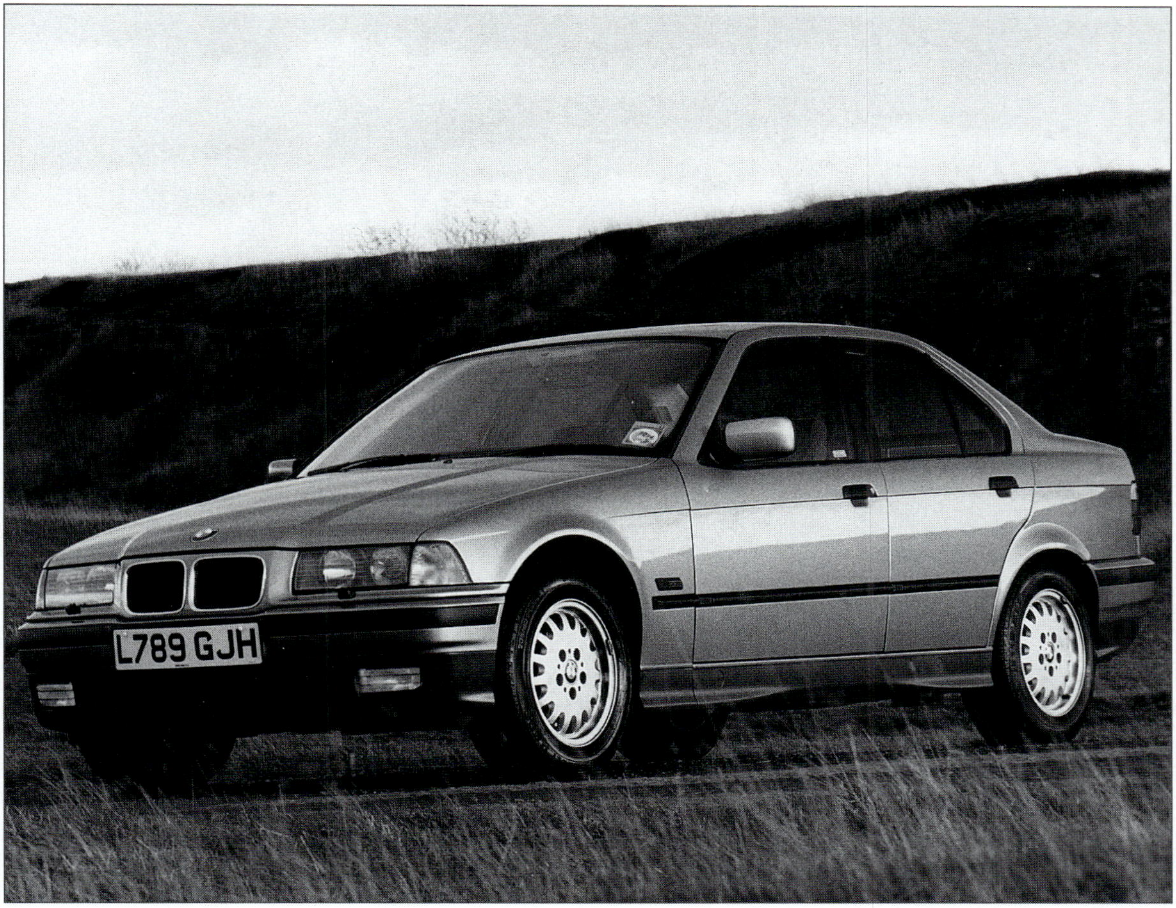

This mid-series E36 saloon shows how BMW family resemblances converged in the 1990s. Efforts were made to reduce drag, for example with those enclosed headlights, successfully as the drag coefficient was claimed to be 0.29. The 325i was the top model, comfortable, with multi-link rear suspension incorporating a central longitudinal link which put to rest a reputation for lift-off oversteer, and quick, with the 2.5-litre M50 six-cylinder engine giving a 140 mph (225 kmh) top speed. The 316i entry model was a 120 mph (193 kmh) car, and had outward features such as quad headlights in common with higher-specification types. (As with some other BMWs, it was not uncommon for chrome numbers on a tail to be changed, and neighbours presumably impressed.)

Opposite: An important newcomer to the E36 3 Series range was the coupé, here posed with forerunners from a 700 coupé to its immediate E30 predecessor. This is a 1994 car; 319iS, 320i and 325i were introduced for that year. The rather ordinary wheels suggest that it is a 318iS, which had the 1.8-litre four-cylinder engine in 140 bhp form (that set it apart from the 318i saloon). Critics found it a little subdued, but BMW claimed a 132 mph (211 kmh) top speed, which seems good enough for a 1.8-litre car. . . . Some 470,000 of these coupés were to be sold before a successor came along in 1999.

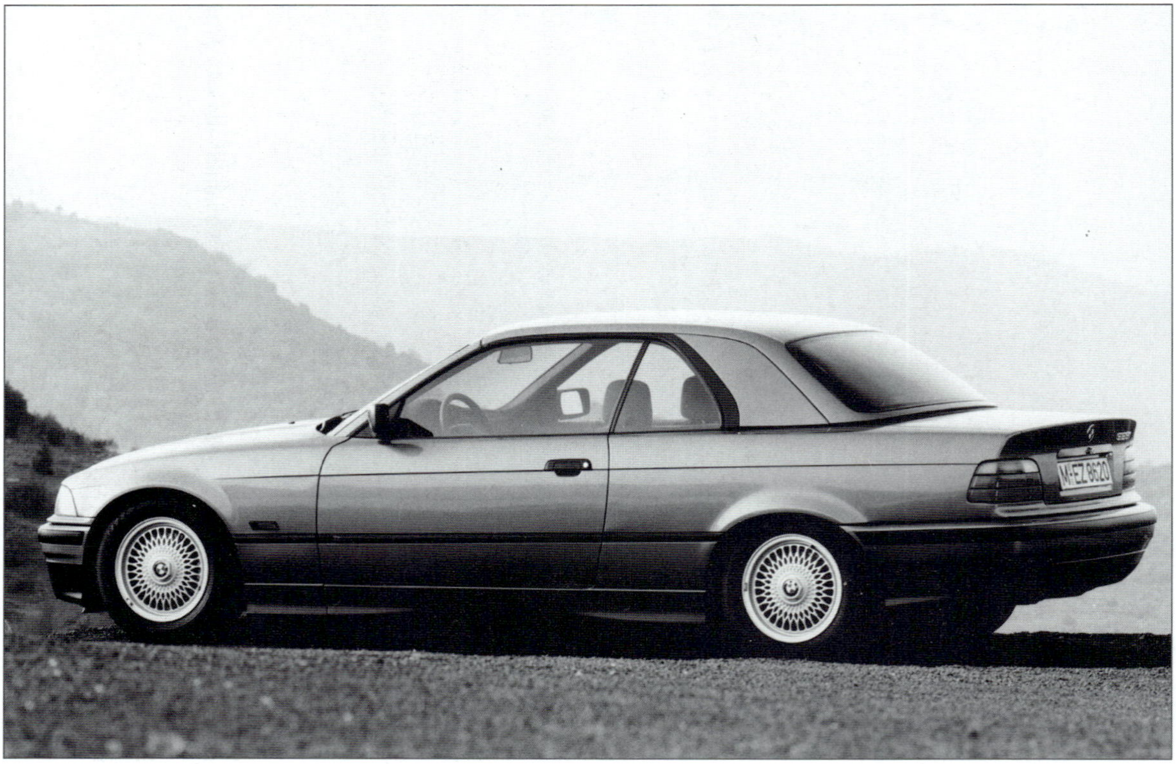

There was naturally a convertible in the new range, and it was most handsome. The optional aluminium hard top fitted to this car gave it attractive coupé lines. Unlike earlier cabriolets, which were essentially chopped saloons, this 1993 version was planned as an open car from the start. Stiffening inevitably meant increased weight, of the order of 270 lb (123 kg), which marginally cut the top speed.

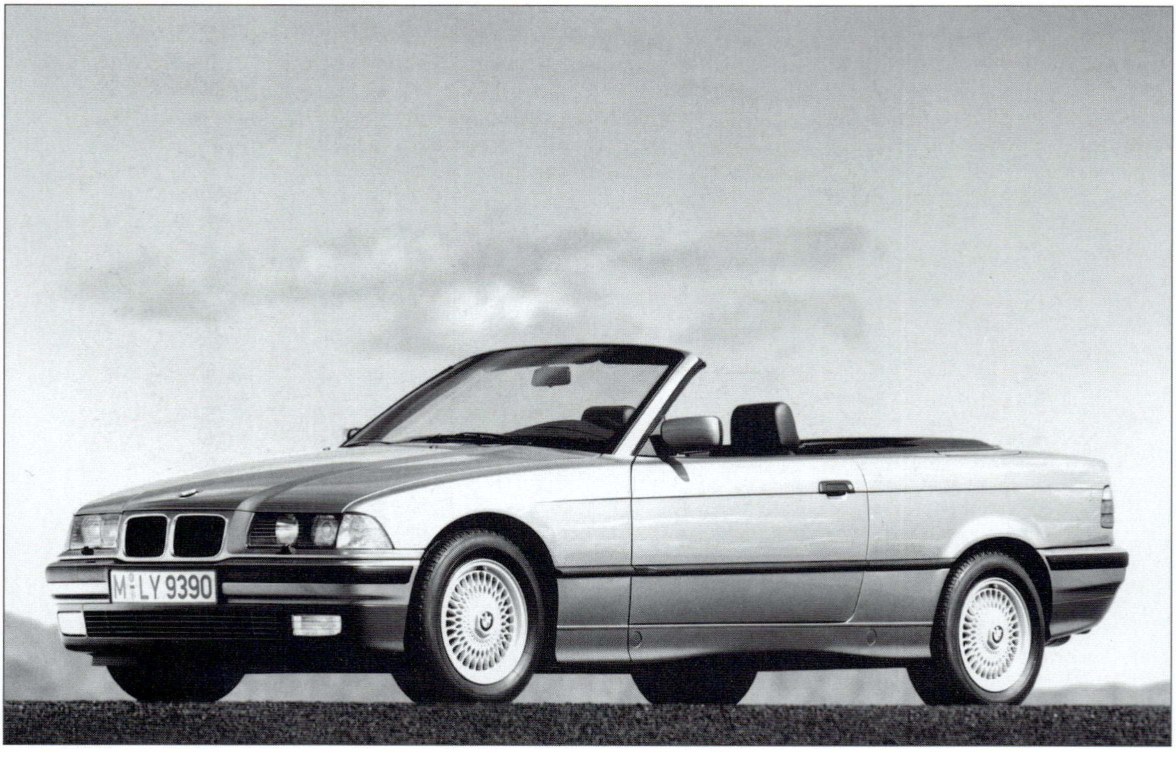

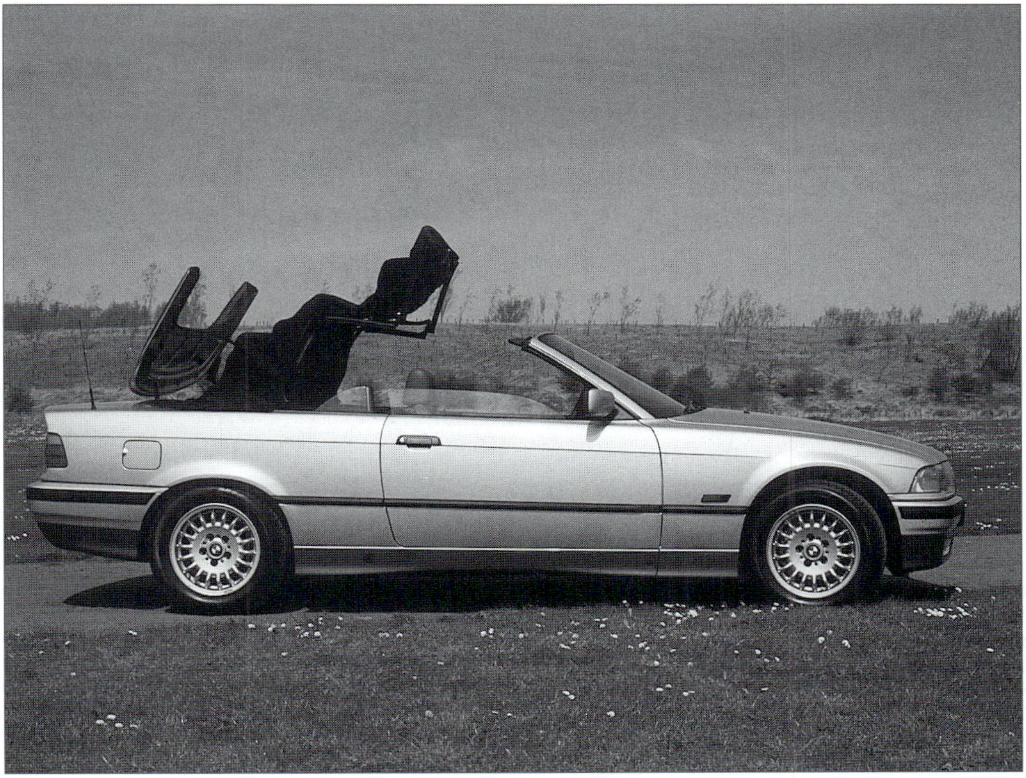

The electrically operated soft top caught in mid-sequence, with the lid of the stowage compartment ready to close down on it. Operation was simple, and all the frameless windows lowered out of sight at the touch of one button. A roll-over bar behind the rear seats operated instantly if sensors detected an imminent roll. This side shot also emphasizes the clean line through to the tail.

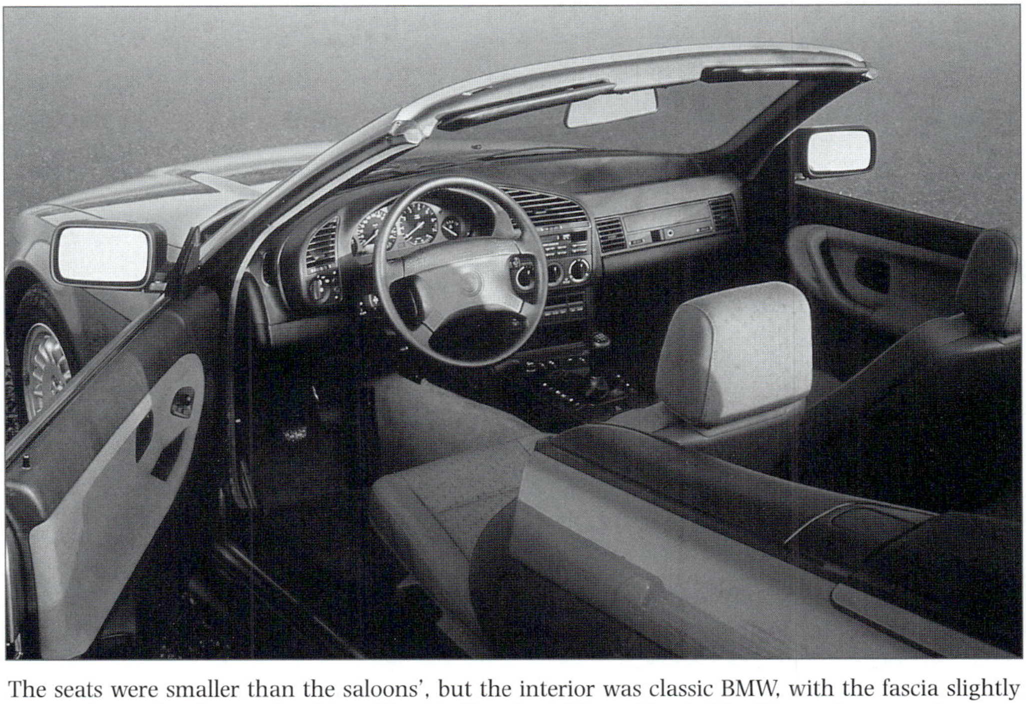

The seats were smaller than the saloons', but the interior was classic BMW, with the fascia slightly wrapped round to favour the driver. Electrically adjustable and heated leather seats were extra.

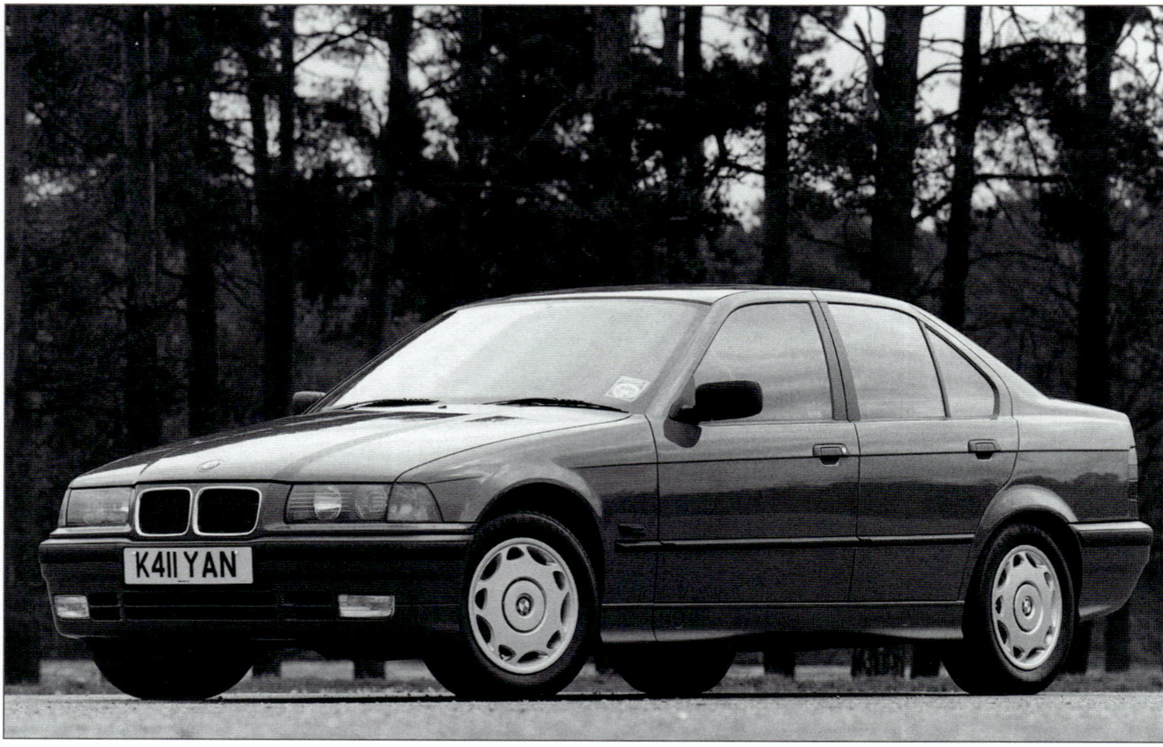

By the mid-1990s BMW's Austrian plant at Steyr had built more than 400,000 six-cylinder diesel engines, which were used in models such as this 1993 325d and the sportier 325 tds. Their respective 2.5-litre engines were rated at 115 bhp and 143 bhp and drove through five-speed gearboxes, to give relaxed performance (with 122 mph/195 kmh and 134 mph/214 kmh top speeds). The 325td was rated highly in 'value for money' terms. In 1995 a four-cylinder 1665 cc diesel was added to the range, in the 318tds.

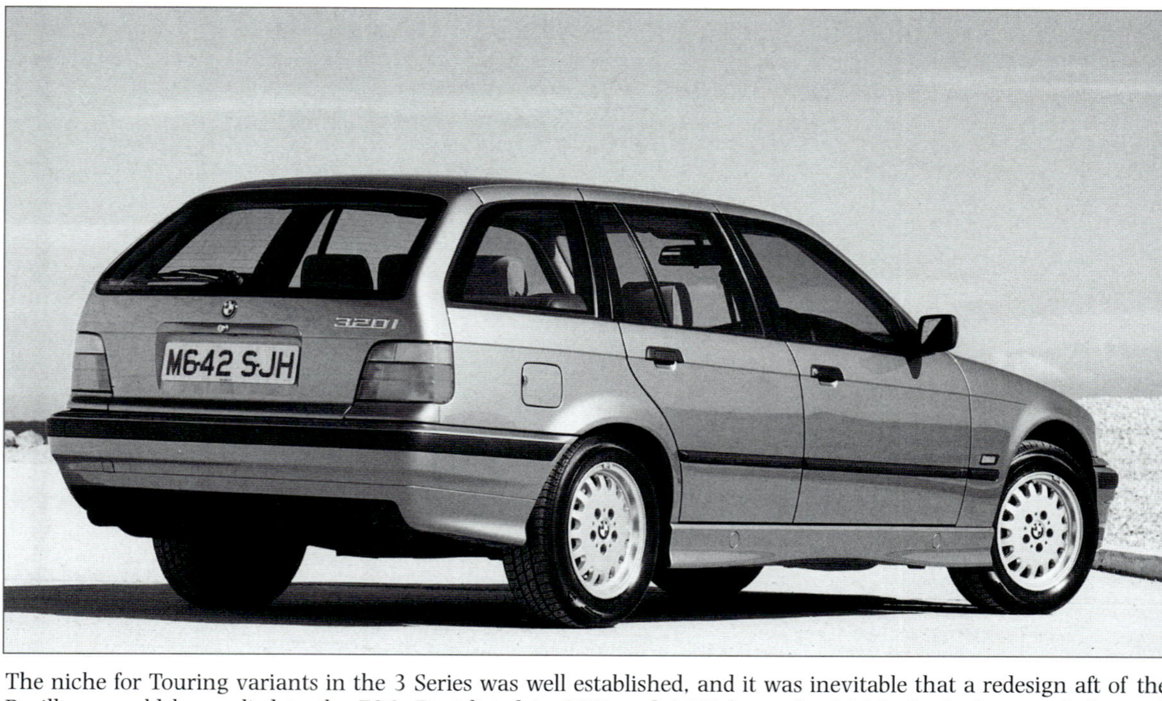

The niche for Touring variants in the 3 Series was well established, and it was inevitable that a redesign aft of the B pillars would be applied to the E36. First listed in 320i and 328i forms for 1995, the body was sleeker and longer than its predecessor, with just a little more load space, and better access to it.

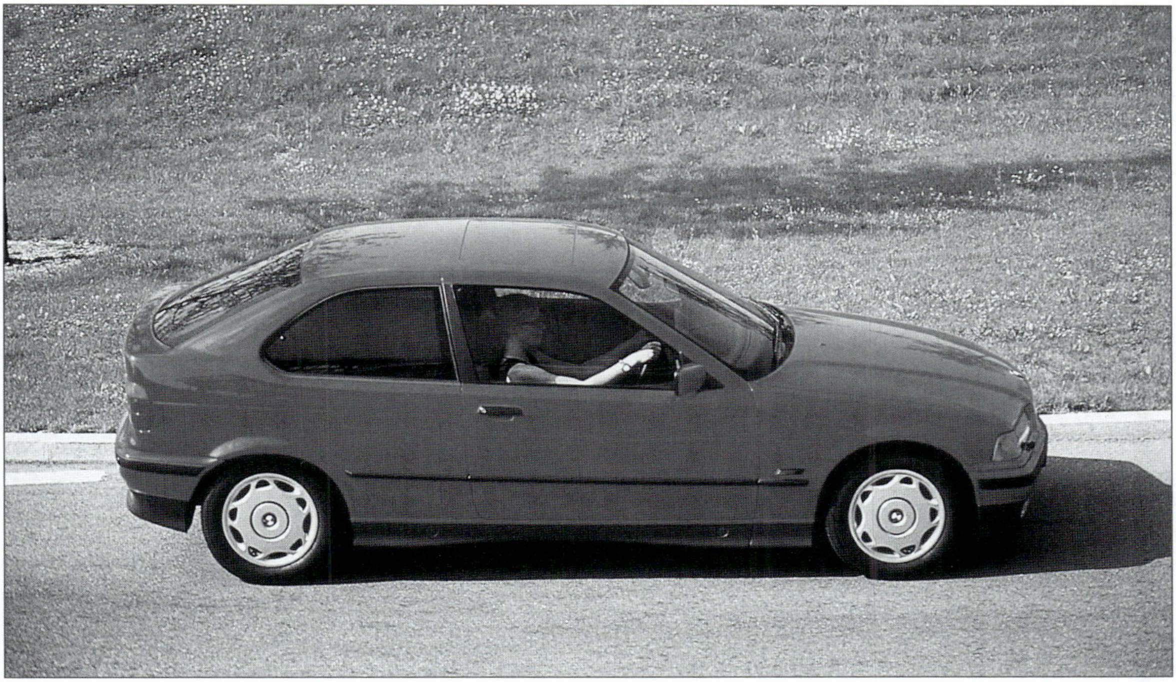

The 3 Series Compact in its first (316i) form could be described as the entry model to the entire BMW range, as the smallest and cheapest saloon. The standard wheelbase was retained, but overall length was reduced by 9 inches, with a very practical three-door body. It was relatively heavy, and performance with the 1.6-litre engine was not exciting; the 318i with 1.8 litres and 140 bhp had a top speed a little shy of 130 mph (208 kmh). The M44 engine came for 1996, in the 318 ti Compact, and by 1998 five Compacts were listed, including the 318tds, the 323ti with a 2.5-litre straight six and in some markets the 316g.

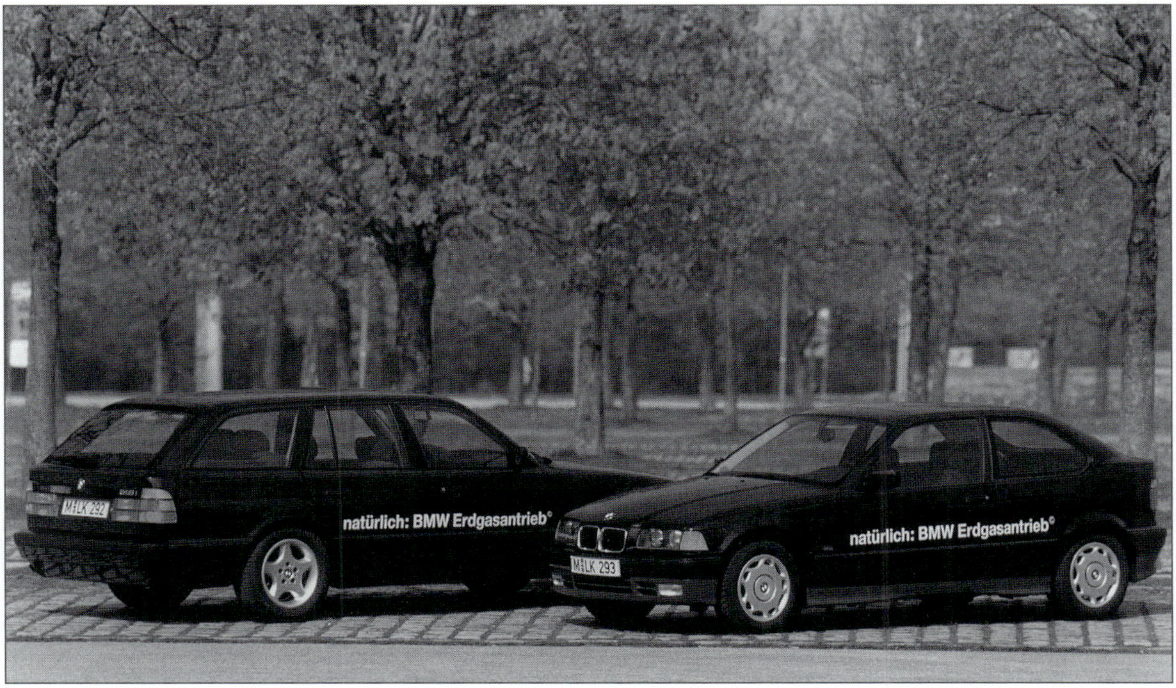

Late in 1995 two 'g' models using natural gas fuel were introduced, initially only in Germany where a refuelling infrastructure had been developed. The 518g Touring and the 316g Compact (right) had a compressed gas tank, and the engines were dual-fuel units, which would also run on petrol.

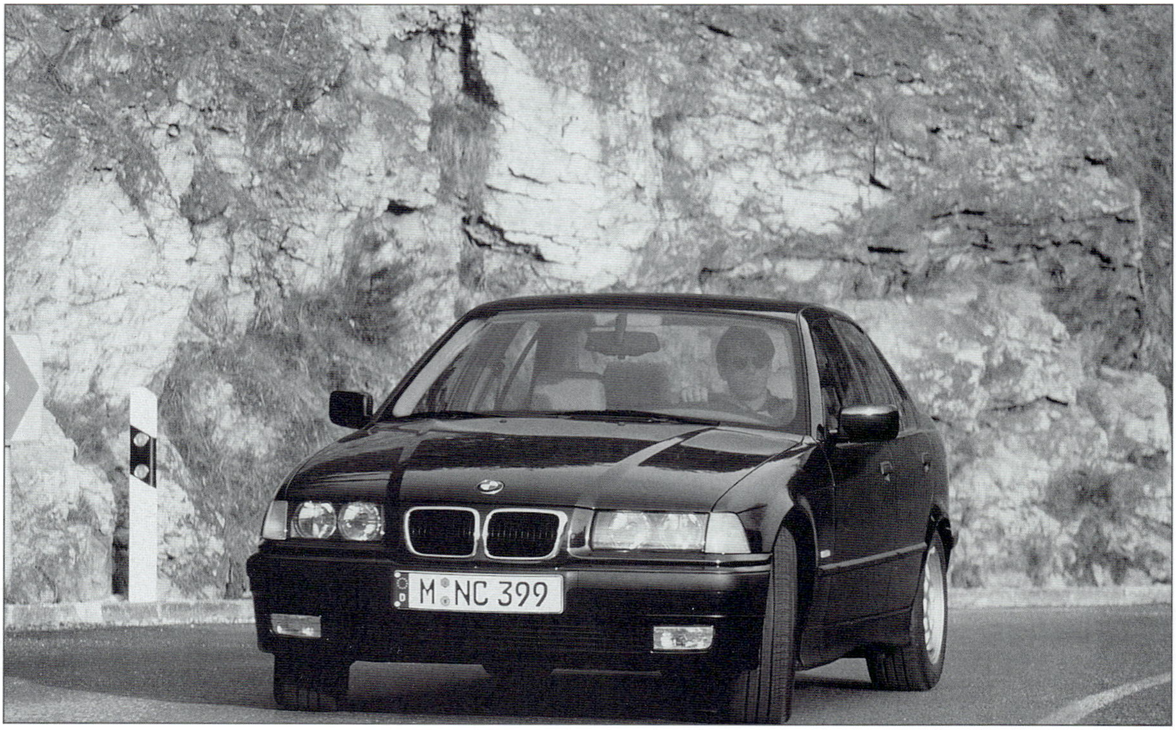

There was a mild facelift for the 1997 model year, when the kidney grille was rounded and the front apron was revised, while there were also worthwhile equipment improvements. The 323i replaced the 320i.

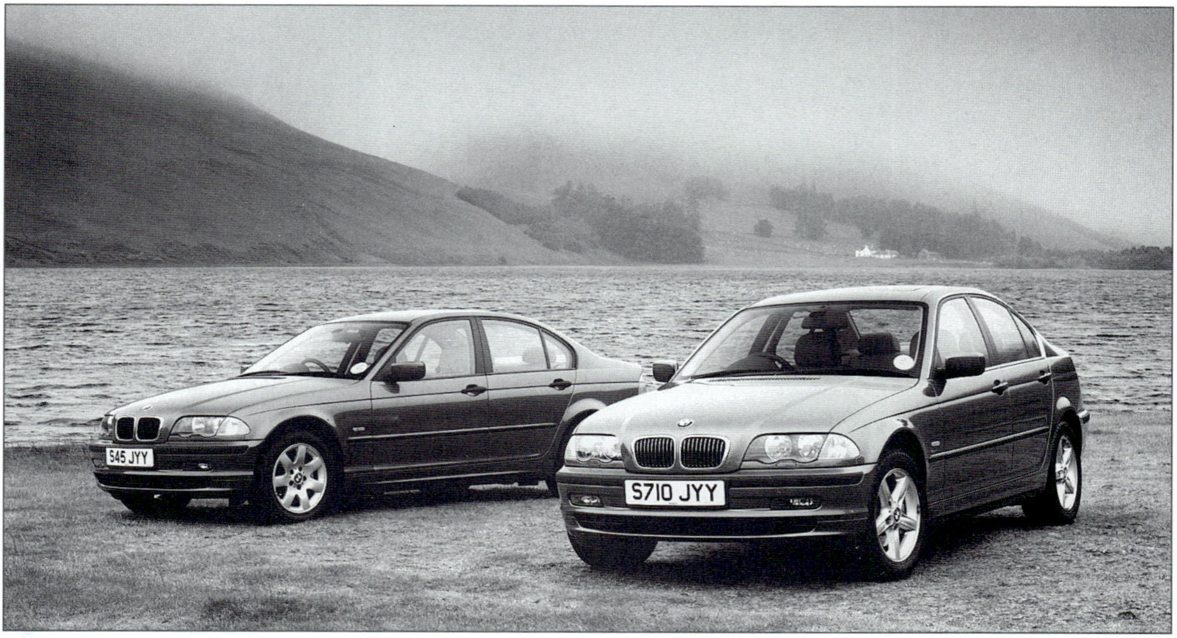

The four-door saloons in the E46 range were announced in 1998, while other types continued unchanged for a while. In appearance the 3 Series was moved towards the 5 Series. The front engine/rear drive configuration was retained, but the platform was new, with suspension on E36 lines. The engine was set well back, and the cars were slightly larger in important dimensions, with wider track and another 1.2 inches in the wheelbase as well as extra height making for increased interior space. Chrome around the side windows, plus the optional wide alloy wheels, distinguish the nearer car of the pair as one of the two six-cylinder models; the other has the 1.9-litre four-cylinder engine.

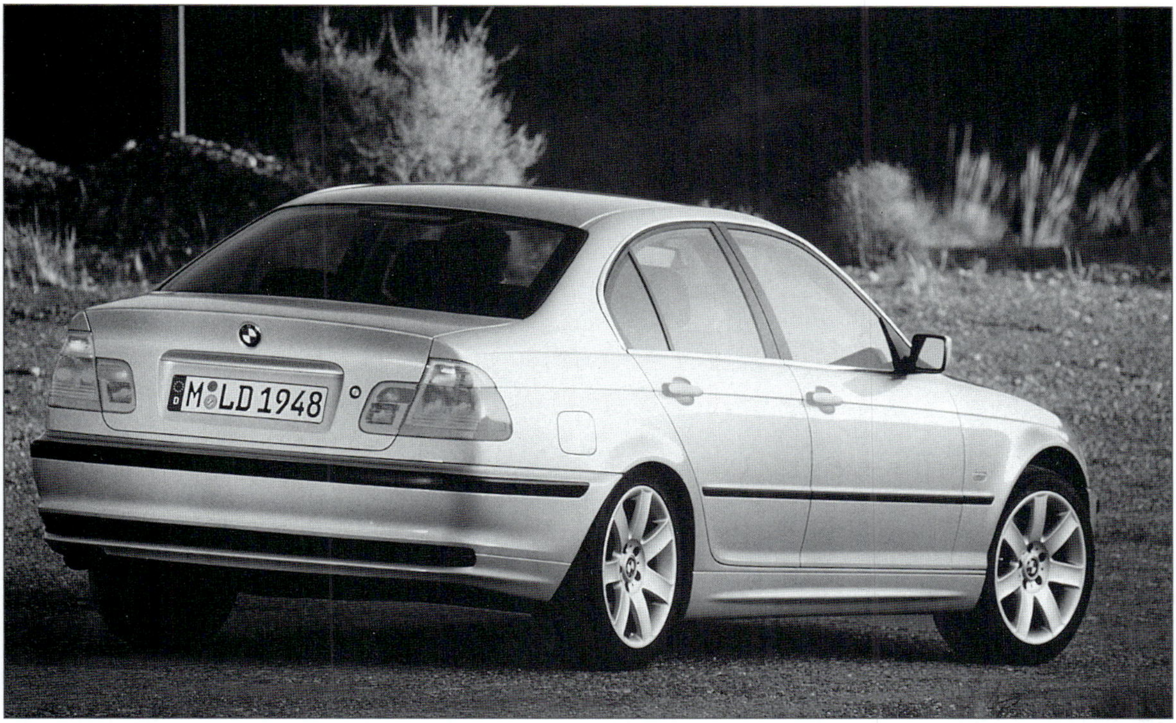

At the rear of the E46 the overhand was minimal, and drag figures were improved, at 0.27–0.29. The grille lifted with the bonnet. The smallest petrol engine was dropped, and a new 1.9-litre diesel in the 320d replaced the 1.7-litre unit.

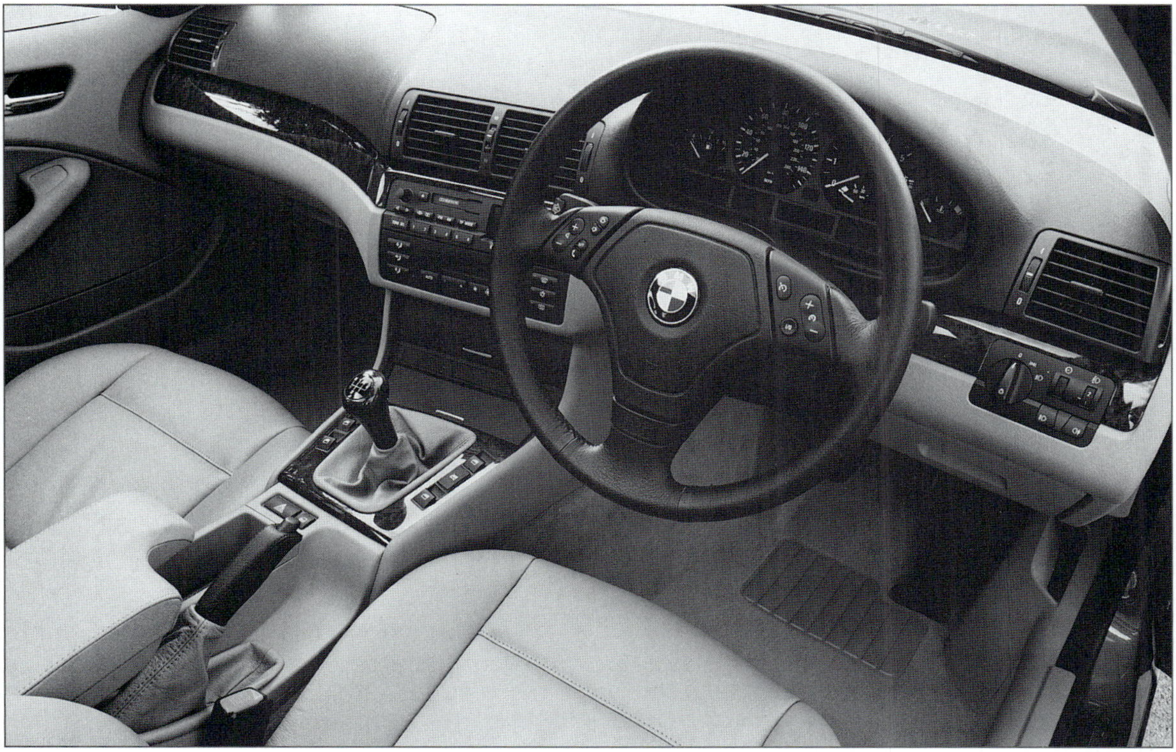

The E46 cockpit was not unfamiliar, as ever with the essential instruments clearly displayed, in a well-moulded fascia. This car had a high trim level, and its steering wheel had multi-function switches. It also has the five-speed manual transmission; a Steptronic automatic 'box was available. Six air bags were a standard fitting.

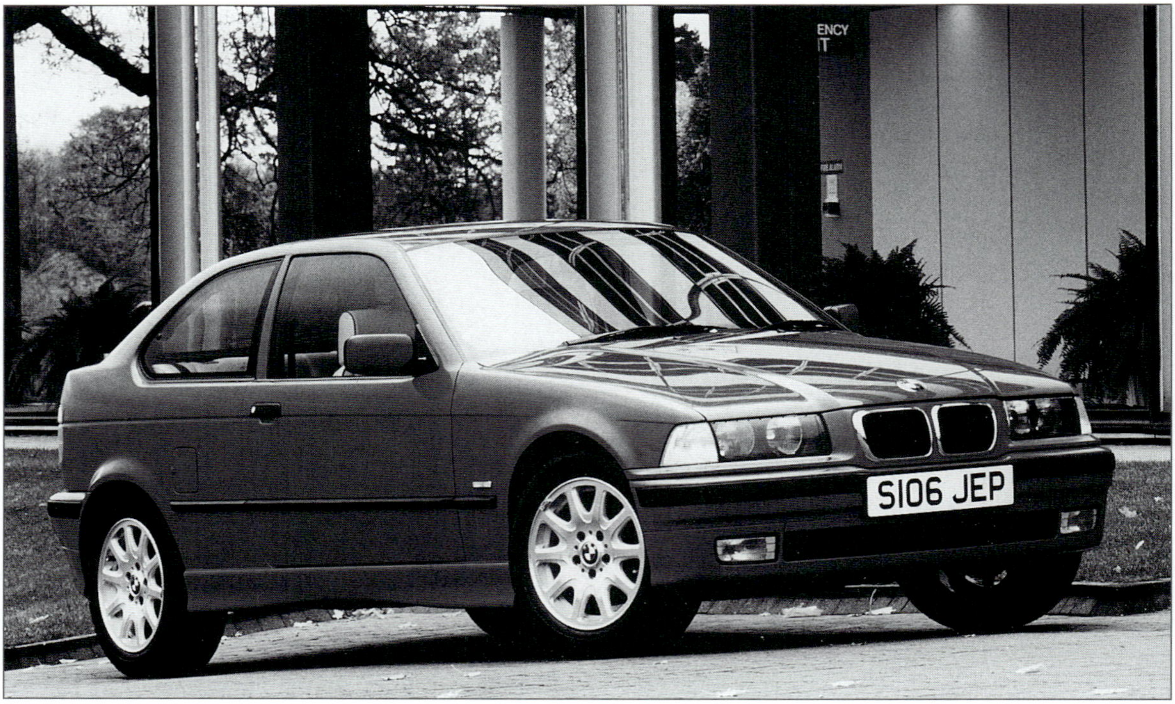

The 1.9-litre Compact entry model was announced early in 1999. With its engine tuned to give better torque as well as 105 bhp, coupled a 119 mph (190 kmh) top speed with quicker acceleration and reasonable fuel consumption.

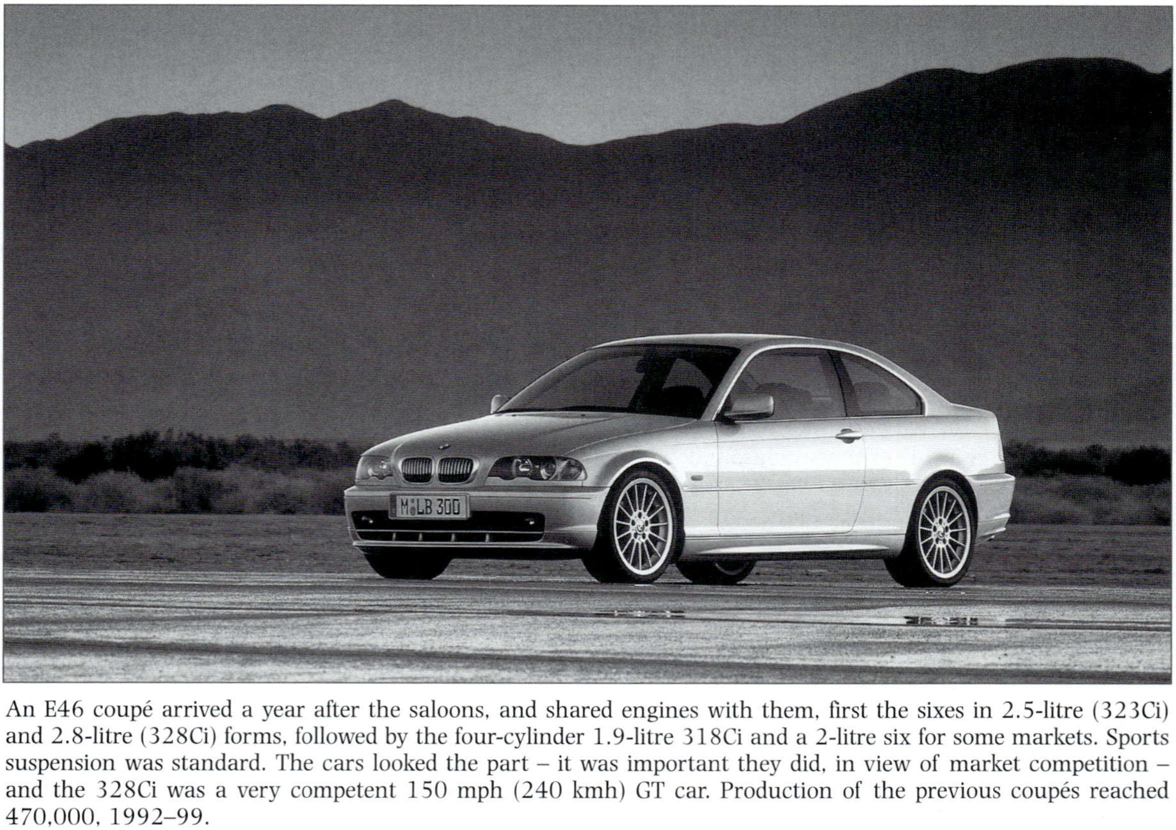

An E46 coupé arrived a year after the saloons, and shared engines with them, first the sixes in 2.5-litre (323Ci) and 2.8-litre (328Ci) forms, followed by the four-cylinder 1.9-litre 318Ci and a 2-litre six for some markets. Sports suspension was standard. The cars looked the part – it was important they did, in view of market competition – and the 328Ci was a very competent 150 mph (240 kmh) GT car. Production of the previous coupés reached 470,000, 1992–99.

CHAPTER FIVE

5 SERIES

'Arguably the best range of cars in the real world' – *Autocar*, March 1999. Chronologically, the 5 Series launched in 1972 was the first of the principal modern ranges, and it was to be continued through a process of evolutionary development into the twenty-first century. The core has generally been middle-ground saloons with strong 'traditional' appeal – high-class executive cars with sporting qualities – but since the second generation types there have also been high-performance versions, 'super saloons' which were to have supercar top speed potential in the 1990s . . . and provide four comfortable seats and sensible luggage space.

The E21 range was launched with two models when the Olympic Games were held in Munich, and these were built at the one-time Glas plant at Dingolfing. The 2-litre four-cylinder engine was fitted in the 520 and the fuel-injection 520i. The first six-cylinder model, the 2.5-litre 525, came in 1973, the 2.8-litre six came in the 528 in 1975 and the 2-litre 'small six' in the 520 in 1977. Meanwhile, a 1.8-litre four-cylinder unit had been offered as a sop to a fuel crisis in 1974, and at the other extreme there was the 530i with a bored-out 3-litre six. This was the first BMW introduced specifically for the US market, and the extra capacity was an attempt to offset power loss when the sapping Federal emissions equipment was added.

Throughout, the suspension introduced with the '*Neue Klasse*' was used, five-speed manual or ZF automatic transmissions were offered, the interiors of the four-door, four-seat bodies were generally tasteful and comfortable, with gradated equipment and long lists of options.

When production was less than a thousand short of 700,000, the E12 5 Series gave way to the E28s in 1981, and these ran to 1987. Second-generation cars had the same bodies, with revisions at nose and tail, and minor styling changes. The suspension geometry was revised, and the earlier cars' reputation for tail happy handling was ironed out. There was a four-cylinder model (518), and five with six-cylinder engines from 1.8 to 2.8 litres, plus a diesel and the 2.7-litre 525e coming in 1983. The 525e (e for *eta*, efficiency) was presented as a fuel-efficient and economical model, deliberately softer in an effort to widen the marque's appeal in the USA.

The 528i and 528iSE topped the 5 Series range as BMW rang the power unit changes, until the 535i was introduced in 1984. Like the M535i in 1980–81 (see Chapter Nine) this was a Motorsport-developed variant, with the 3.5-litre engine but not the aerodynamic add-ons of the M535i.

The total production of second-generation 5 Series cars was 722,327, as the E34s were introduced, with five basic models. In time, combinations meant that eighteen types would be available. Less boxy lines gave subtle aerodynamic improvements, and coupled with lower weights and some mechanical refinement, the new 5 Series cars were more economical – the term might be relative, but nevertheless an asset in energy-aware times! Disc brakes and ABS were standard. A little more interior space was contrived.

Among other changes, new four-valve engines replaced the 'small six' in 1990, and were refined for 1992, while the 1991 diesel was a turbocharged and inter-cooled unit. Two models with six-cylinder engines (530i and 535i) gave way to the 530i and 540i and

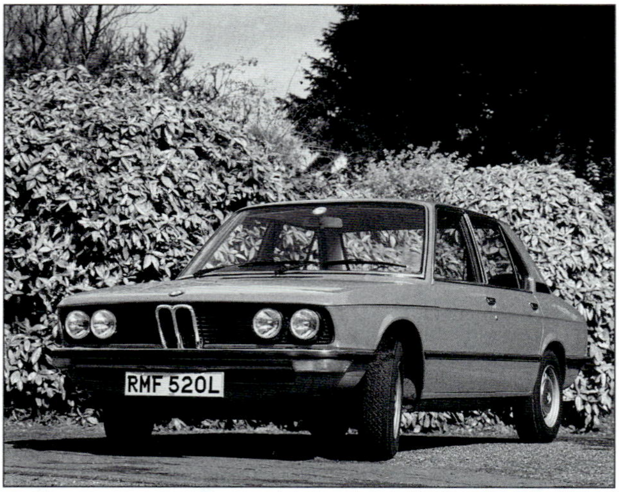

By later standards, the first cars of the new phase in BMW history appeared staid, upright and even angular. But this 520 was praised for its accommodation and comfort, handling and transmission. Outright performance with the aluminium-head 1990 cc four-cylinder engine was not so highly rated, a 106 mph (171 kmh) top speed being regarded as below par for the class.

32-valve alloy V8s in 3- and 4-litre forms in 1992. Both models had five-speed automatic transmissions, whereas earlier types had five-speed manual or four-speed automatic options.

To a degree, the 540i did not become the 5 Series flagship model, as there had been a 'new M5' since 1990, with a 3.5-litre stright six rated at 315 bhp compared with the 4-litre V8's 286 bhp. However, the V8 cars were ultra-smooth cruisers, the M5 a stiffer sporting machine. In any case, it lasted only to the 1993 model year, in part as the V8 cars were felt to be appropriate range-topping variants.

There were Touring models with a twin sunroof arrangement, but this novelty was quietly dropped. (Fewer customers even wanted a normal sunroof as air conditioning became more popular.) Incidentally, a handful of M5s had Touring bodies, and performance that might have surprised other autobahn users.

A fourth generation was introduced in 1996, as BMW sought to build on sales of 1.4 million units, with a substantial development investment, although the E39 cars were evolutionary in many respects. The saloons now had a near-coupé appearance, and the bodies were a little longer and wider, and slightly higher. Together with the attention paid to under-car aerodynamics, this new shape made for considerable reductions in drag. Although overall weight was only fractionally reduced, it was at least kept under control, partly through extensive use of aluminium, in suspension members (where it was claimed to be a world first) and in cross bracing and side impact bars.

The launch models had six-cylinder engines, including a diesel, with production of the V8 types starting later in 1996. There was a 525iX four-wheel-drive variant, with the 2.5-litre straight six, and a 518i basic model was to come with a new four-cylinder engine. Touring types were introduced for 1997, and shared the saloon specifications in all save one engine option. By 1999 both types could be fitted with a new 3-litre turbocharged diesel.

At the end of the decade the saloon range comprised seven models, and only the 535i with a 3.5-litre V8 was missing from the Touring line up. Once again, an M type was perhaps the flagship, in the shape of the new M5 that had come in 1998 with a 4.4-litre V8. Like the 540i, its top speed was limited to 155 mph (250 kmh), so in that respect discussion was academic. In March 1999 *Autocar* compared rivals from four countries in this class, the Alfa Romeo 166, BMW 528i, Jaguar S-type and Lexus GS300, rating the BMW above its rivals on grounds of 'brilliant performance and strong economy . . . sweetness and refinement . . . superb build quality', which made it a 'peerless executive-cum-sporting saloon'. That was just one positive outcome of BMW's battle to maintain market position in this important sector, where the competition was more intense than ever.

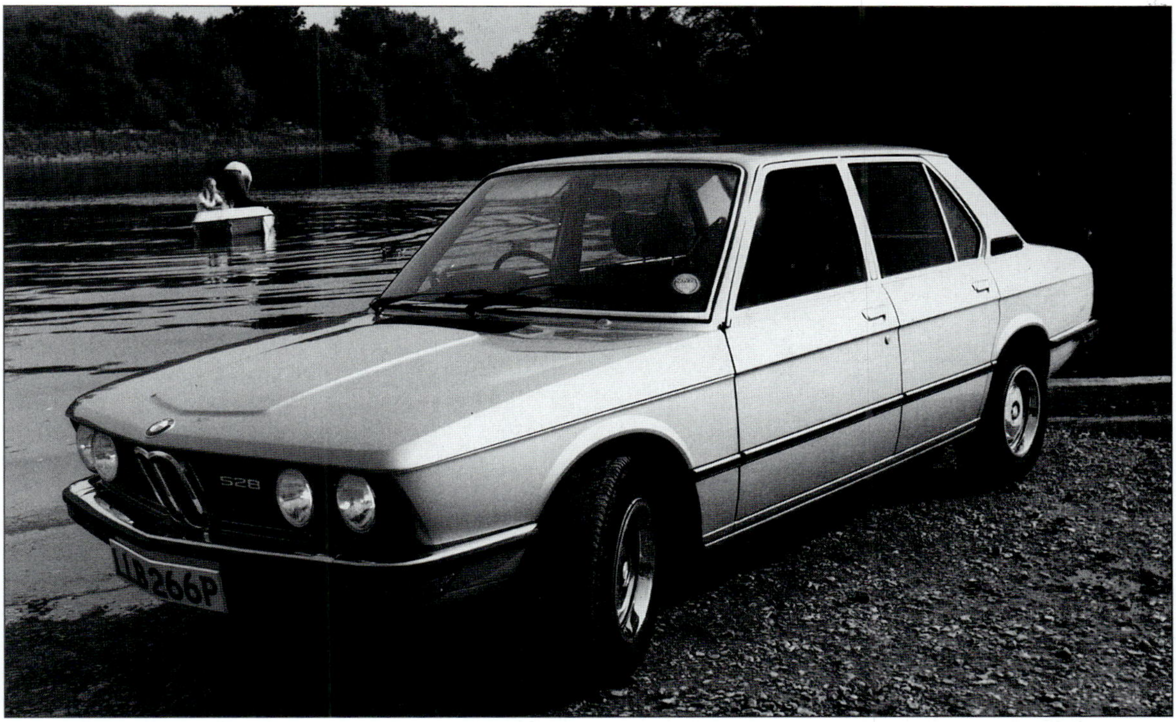

The top of the range in 1975 was the 528, with a 2788 cc straight six. The six-cylinder types (525 and 528/528i) could be distinguished from the four-cylinder types (518, 520 and 520i) by the modestly raised centre section on the bonnets.

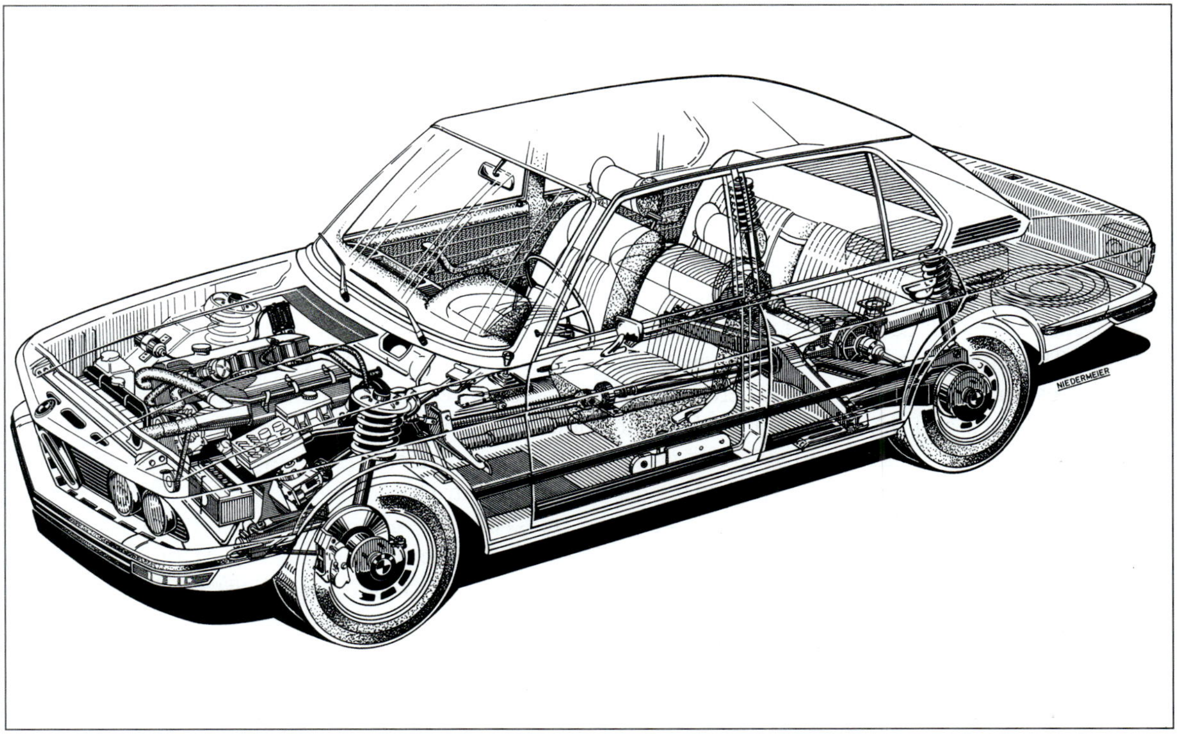

The E12 cars were generally conventional, with MacPherson strut front suspension and semi-trailing arms at the rear. The engine bay was fairly full, and in the foreground a transparent box housed fuses, relays and a service diagnostic unit. Interior space was acceptable for the class, and luggage space generous.

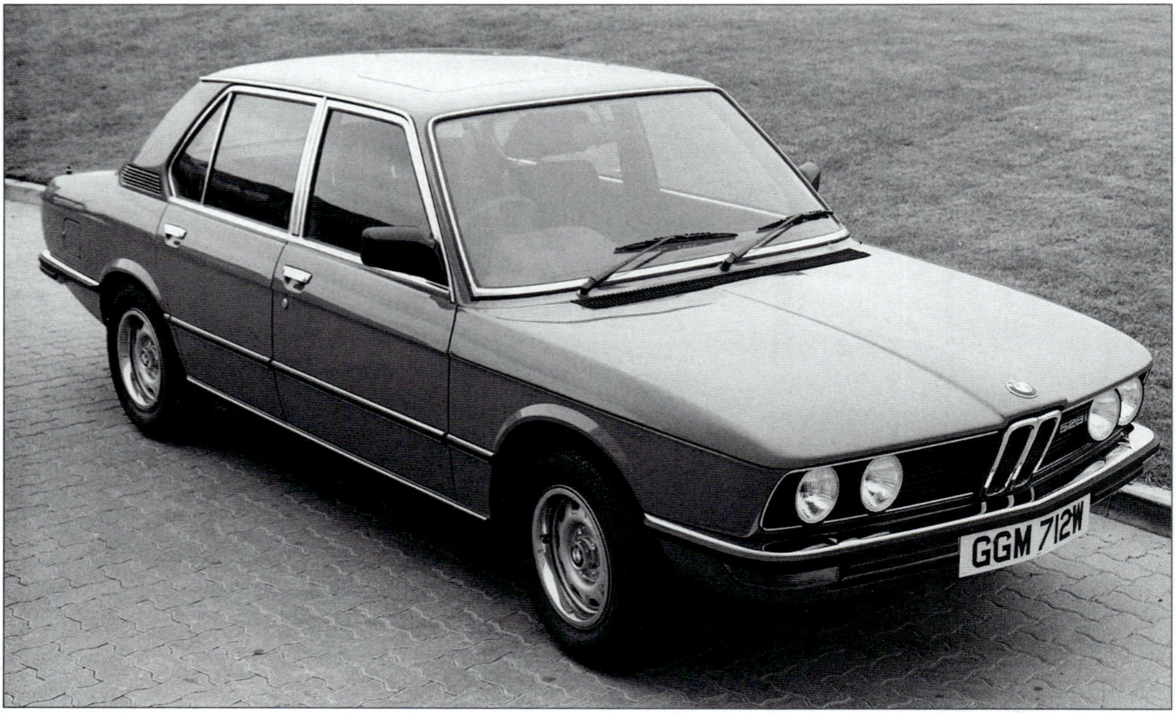

There was a minor facelift in 1976, and then no outward change through the rest of the life of the E12 cars. A slight 'peak' above the kidney grille was noticeable, while the six-cylinder cars had wider wheels (and this 528i had twin exhausts). In 1977–81 the 2.8-litre fuel-injected engine made this a 128 mph (205 kmh) car, and that performance was closely matched by the US market 530i. However, the 2985 cc version of the engine in that car gained a reputation for fragility, with cracking cylinder heads.

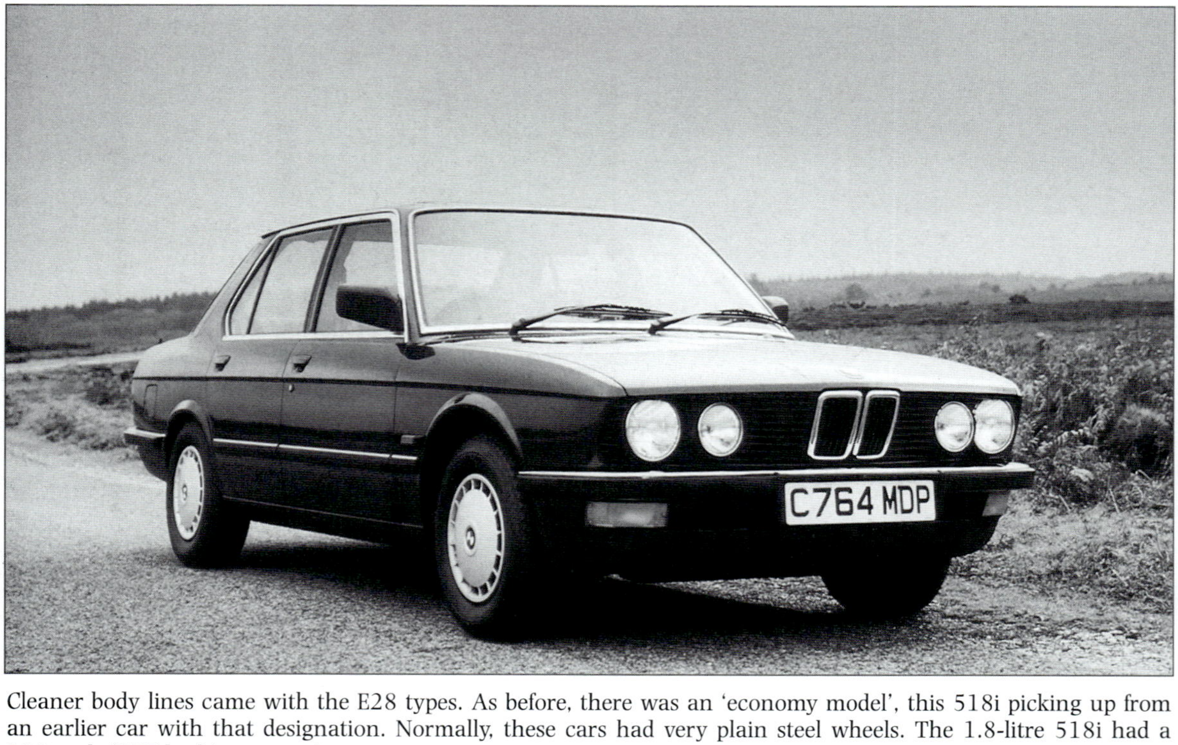

Cleaner body lines came with the E28 types. As before, there was an 'economy model', this 518i picking up from an earlier car with that designation. Normally, these cars had very plain steel wheels. The 1.8-litre 518i had a 107 mph (175 kmh) top speed.

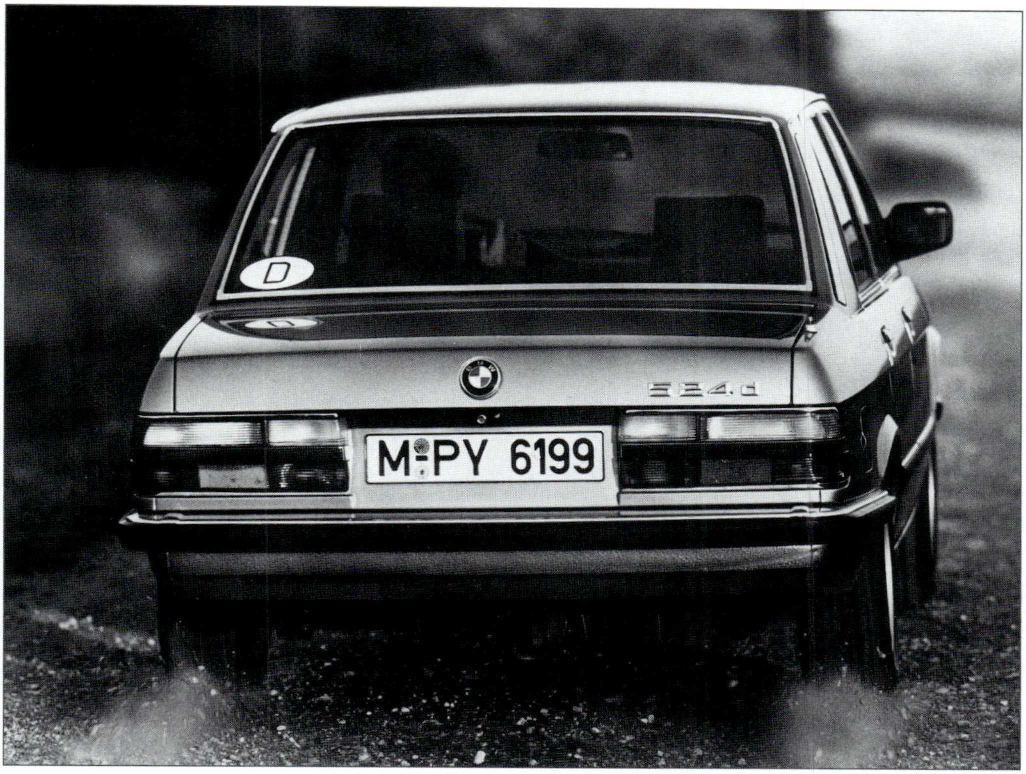

A diesel engine came to the 5 Series in the mid-1980s, a turbocharged 2443 cc straight six rated at 115 bhp which gave this 524d a claimed top speed of 112 mph (180 kmh), or slightly less with the optional ZF four-speed automatic transmission. The diesel was not available in all markets. Apart from the 'd' designation, this view shows the neatly integrated rear end of the E28.

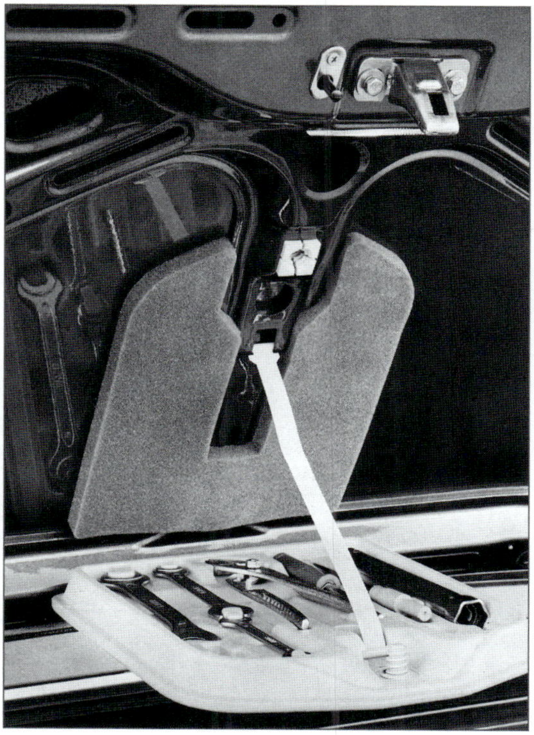

A tool kit housed in the boot lid attracted attention in many BMW models, including this E28 5 Series car. When it dropped down it formed a handy tool tray, held by a stay.

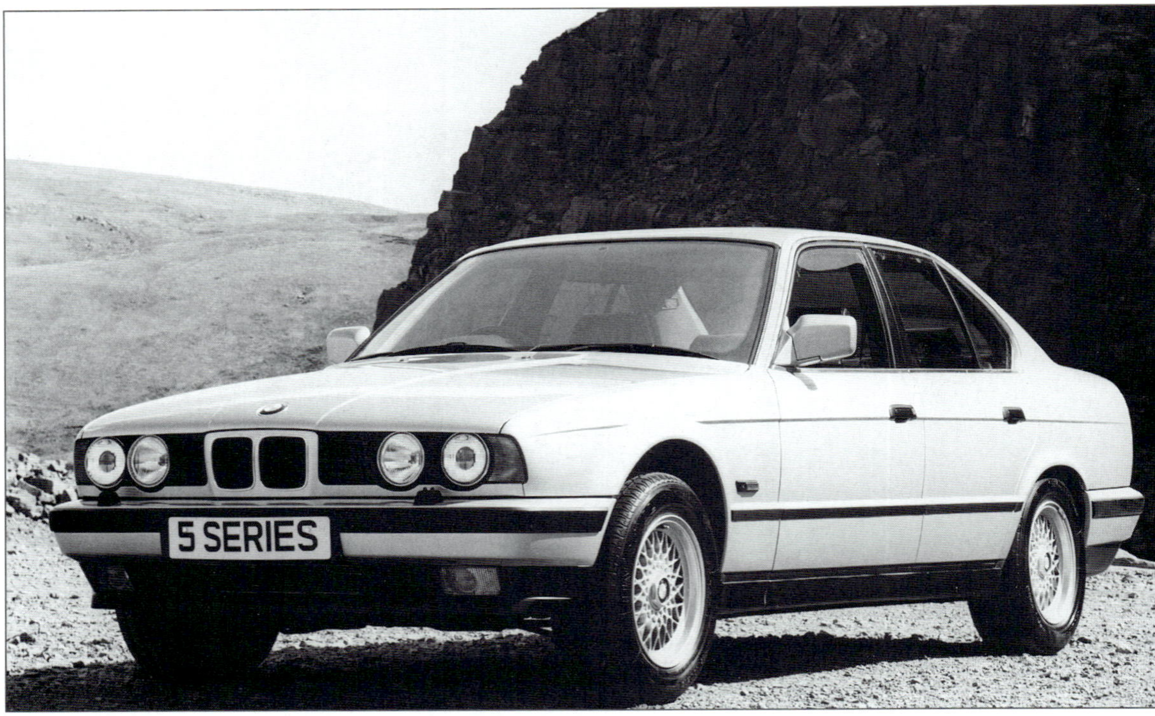

The 5 Series body was completely restyled for the E34 saloons that came at the beginning of 1988. Appearances were in keeping with the late 1980s, and related more closely to the 7 Series, while the drag figure was improved to 0.30–0.32, front/rear aerodynamic balance was better, and so was stability in crosswinds. The new body was also stiffer. Detail fittings varied according to specification, and the top models were fitted with speed-dependent windscreen wipers developed for the 7 Series.

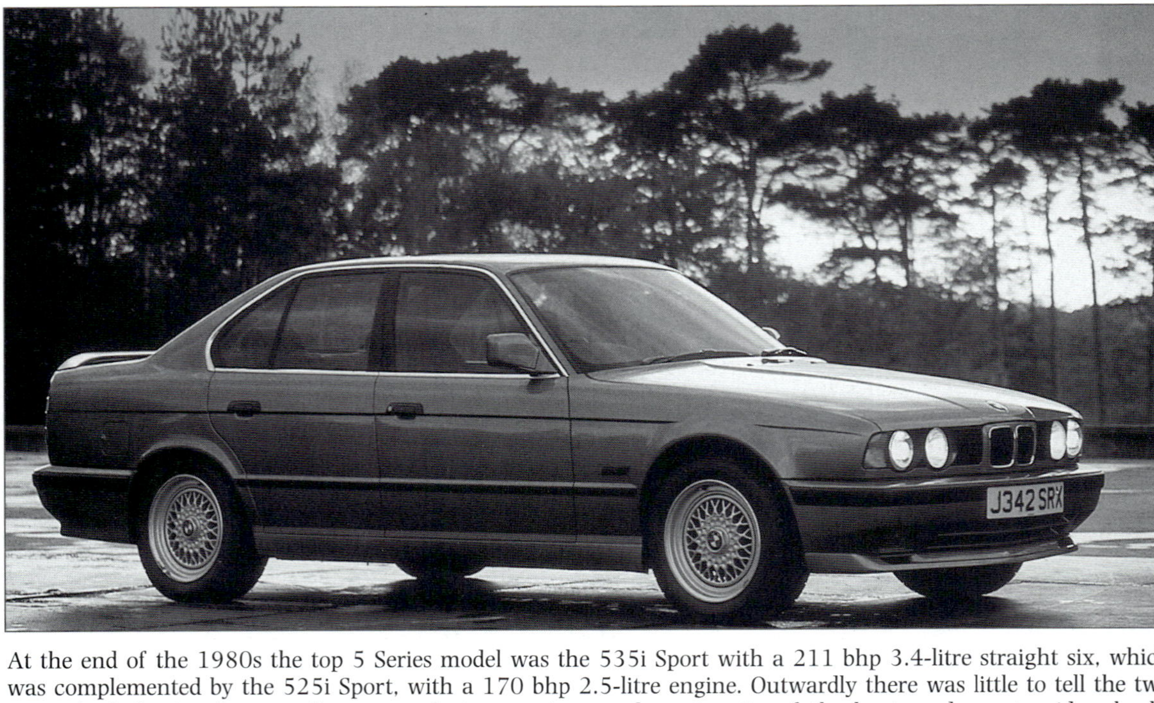

At the end of the 1980s the top 5 Series model was the 535i Sport with a 211 bhp 3.4-litre straight six, which was complemented by the 525i Sport, with a 170 bhp 2.5-litre engine. Outwardly there was little to tell the two apart, both having front spoilers, a modest rear wing on the extremity of the boot, and squat, wide wheels. Acceleration with the five-speed manual gearbox was marginally better than for cars with the otherwise excellent four-speed automatic, and both types could exceed 140 mph (225 kmh).

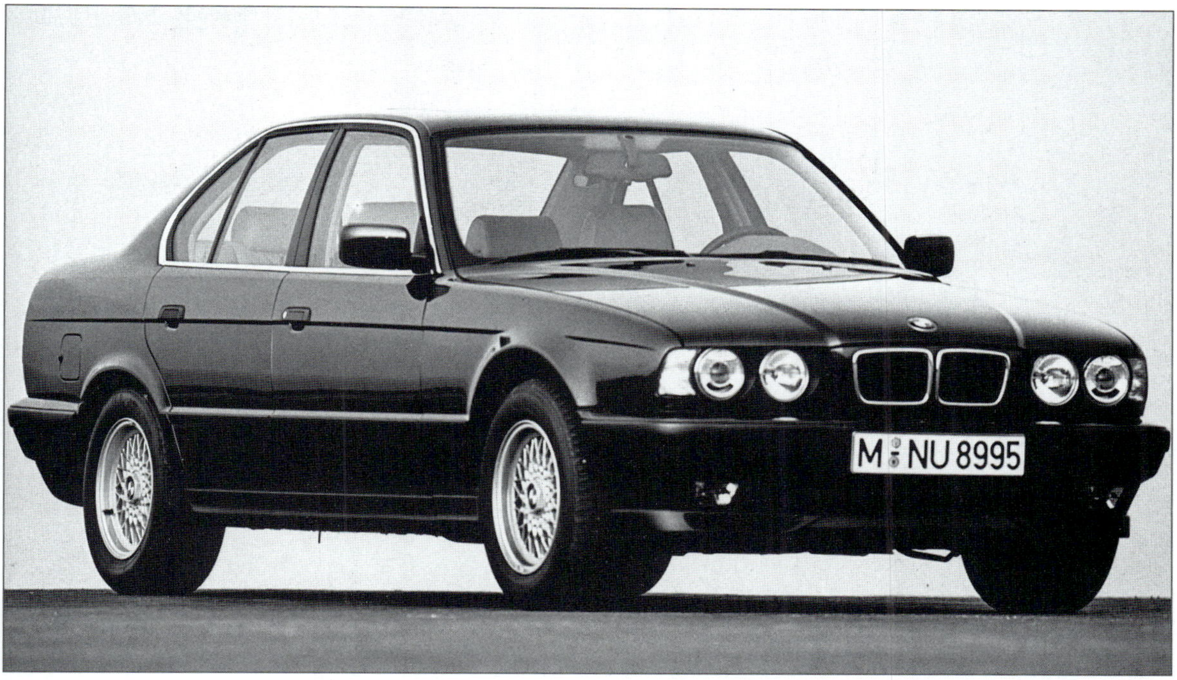

A fundamental change might have been expected to be accompanied by a new E number, but the introduction of a V8 – already seen in the 740i – apparently did not call for one. The larger of the V8s (3982 cc) in this 540i was rated at 286 bhp, which made the ASC traction control system desirable in this car. The 540i was introduced with a ZF five-speed automatic transmission, with a six-speed manual gearbox to follow. The other V8 car, the 530i, had a five-speed manual alternative.

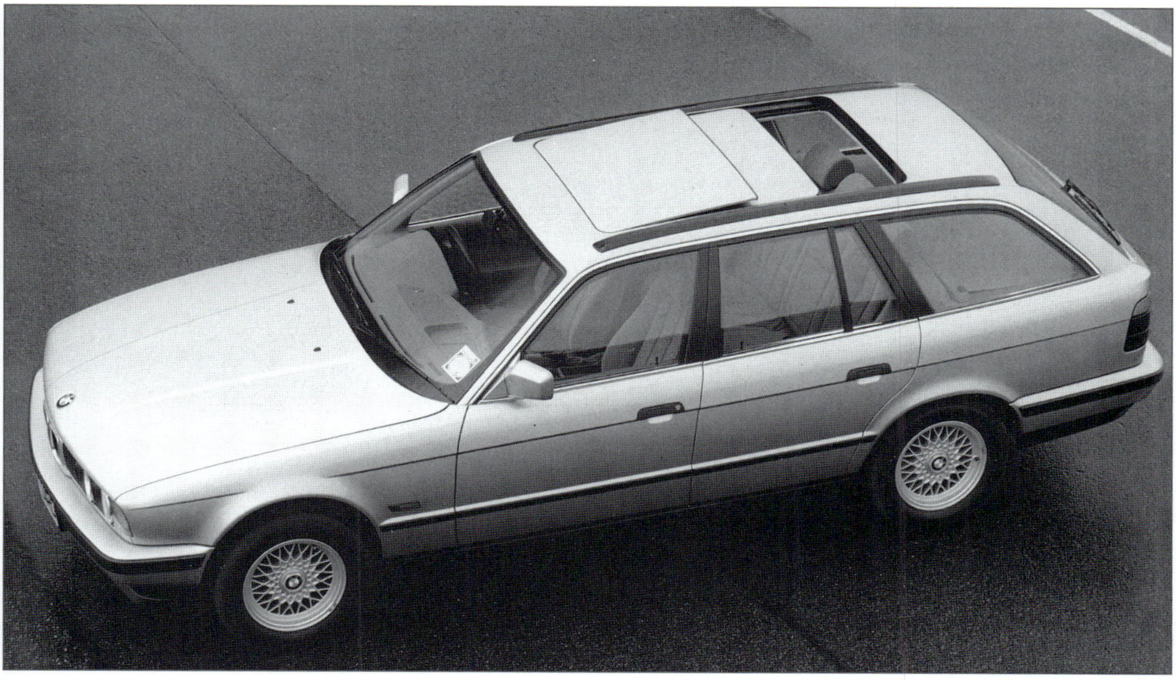

For 1992 Touring bodies were introduced, in 520i and 525i forms, shortly followed by the 525iX four-wheel-drive version in some markets. These have been described as 'executive estate cars' – but not by BMW! – and with rear seats folding to give a flat floor they came closer to fulfilling the estate car role than the 3 Series Touring models. An oddity was the double sunroof, which allowed for the front section open, the rear section open, or both partly open. This novelty was not repeated on other models.

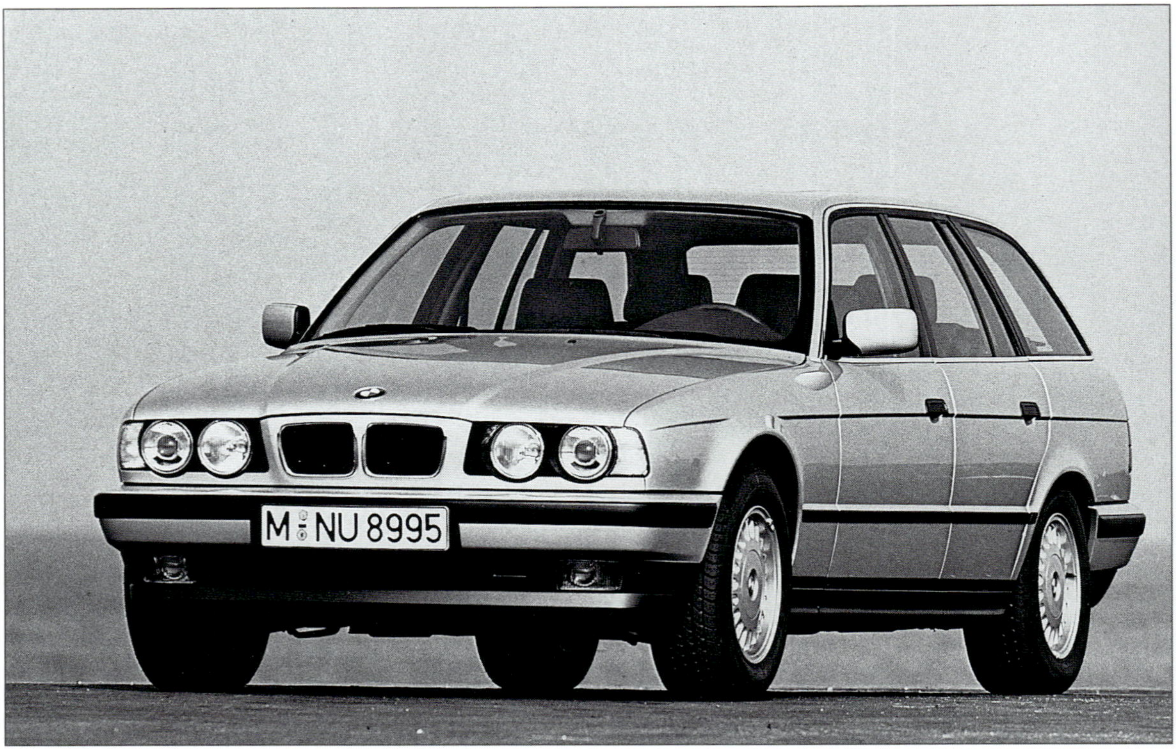

The first 5 Series Touring models were soon joined by the 530i and 540i with V8 engines, bringing higher levels of performance and sophistication, for example in the Adaptive Gearbox System which matched selection to individual driving habits, and had winter, hill-climbing and touring programmes. Interiors were furnished to high standards. Performance matched the equivalent saloons, the 530i illustrated being capable of 140 mph (225 kmh).

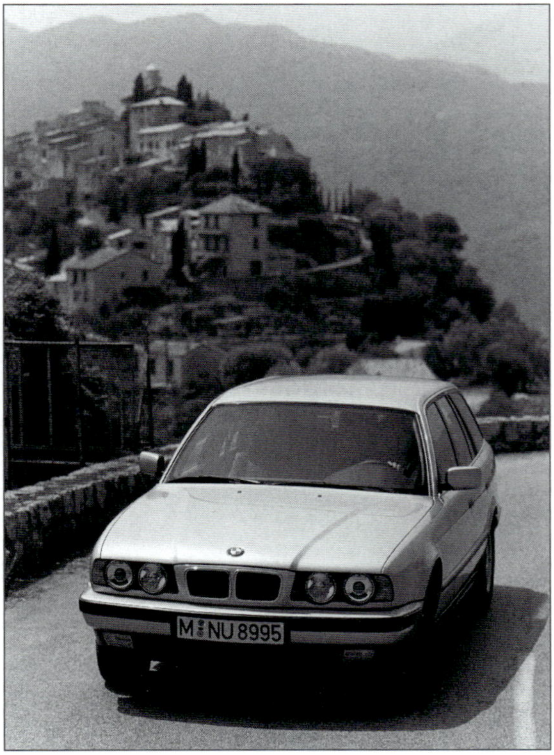

A pleasing shot of a 1994 5 Series Touring six-cylinder model. By that time the subtle styling changes at the front introduced with the V8 cars had been carried across the range, noticeably the wider kidney grille.

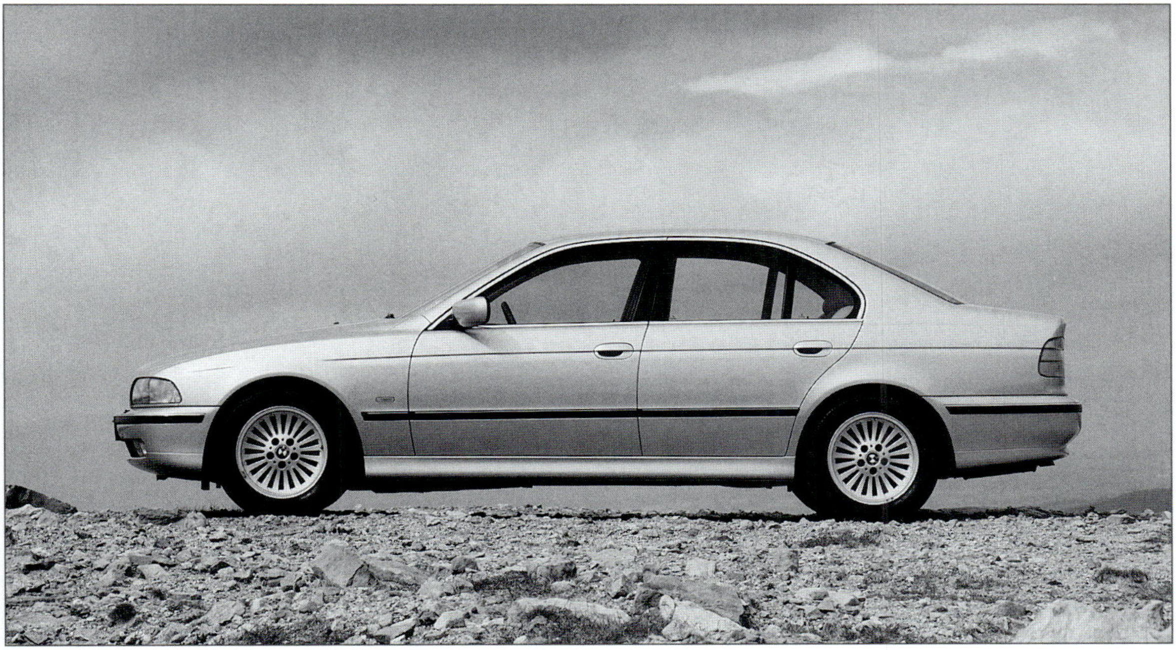

The fourth-generation 5 Series, launched at the 1995 Frankfurt Motor Show, was a formidable challenge to Mercedes's E class. These E39 5 Series did not look larger than the E34 models, but they were a little larger, and the overall lines gave an impression of efficiency – this side elevation shot of a 540i says everything about its clean profile. This version came a little later than the 520i, 523i, 528i and the 525tds diesel.

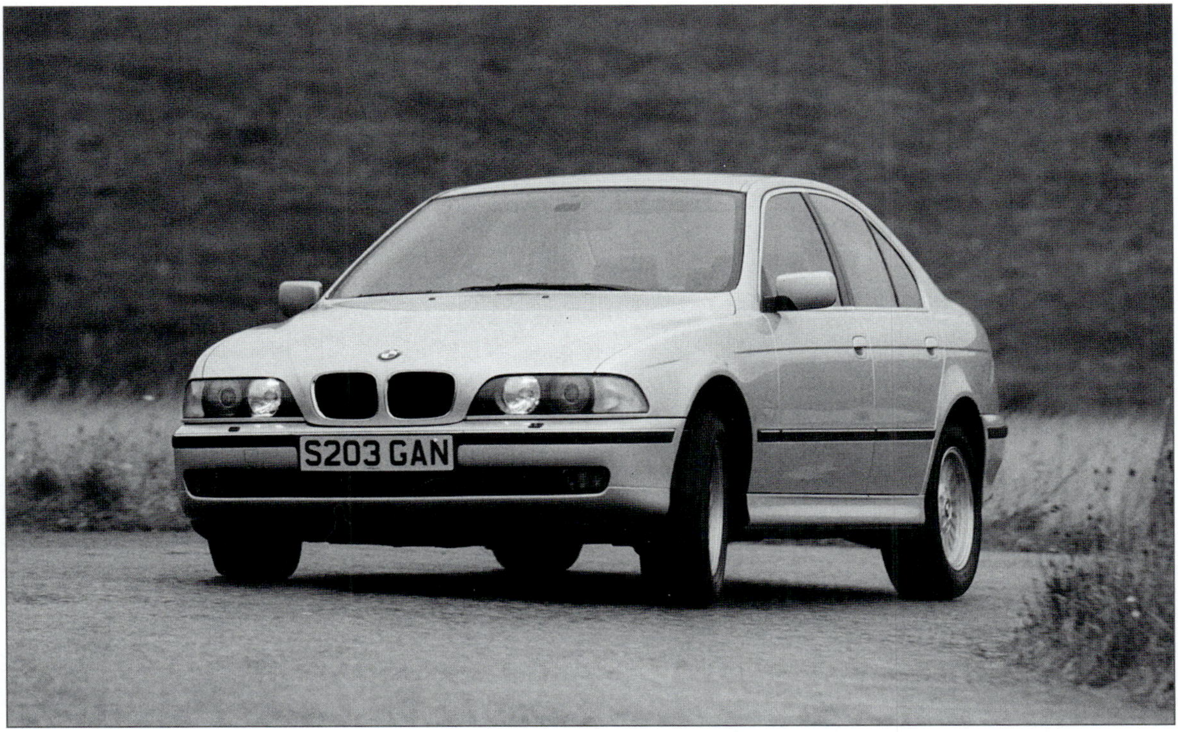

The nose picked up styling cues from other ranges, and had a sweep to the bonnet not found in earlier 5 Series models. The windscreen and rear window were steeply raked, and the drag coefficients ranged from 0.27 (520i) to 0.30 (525tds). The core model was the 528iSE, which had a 2793 cc 193 bhp straight six, and reached 147 mph (236 kmh) in manual transmission form.

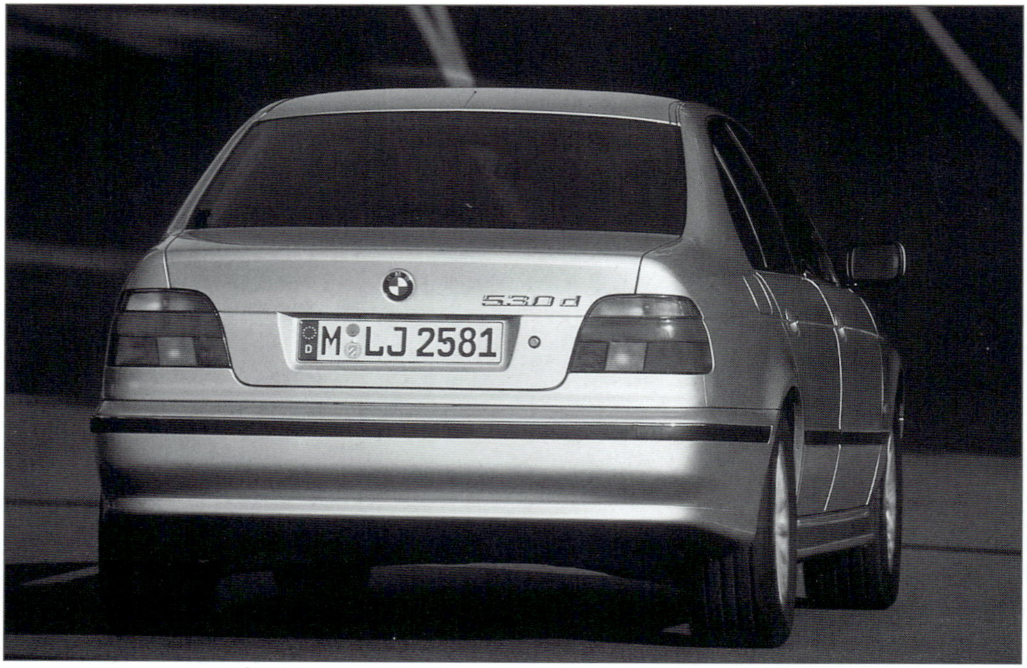

The rear quarters were equally clean, and this shot compares with the 524d of some ten years earlier (page 85): this is a 530d, a direct successor. It was capable of 138 mph (222 kmh).

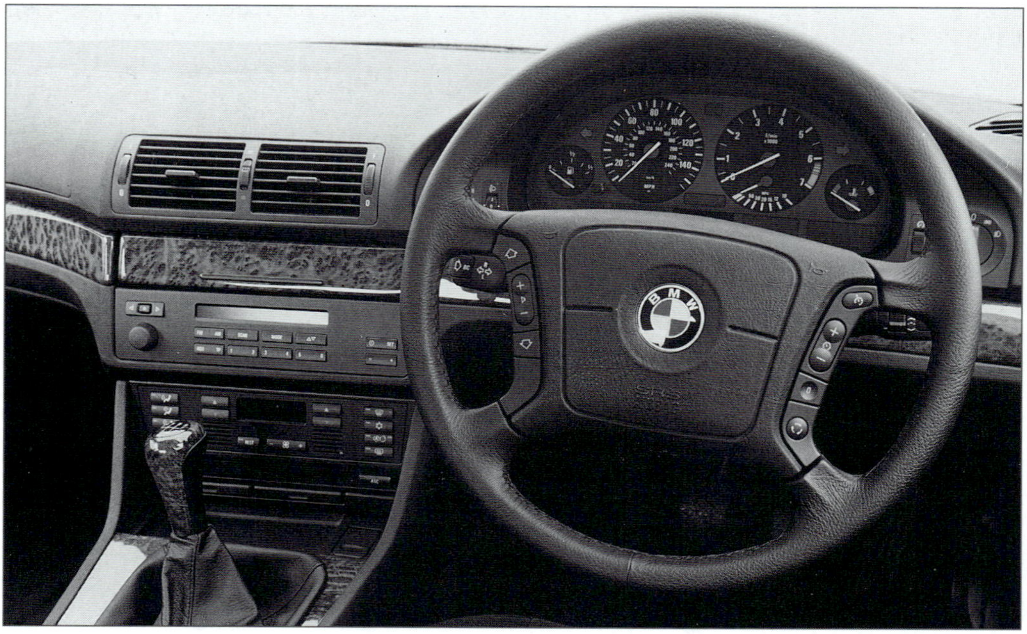

Late-1990s 5 Series fascia, with the optional multi-function steering wheel first seen on 7 Series cars. Somehow, BMW never seem to mention the primary function, steering, while pointing out the finger-tip operation of radio, cassette, CD, cruise control and air recirculation. The instruments are again admirably clear and to the point, and on SE models information from the computer and check controls was also displayed in front of the driver. This car has the five-speed manual gearbox that went with the straight sixes; automatics were also available, on the SE with the Steptronic clutchless manual over-ride alternative.

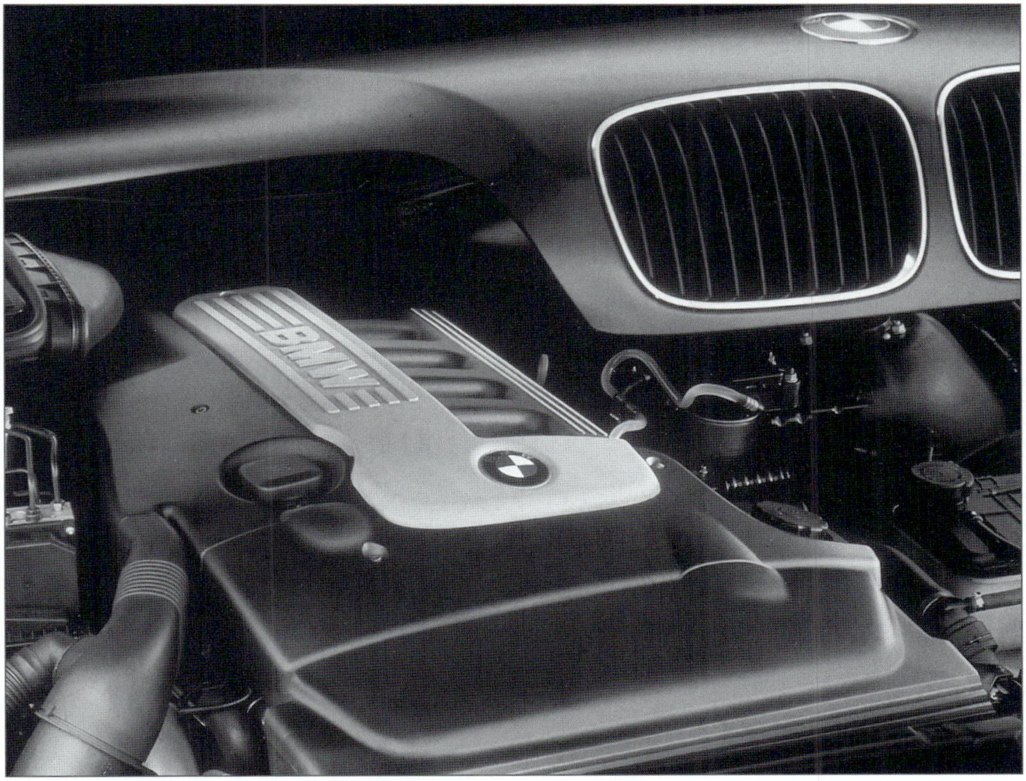

Under-bonnet impressions were of a full engine bay, and neatness. This is actually a diesel, and the photograph incidentally shows how the kidney grille lifted with the bonnet on the E39 models.

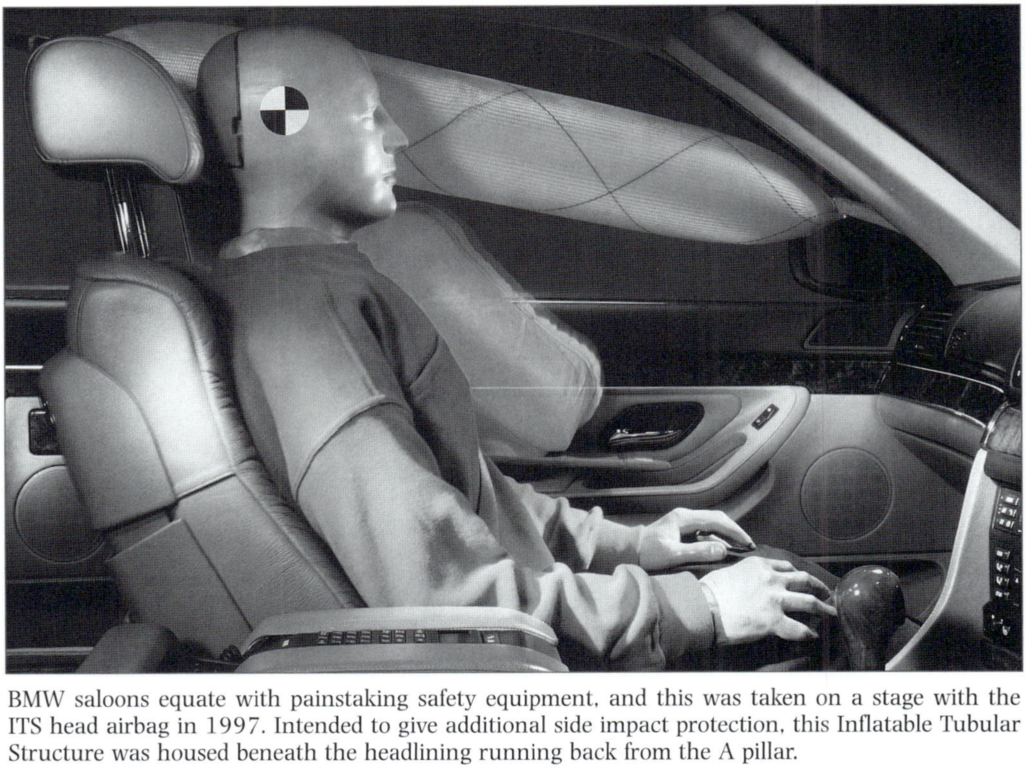

BMW saloons equate with painstaking safety equipment, and this was taken on a stage with the ITS head airbag in 1997. Intended to give additional side impact protection, this Inflatable Tubular Structure was housed beneath the headlining running back from the A pillar.

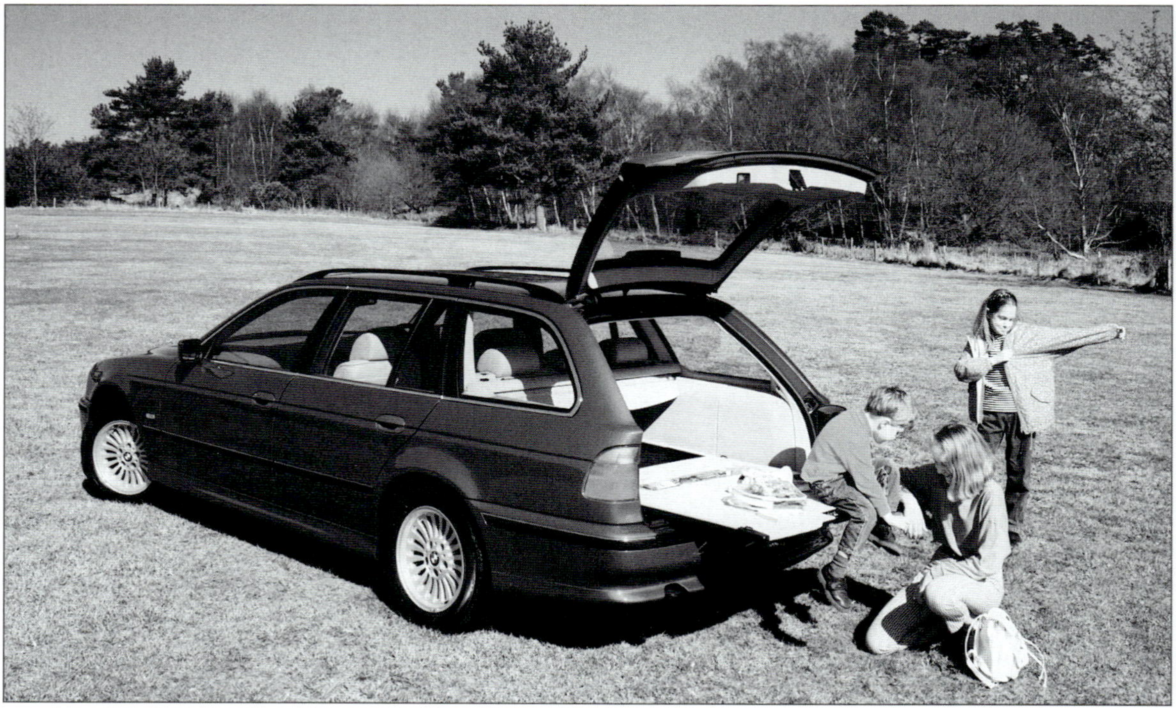

The E39 Touring range was completed in 1997 when the 540i was announced in the autumn. These were still regarded as touring cars with very generous luggage compartments, although the load area was increased again. (The struts of the rear suspension were almost horizontal.) An option on all Touring models was a retractable luggage compartment floor, shown here in the extended position, and bearing a weight. It extended more than 23 inches beyond the tailgate, and was simple and practical, for loading awkward or heavy objects, as well as for picnics.

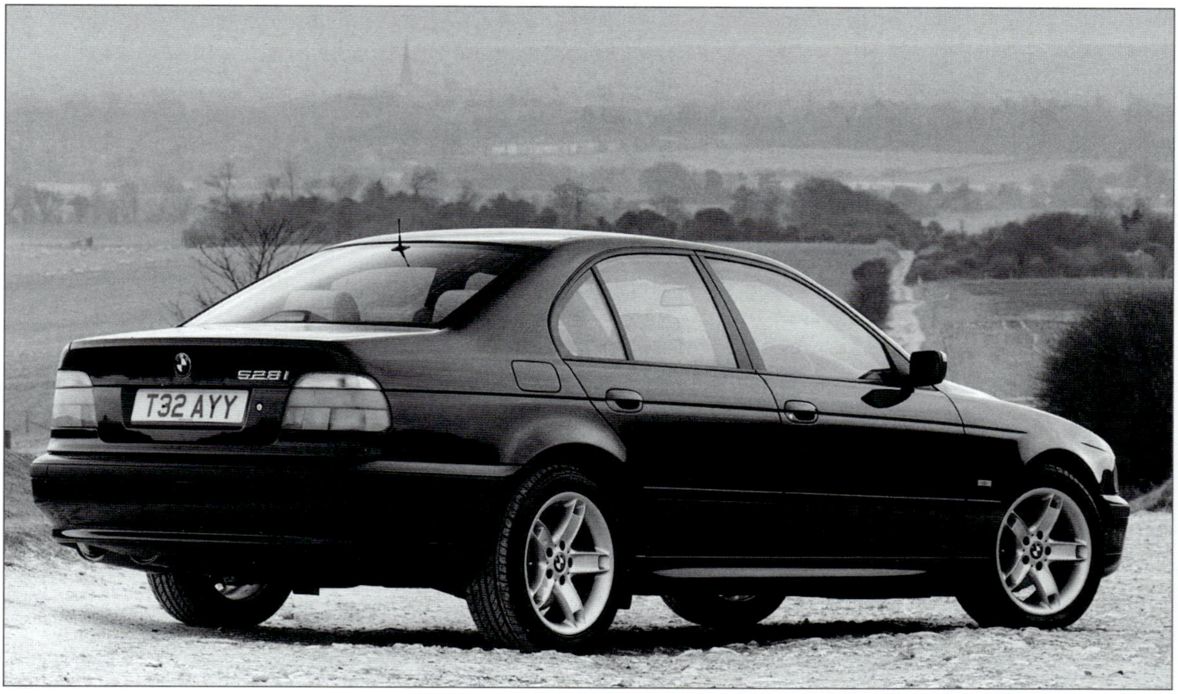

A Sport option for the six-cylinder petrol-engined models was extended to the diesel 530d in 1999. Outwardly, these cars could be recognized by distinctive five-spoke ('star spoke') alloy wheels and black door trim.

CHAPTER SIX

SUPREME COUPÉS

While the coupés with origins in the 1960s ruled European touring car racing, BMW faced the challenge of replacing the CS types. It was met in the 6 Series, which were road cars – true Grand Touring cars but not circuit cars. These were introduced in 1976, then BMW moved to a higher plane with the technically advanced 8 Series coupés in 1990.

In the 6 Series (E30) the coupé concept was refined, to give immensely satisfying mile-eaters, first in the 630CS and CSi introduced at the Geneva Motor Show in 1976. These borrowed from the 5 Series and the then yet-to-be-announced 7 Series, and the construction was familiar. However, the chassis was new and carried particularly shapely bodywork designed by Paul Bracq. This lasted well for a decade before it began to appear dated, and the only really obvious modification during that period was the addition of nose and tail spoilers in 1978. The cockpit was refreshingly new, but the coupé body led to rear seat limitations (again); BMW claimed that research had suggested the cars would seldom be expected to carry four adults. Luggage space was as generous as in previous coupés, appropriately for a grand touring car.

These were relatively heavy cars (the CSi was almost 100 kg heavier than its early 1970s counterpart) and performance did not quite match up, at first. The 3-litre straight six in the 1976 CS was rated at 185 bhp, and a 3.2-litre version in the CSi gave 200 bhp – a little less than the 1971 CSi, while the emissions-strangled units in US-market cars fell well short.

The CSi for 1978 had a 3.5-litre engine, and this brought a sporting element into the 6 Series repertoire. Then the twin-cam 286 bhp engine that came from the previous M5 was fitted in an 'M' model, the M635CSi, to bring a four-seater supercar into the range.

There had been early production problems, but these were settled as the work was concentrated at Dingolfing in 1977. BMW's usual policy of detail enhancements was followed, for example with computerized fuel-injection management introduced in 1982. Early cars had a Getrag four-speed manual gearbox or a three-speed ZF automatic transmission; later a five-speed manual box was listed.

BMW looked to cabriolet possibilities in the early 1980s, but reluctantly conceded that the investment (in structural revisions) could not be justified. Independent open conversions were never approved by the company. Sales of the production coupé were not exciting, and in the second half of the 1980s the range was cut back to two versions, as the 6 Series just kept a presence in market sector until the 8 Series arrived in 1989.

The E31 coupés were as up-to-the-minute as the 6 Series was beginning to date, and complemented the 7 Series at the top of the BMW line-up for the 1990s. Very little derived from other models – principally, the engine and the automatic transmission were contributed by the 750i, together with the front suspension. The body looked slippery, rakish even, and the drag coefficient was an excellent 0.29. Generally, the body was regarded elegant, but Klaus Kapitza's styling was not without critics.

The traditional kidney grille was naturally perpetuated; so was the kink in the C pillar. The retractable headlights were reminiscent of the M1, and did not extend far enough to mar the overall nose lines when they were in use. In another neat detail, the frameless door glass slid down fractionally as a door was closed, then resealed. Boot space was

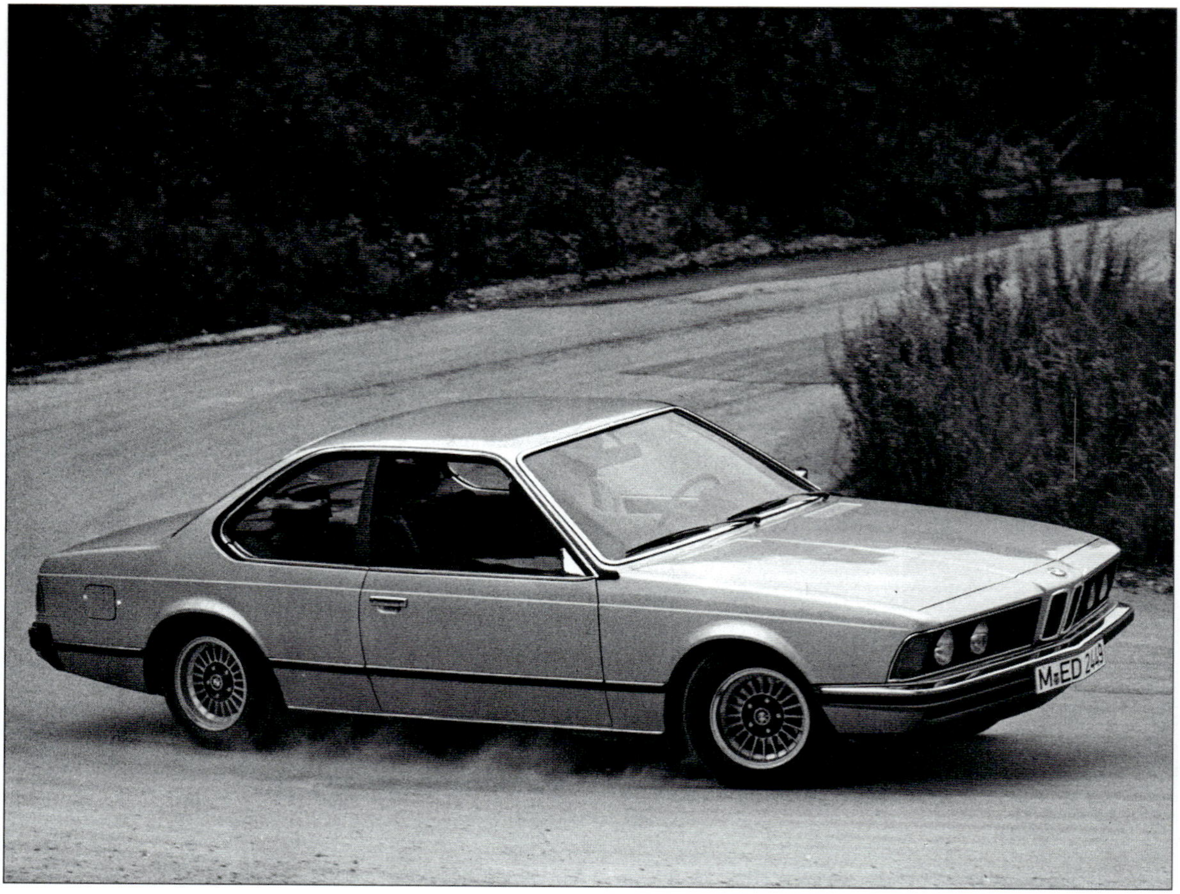

A familiar photograph shows the early 6 Series coupé to advantage. The body lines on this 630CS are clean, with no aerodynamic appendages, there is a large glass area and the bonnet is wide.

good, and as far as people accommodation was concerned, this was another 2+2 rather than a full four-seater. The 850 was a little shorter than the 6 Series cars, and lower, but considerably wider and on a longer wheelbase, and it was heavier.

The launch engine was the V12 from the 750i, driving through a ZF four-speed automatic transmission, with the option of a six-speed manual Getrag gearbox, perhaps hardly called for as the engine had abundant torque that made six speeds an embellishment. An ungoverned 850 would probably exceed 180 mph (290 kmh), but since the 1980s BMW has not made extravagant top speed claims and has stood by the general German 250 kmh (155 mph) limit.

The MacPherson strut front suspension – 'constant-velocity joint spring strut suspension' – was complemented by a system BMW described as 'integral' at the rear. In this complex multi-link system, Automatic Stability Control (ASC) was standard on automatic cars, obviating differences in wheel speeds; manual cars had ASC+T (T for traction), which applied an appropriate rear brake to slow a spinning wheel and thus maintain optimum traction. It was followed by Dynamic Stability Control, which went beyond wheel spin, and reduced power as soon as it detected a loss of stability due to excess understeer or oversteer. Then there was Active Rear Axle Kinematics (ARK), which allowed a tiny degree of steering at the rear. This complexity did not all come at

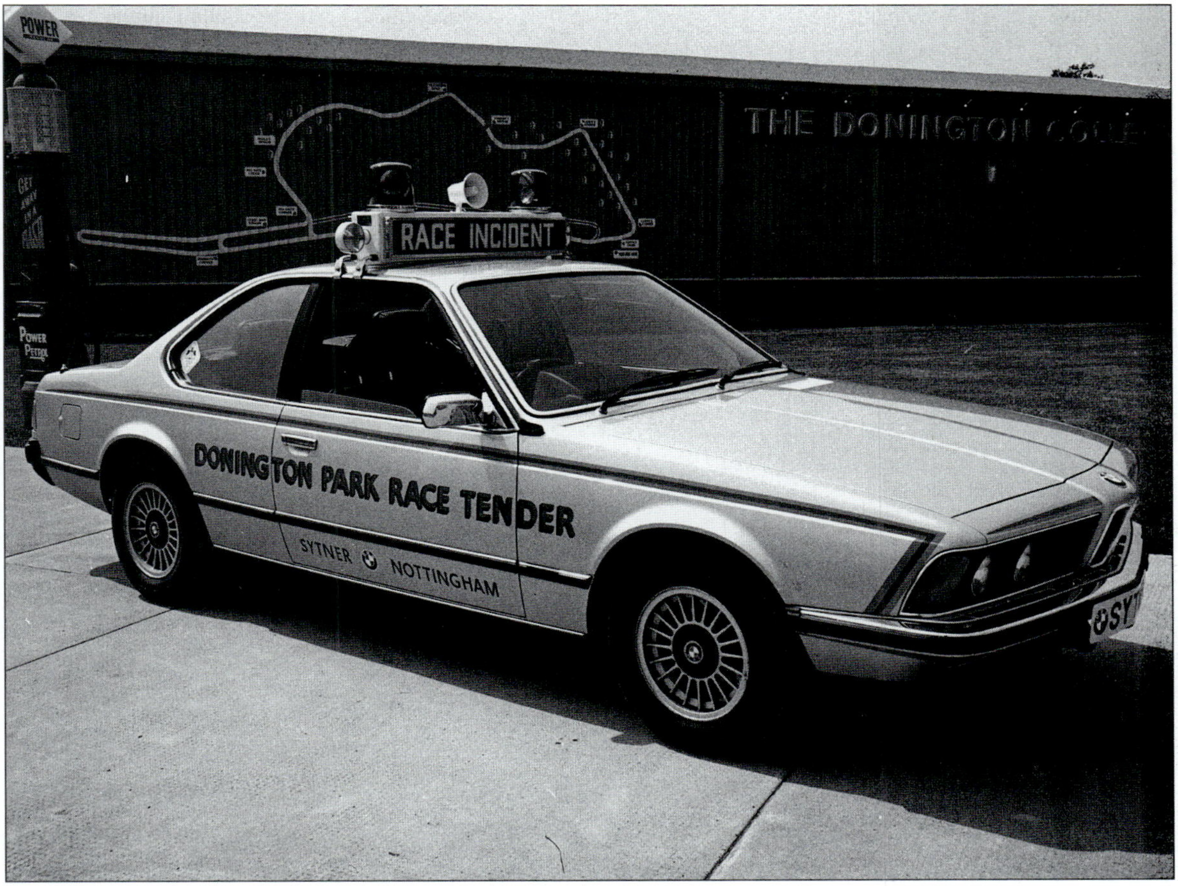

A rather special 633CSi, extensively modified as a safety car for the Donington Park circuit. Fire fighting equipment took the place of the rear seats, and self-levelling suspension was fitted to cope with its weight.

once, but it made for a car that was forgiving of driver lapses. By the way, there were five suspension settings, from Comfort to Sport.

An 850CSi came in 1992, with a 5.6-litre V12, and a drive-by-wire electronic accelerator system. Additions such as the front spoiler might have qualified it for an 'M' designation, but the possibility of such a car in the 8 Series had been dismissed the previous year. Features such as that spoiler and a rear diffuser were carried over to the 840Ci Sport when the 850CSi was discontinued in 1996.

The range was broadened late in 1993 with the 840Ci, which used the V8 first seen in the 7 Series. This model had a five-speed automatic gearbox, with three driving programmes to suit road conditions and driving style. (A six-speed manual 'box was offered in 1994.) The 840Ci Sport had a 4.4-litre version of the V8.

The V12 cars continued in production through the 1990s, although never in large numbers. In the second half of the decade the V8 cars were felt appropriate – and adequate – in some markets, such as the UK; the cars listed by the British company for 1999 were the 4.4-litre 840Ci and the 5.4-litre 850Ci, both 155 mph coupés.

In most respects the 8 Series cars were exemplary, equipped to very high standards. They were not sports cars, but they were outstanding GT cars for rapid long-distance journeys, which wholly justified the BMW Power and Dynamics slogans.

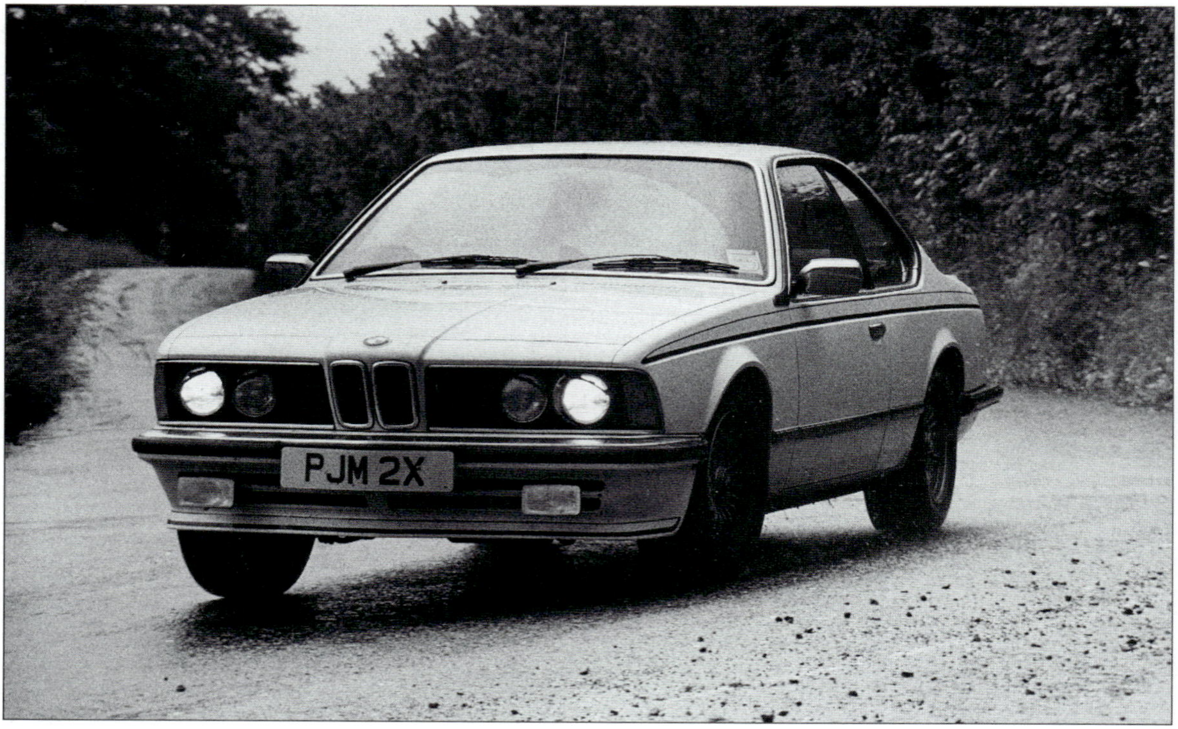

The first high-performance 6 Series variant was the 635CSi, with a 3.5-litre 218 bhp straight six which used the block of the M1 unit, a lowered suspension and ABS, more sophisticated electronics and revised trim. Its top speed was within a whisker of 140 mph (225 kmh).

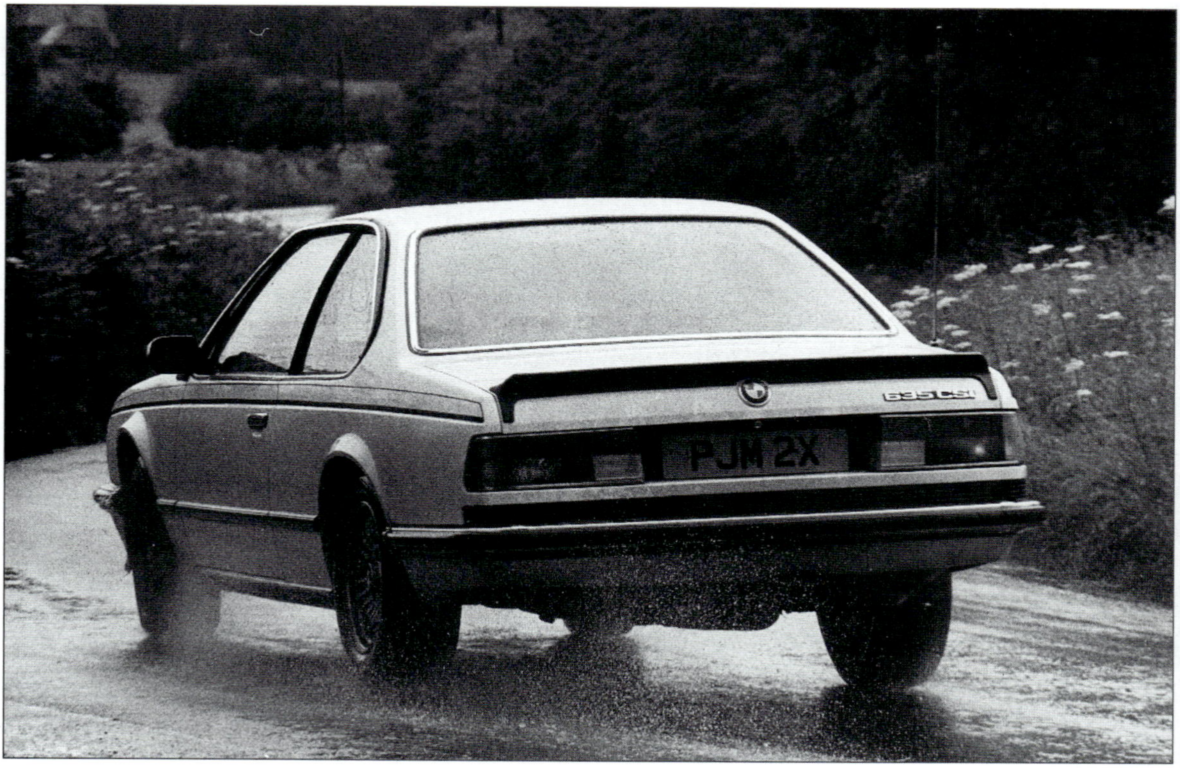

Road stance with the revised suspension, and a rear spoiler, backed up identification with chrome letters. All-round visibility was first-class, and handling on wet and winding roads inspired confidence.

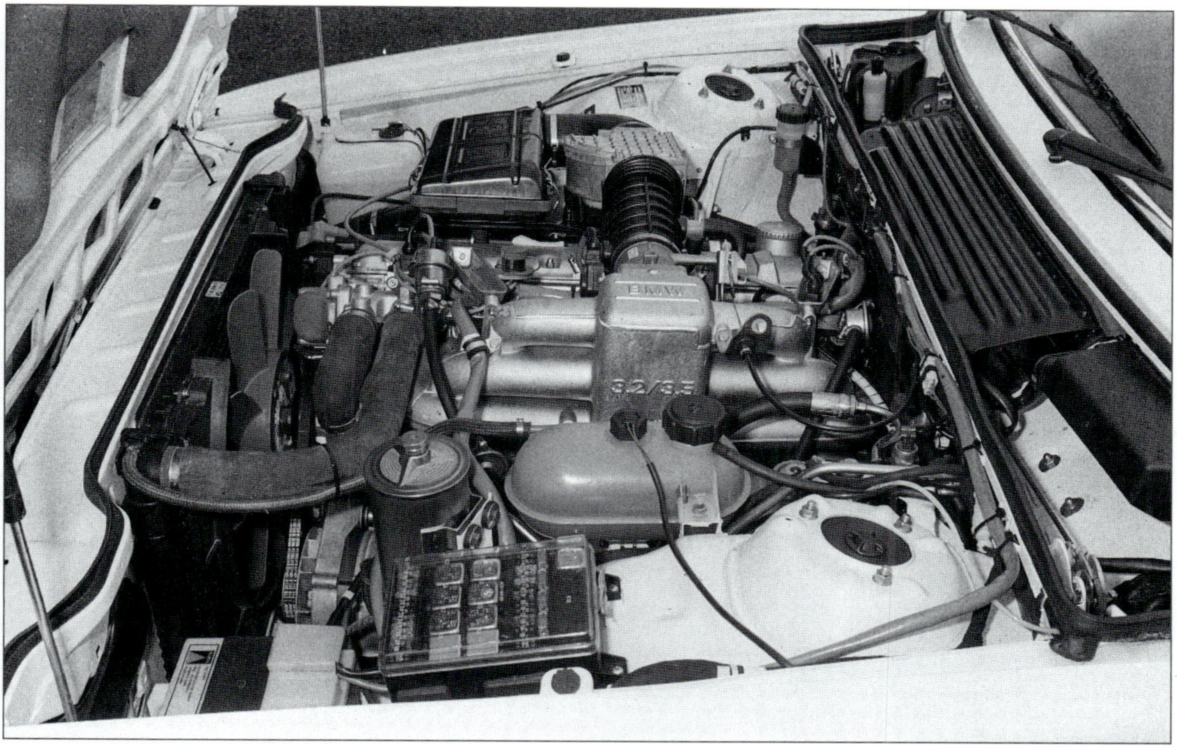

There was little space under the bonnet, but accessibility was good. The Bosch Motronic electronic fuel injection equipment dominates the engine. In this car that was the 'small-bore' 3.4-litre unit, in which 23 cc was lost.

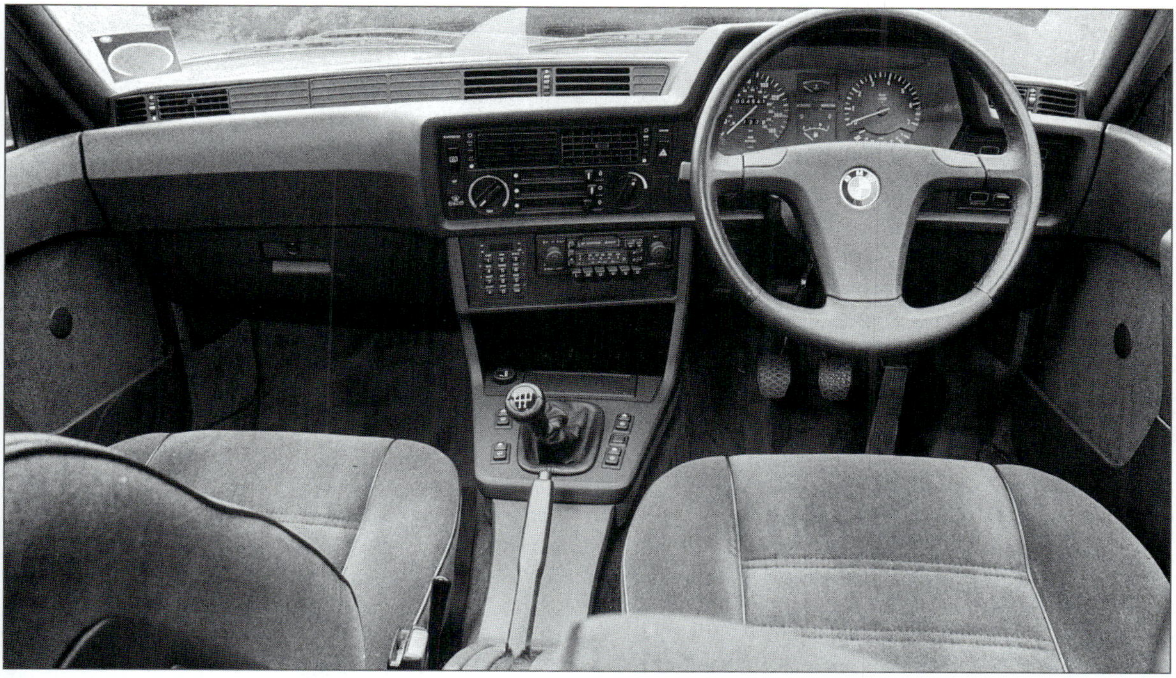

There was no ornamentation on the fascia, and as usual the instruments were strictly functional. The large dials are the speedometer (left) and rev counter, with an 'econometer' in its lower segment. Water temperature and fuel gauges are in the centre. To the left of the radio is a function panel, giving outside temperature, average speed, fuel consumption, journey time, and so on. The display included a check control, whereby functions could be tested when a button was depressed.

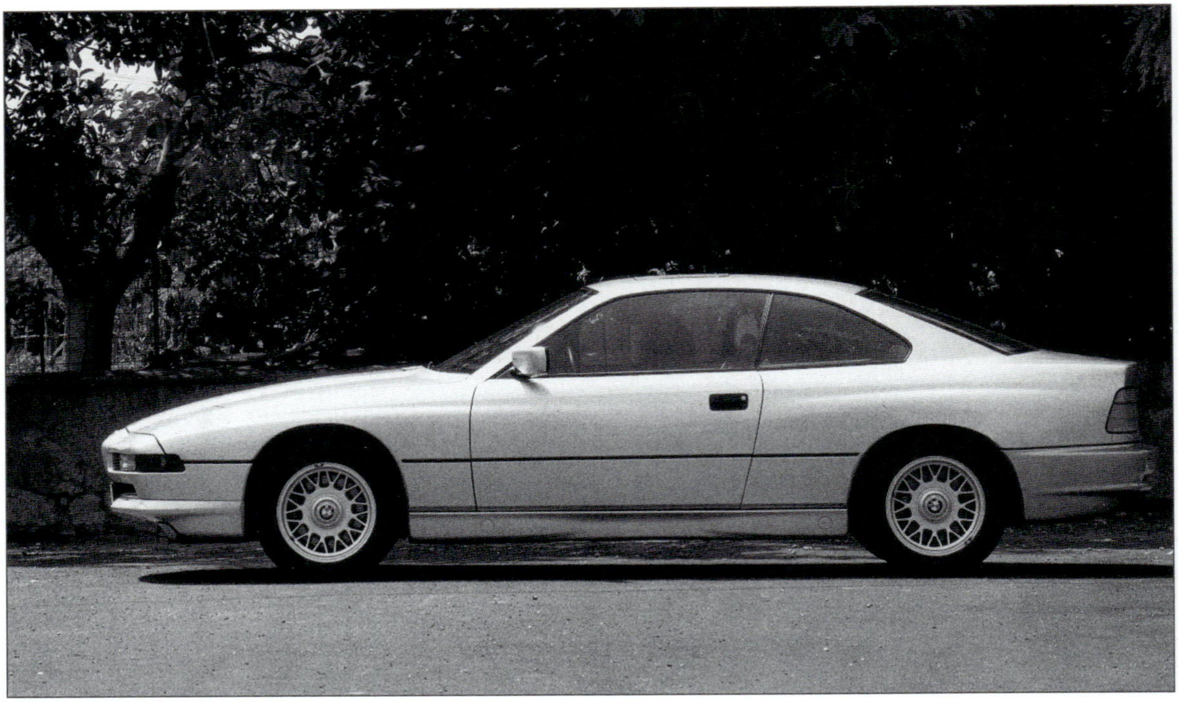

In side elevation, the 8 Series cars were notably clean (above), rising with a gentle wedge line from nose to tail. The nose (below) naturally included a kidney grille, and the appearance of aerodynamic efficiency was borne out in practice.

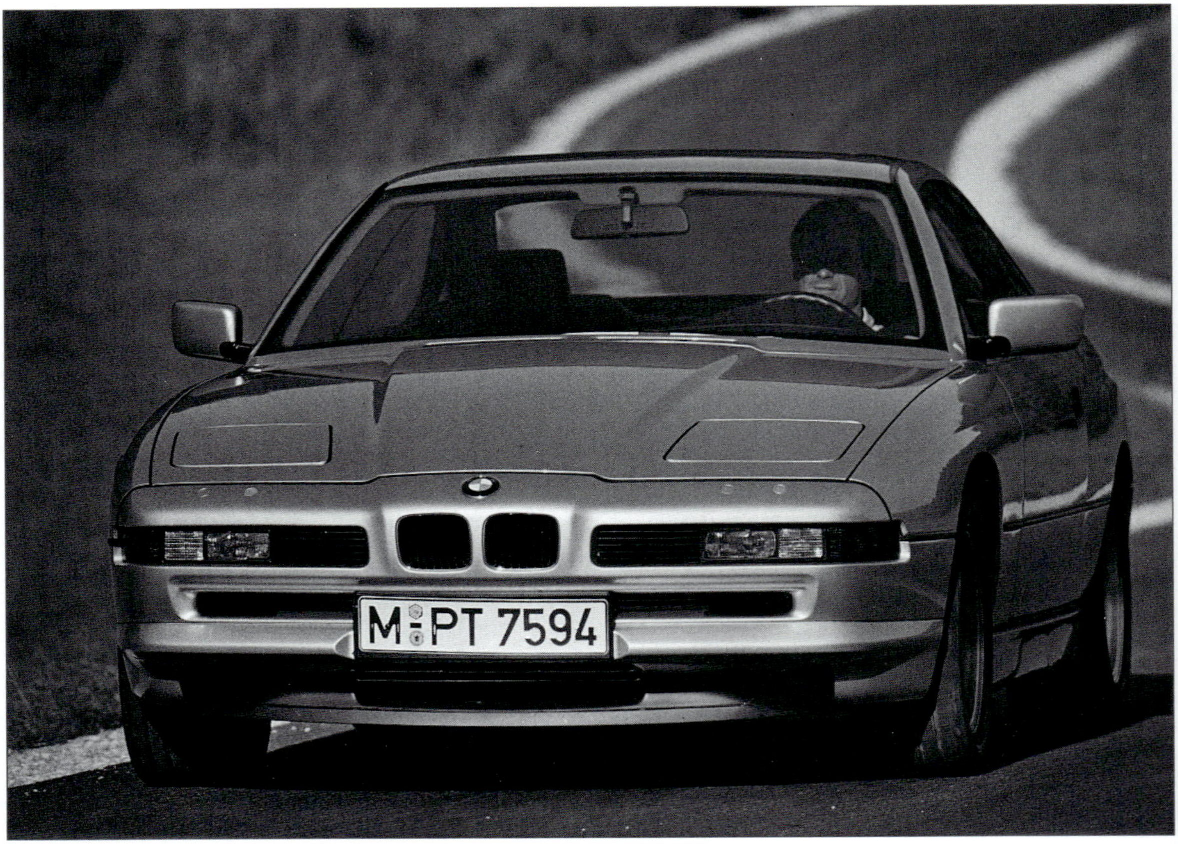

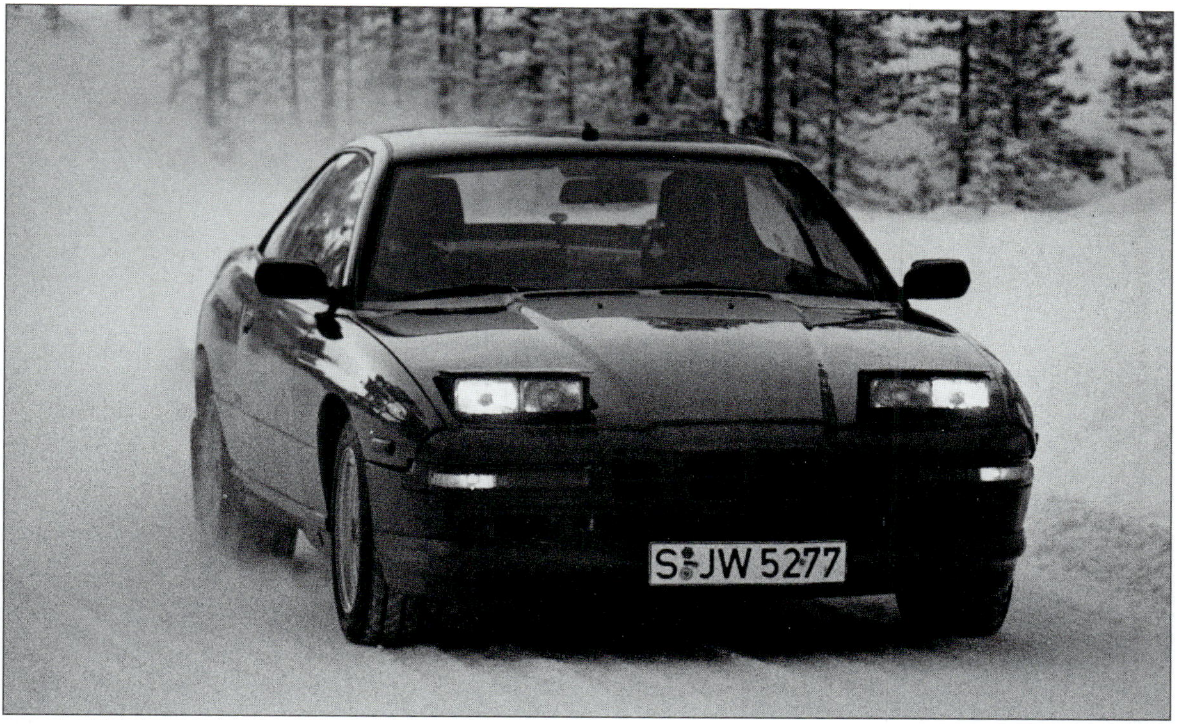

The pre-production 850i seen here during winter tests (above) shows that even raised headlights did not spoil the lines. The tail (below) was equally clean, with no afterthoughts, overall or in details. This 850CSi carries no model identification – an anonymity affectation that is not uncommon on modern BMWs, but the wide wheels and the tail pipes will tell other drivers that they have been passed by something special. . . .

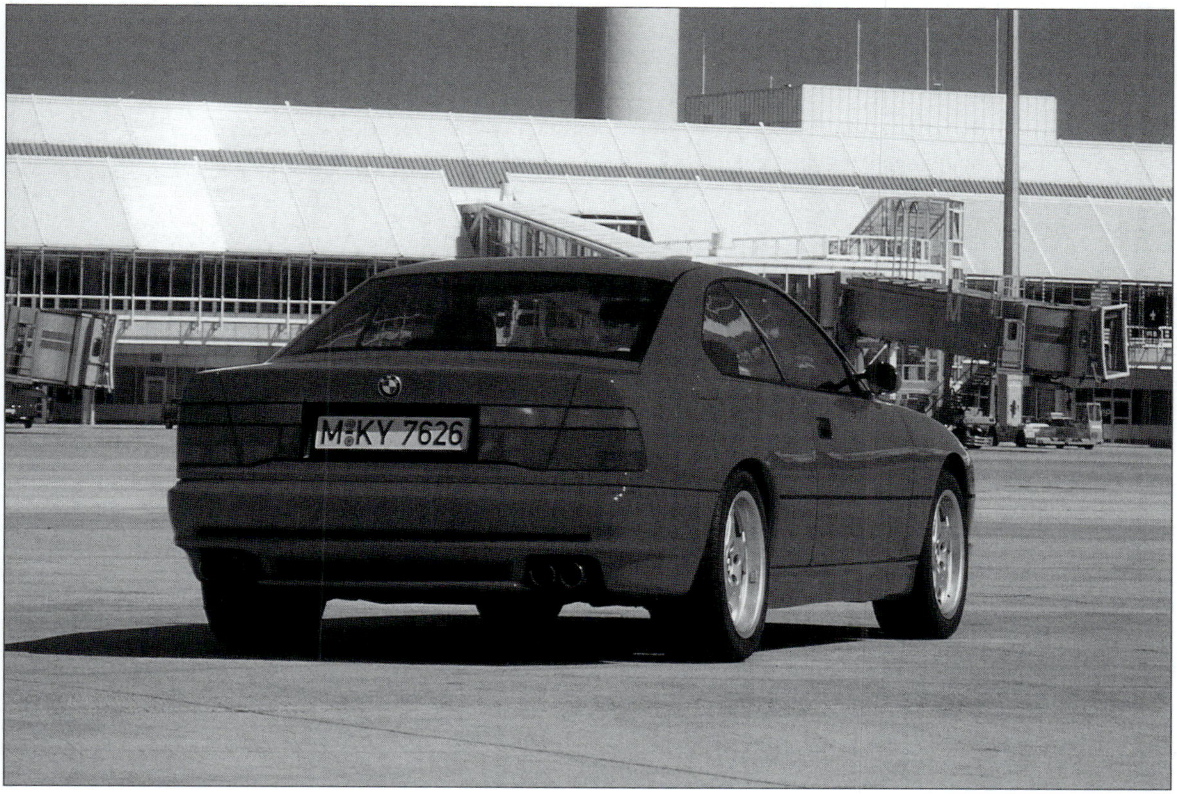

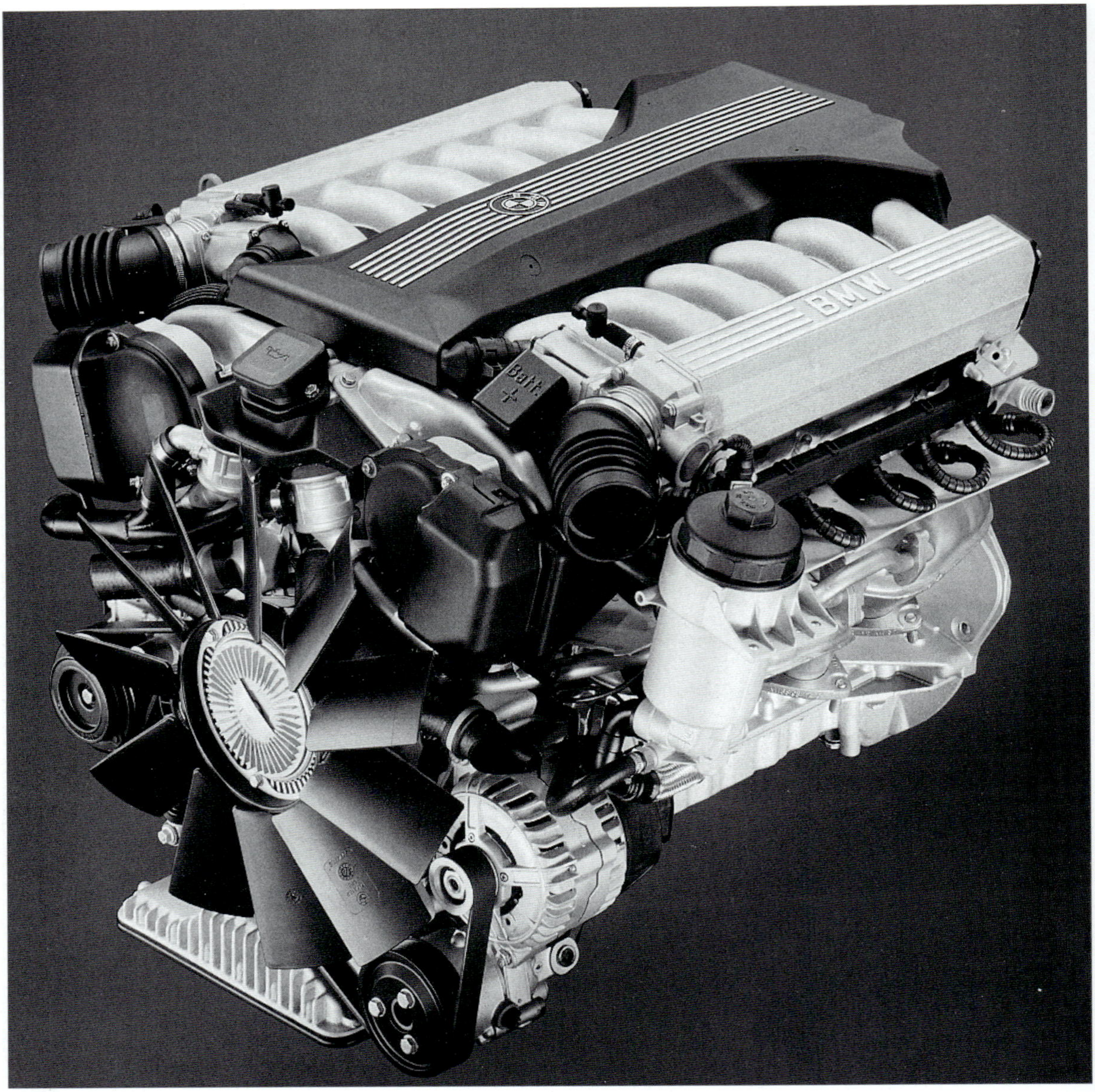

The V12 in its revised 5.4-litre 1995 form was another very tidy power unit. The intake pipes fed air to the opposite bank of cylinders, in the interests of efficient breathing. The block and heads were in aluminium alloy, with the pistons running directly in the block. The V12 was rated at 326 bhp, and drove through a new five-speed automatic transmission. V12 8 Series cars had been discontinued in Britain in autumn 1994.

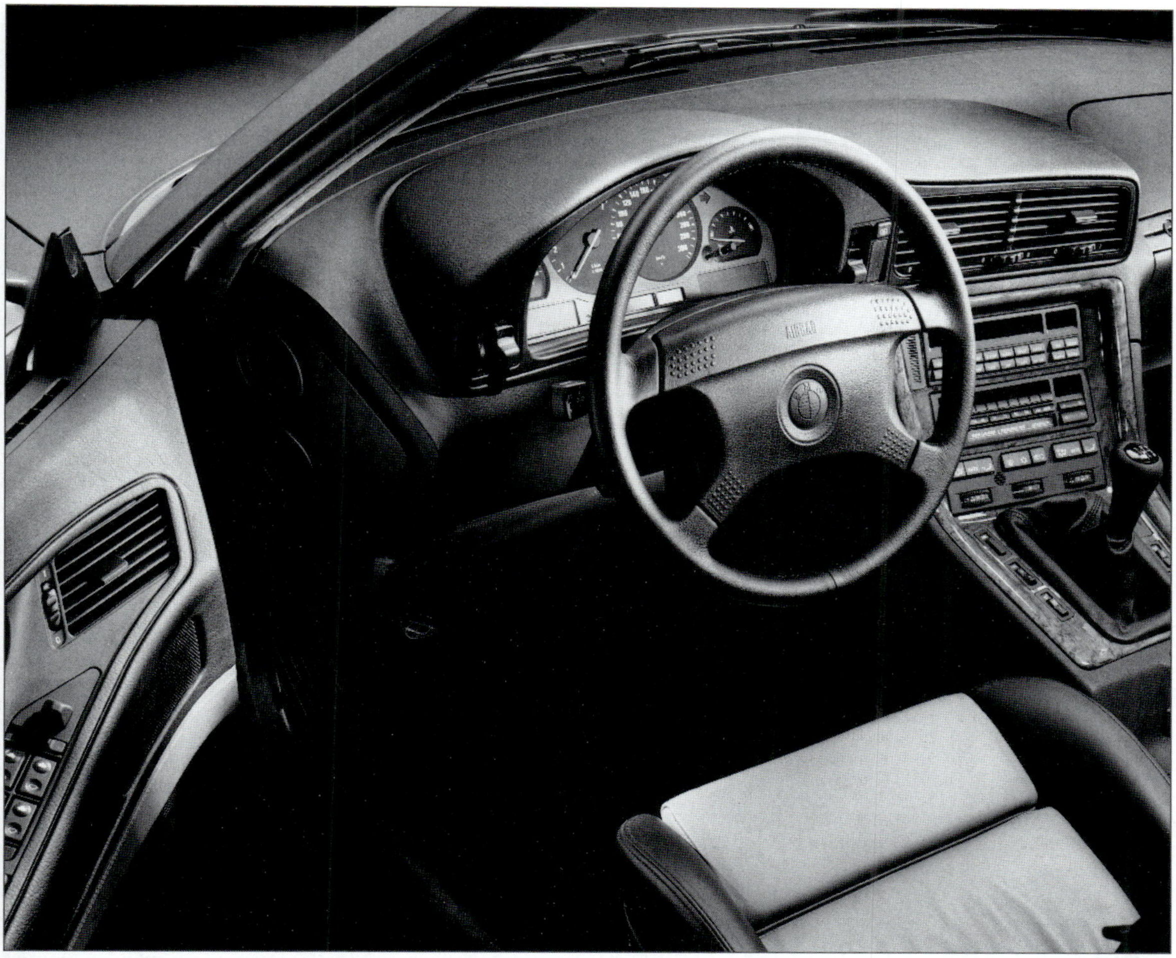

The 8 Series fascia had all the 'family' traits, with only the necessary dials directly in front of the driver (and to have red numbers and indicator needles by the mid-1990s). The speedometer overlapping the rev counter was a styling quirk that was not always appreciated. As usual, the secondary controls were angled towards the driver. This 850CSi has two-tone leather seats; the driver's seat has a memory, and would self-adjust to a regular driver's preferred position.

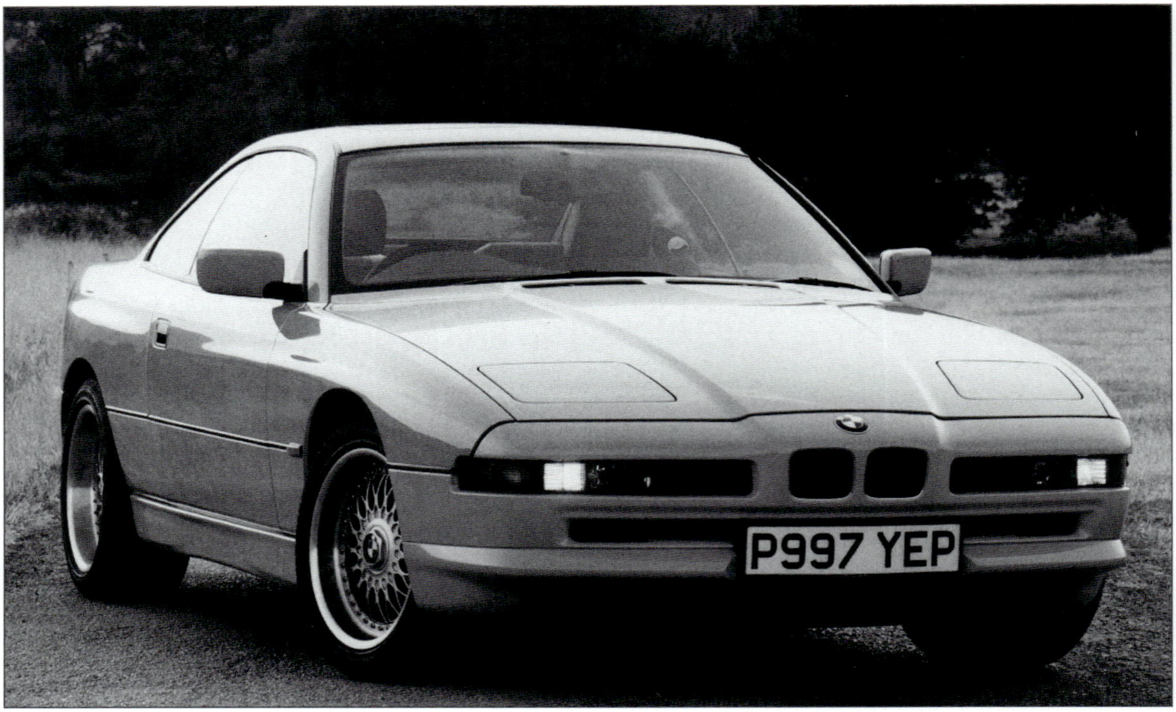

Outwardly, there was little to distinguish the cars with V8 engines, in this case a 1996 840Ci with the 4.4-litre four-valve unit which gave it a governed top speed of 155 mph (250 kmh). Among its features was a chemically treated water-repellent windscreen.

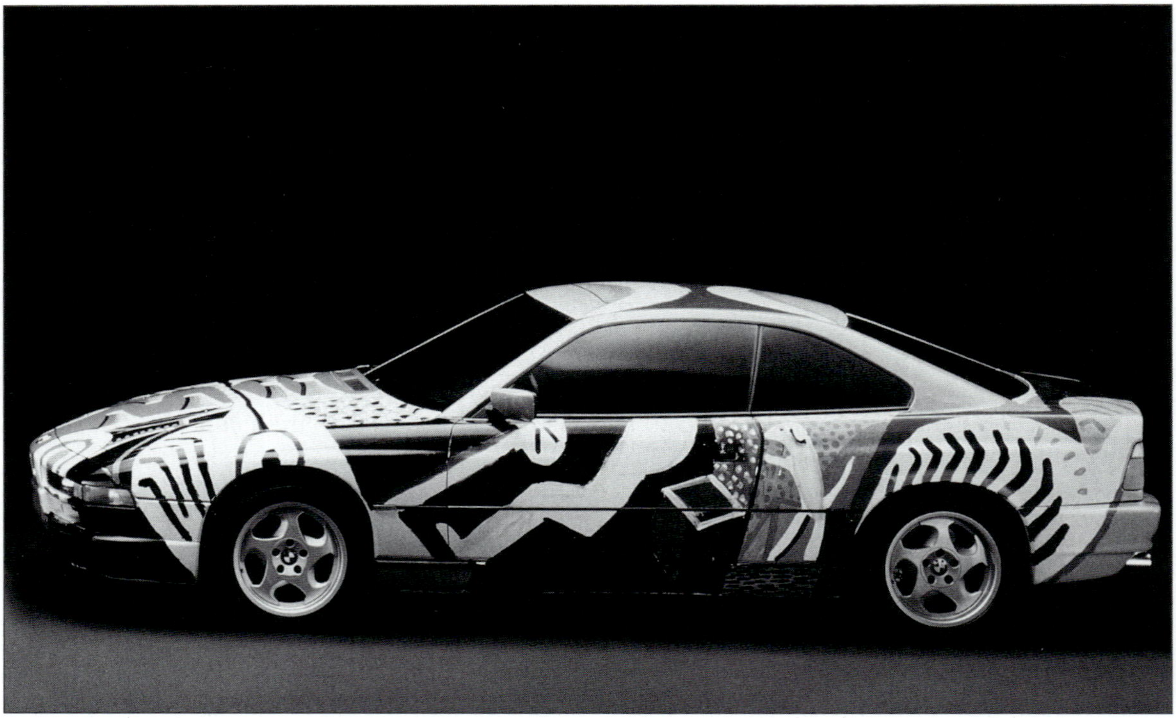

In 1975 Herve Poulain's 3.0CSL entry at Le Mans was run with an odd graffiti colour scheme by Alexander Calder, and that inspired the 'BMW Art Car Collection'. The decorated cars certainly attracted attention, and as some were exhibited in leading galleries and museums, they presumably attracted the approval of the art world. The fourteenth art car was this 1995 850CSi, painted by David Hockney.

CHAPTER SEVEN

7 SERIES

The large BMWs introduced in 1977 revived the line of elegant luxury saloons, and were conservative cars with strong family resemblances to their smaller contemporaries. They were also challenges to Mercedes-Benz – the first in this sector for more than two decades – and Jaguar, while in years to come 7 Series cars would have to face competitors from marques such as Audi and Lexus. They were orthodox, but the design was to prove versatile, and equipment levels were high – as, indeed, they needed to be in this class.

The first E23s were bigger than previous BMWs, and at first they were not particularly sprightly. When the inevitable development got under way, they were to mature as superb luxury cars with outstanding performance. The series was introduced with three versions, using the straight six – 728 (2.8 litres), 730i (3 litres) and 735i (3.2 litres). These first cars were to be criticized in several respects – handling at high speeds and body roll, fuel consumption and noise – and were considerably reworked under the skin. Production, 1977–79, reached 54,000 units.

Most shortcomings were overcome by 1979, when the base model was still the 728 (but 728i, as all 7 Series cars had fuel injection), with the 732i and 733i (both 3.2 litres) followed by the 735i (3.5 litres) in 1980. There was also the 745i with a KKK turbocharger, for the German market: its designation was assumed to be a clue to an original intention to fit a 4.5-litre V12. That was an automatic transmission car, whereas the others had four-speed Getrag manual gearboxes (728i–733i) or five-speed 'boxes in the 735i, the first BMW with an overdrive fifth gear. Generally, these cars were rated more responsive than the early 7 Series types; the 735i was a 130 mph (210 kmh) car, while the 745i reached 138 mph (220 kmh) with its smaller, but turbocharged, engine.

There was an interim facelift in 1982, when the 735i/735iA got a lighter and fractionally smaller straight six. With market variations – the South African 735i, for example, was effectively the 745i without a turbo, while the 733i was the only model offered in the USA – this range lasted until the E32 marked a major overhaul.

Introduced in 1986, the E32 cars were first listed with the straight six in 3.4-litre form in the 735i and 735iSE (145 mph/230 kmh cars), and then as a 3-litre unit in the 730i. The body was more rounded, and with attention to under-car aerodynamics the drag coefficient was cut to 0.32. Electronics were more abundant, for example with circuits that selected fall-back channels if one failed, and these were the first BMWs to have 'ellipsoidal' headlights.

The first E32 cars were a prelude to the 750i, introduced in 1987 with a 5-litre V12. A 60-degree engine, with two valves per cylinder and a chain-driven overhead camshaft, this was rated at 300 bhp. It was the first mass-production V12 since Jaguar's early in the 1970s (that unit was still in production). BMW used its V12 in the 750i 'sports saloon' and in the range-topping 750iL, a slightly heavier car that some testers found had better ride characteristics, presumably because of its longer wheelbase (L-signified). A ZF four-speed automatic transmission was normal. Electronics controlled even more functions in these cars, from the fascia computer that monitored them, to the 'memory' seats, to engine diagnosis and, this being a German high-performance car, to rein in the engine when road speed reached 250 kmh.

Surprisingly soon, in 1992, another engine was introduced into the 7 Series, a V8. This was another light-alloy unit, but with four valves per cylinder and twin overhead camshafts; in 3-litre form it gave 218 bhp and the 4-litre version was rated at 286 bhp. The smaller V8 gave little immediate power advantage over the 3.5-litre straight six, while the larger V8 led pundits to question the need for the V12.

The exterior lines changed little, the cabin hardly at all. More change did come with the E38 types in 1994. The front suspension was revised, and there was a new multi-link system at the rear (described as 'an integral axle with multi-dimensional suspension'). Reports suggested that the ride was improved with no reduction in the car's agility. The body was wider and higher, and also longer, so that rear seat space was improved.

The V12 750s continued, of course, while revised V8 cars were introduced for 1996. The 735i and the 740i had 3.5-litre and 4.4-litre engines; better torque, rather than more power, was claimed for these, together with improved fuel consumption. That year the AGS Steptronic transmission was brought in for these cars, giving drivers the choice of automatic or manual shifting. This transmission was also available on the new 728i, which had a lighter 24-valve straight six (the previous 728/728i had been run 1977–86, and production reached 103,000). It also had the Automatic Stability Control mentioned in the previous chapter. The fully integrated navigation system, standard on the 750 models in 1996, could be fitted to other 7 Series types – at a price. . . .

Supercharged straight six diesel engines were added to the 7 Series range for 1999 (announced in 1998), in the 725tds (2.5 litres, 143 bhp) and the 730d (3 litres, 184 bhp). The luxury 3.9-litre 740d was not announced until 1999. The 730d and the 740d had Steptronic transmission, but the diesels did not have the Dynamic Stability Control (DSC) fitted to the V8 and V12 cars, and optional on the 728i, and backed up by Dynamic Brake Control (DBC – another of those sets of initials beloved by BMW publicity people).

Although the equipment levels of the 750s was already high, more was added to these top models, with particular attention to anti-theft aspects, while double glazing brought them into line with other leading marques. Then the L7, first intended for 'special markets outside Europe', was made more widely available 'upon special request'. With an extra 10 inches in the wheelbase, it became the prestige model of BMW's prestige series. By the end of the century BMW needed only a really large limousine to complement these fine luxury saloons, to top out its range. But then it was due to inherit Rolls-Royce in 2003. . . .

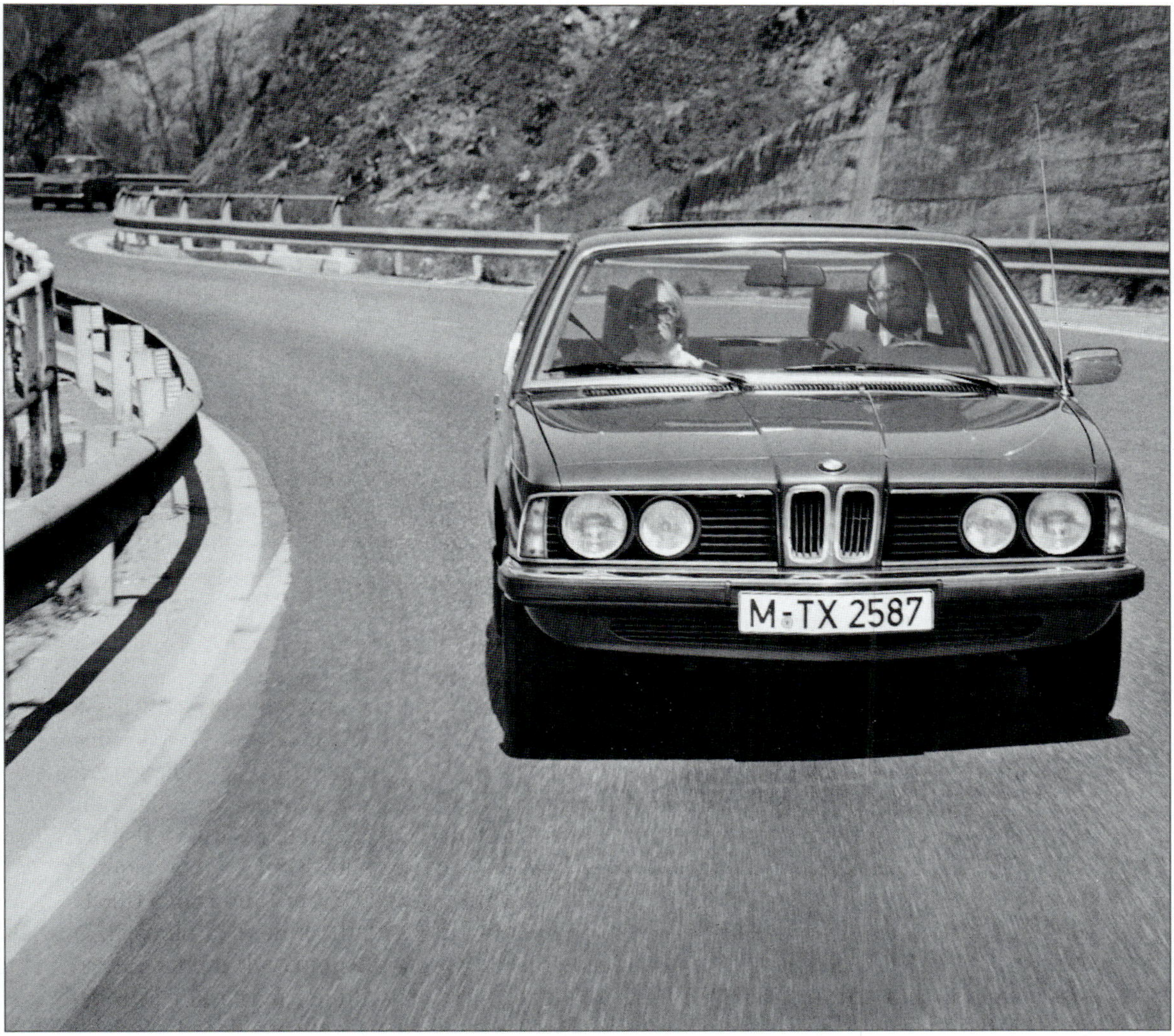

An early 7 Series car, this 730 shows many of the characteristics of BMW's premier line, as it returned to the large-car class in 1977. The styling related it to other BMWs, and its road stance looks confident here. The smallest of the debut range, the 2.8-litre 728, was a 120 mph (193 kmh) car, while this 730 was around 5 mph faster.

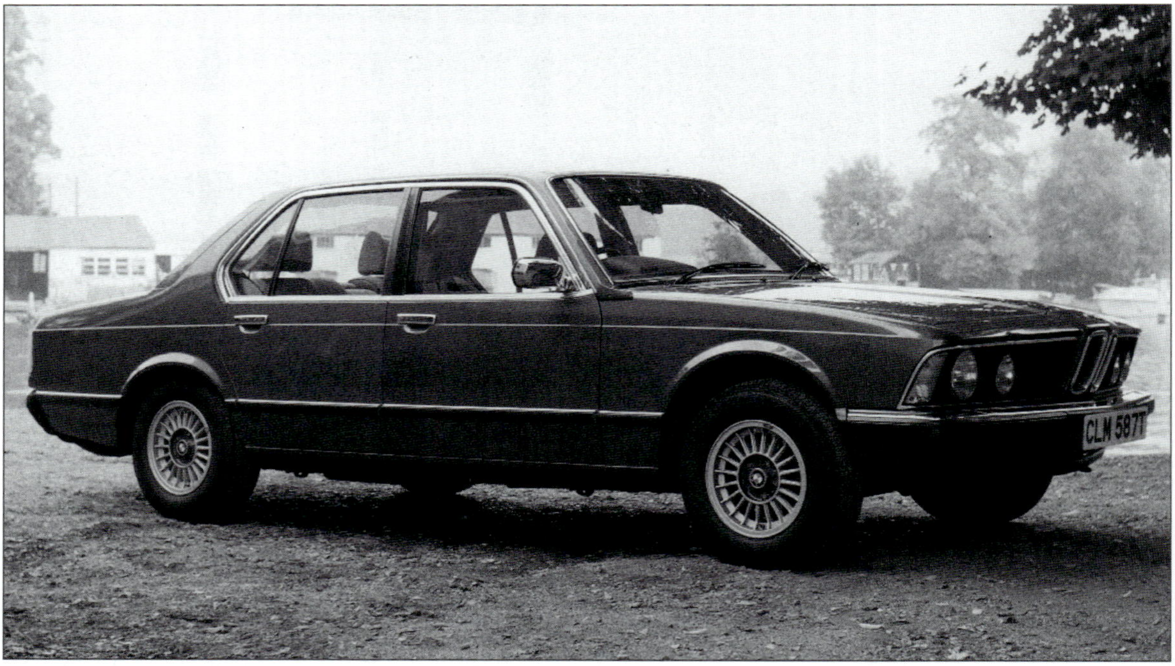

This right-hand-drive 730 shows the well-balanced overall lines, with the trade-mark C pillar and the reasonable glass area in the doors. The alloy wheels are some of the simplest and most attractive types fitted to 7 Series cars.

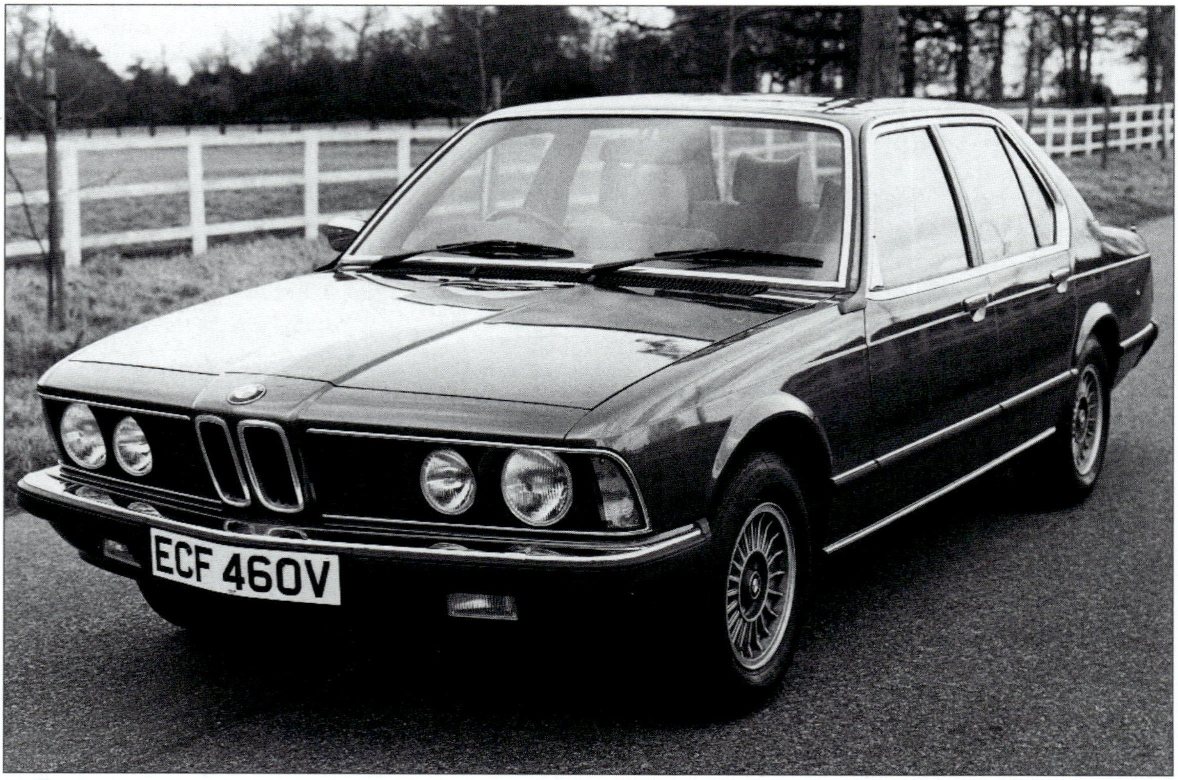

Two years into production, the 7 Series cars appeared little changed, although by this time all had fuel-injected engines. The fog lights suggest that this is a 735i, a 132 mph (211 kmh) car. Standard fittings included items such as a heated driver's mirror and door lock, but an original-equipment radio might have been more welcome. . . .

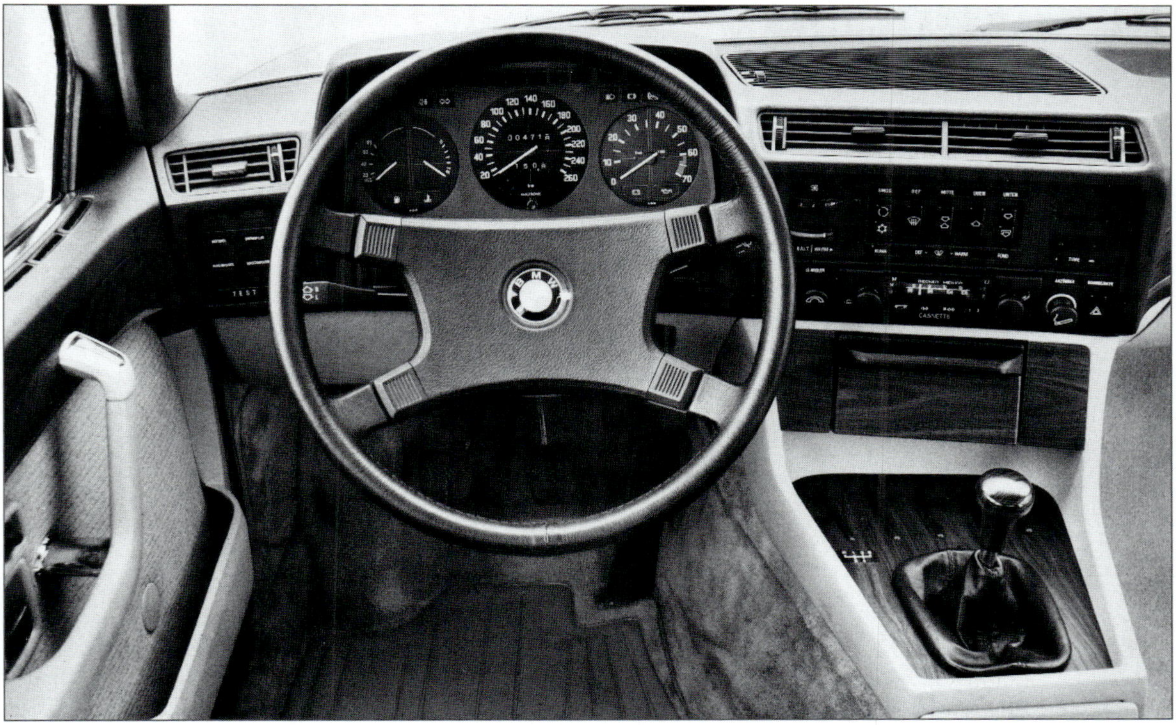

The second-generation 7 Series cars in the late 1970s had straightforward fascias, with clear and easy-to-read instruments. Left to right, through the wheel of this left-hand-drive car: fuel and water temperature, speedometer reading to 250 kmh (on right-hand drive cars the scale reached 160 mph) and rev counter. The entire console houses the radio and complicated heating and ventilation controls. Velour-upholstered seats (below) gave good lateral support, but the window controls between them called for an awkward movement. Door capping was plastic but simulated wood, which seems a trivial economy.

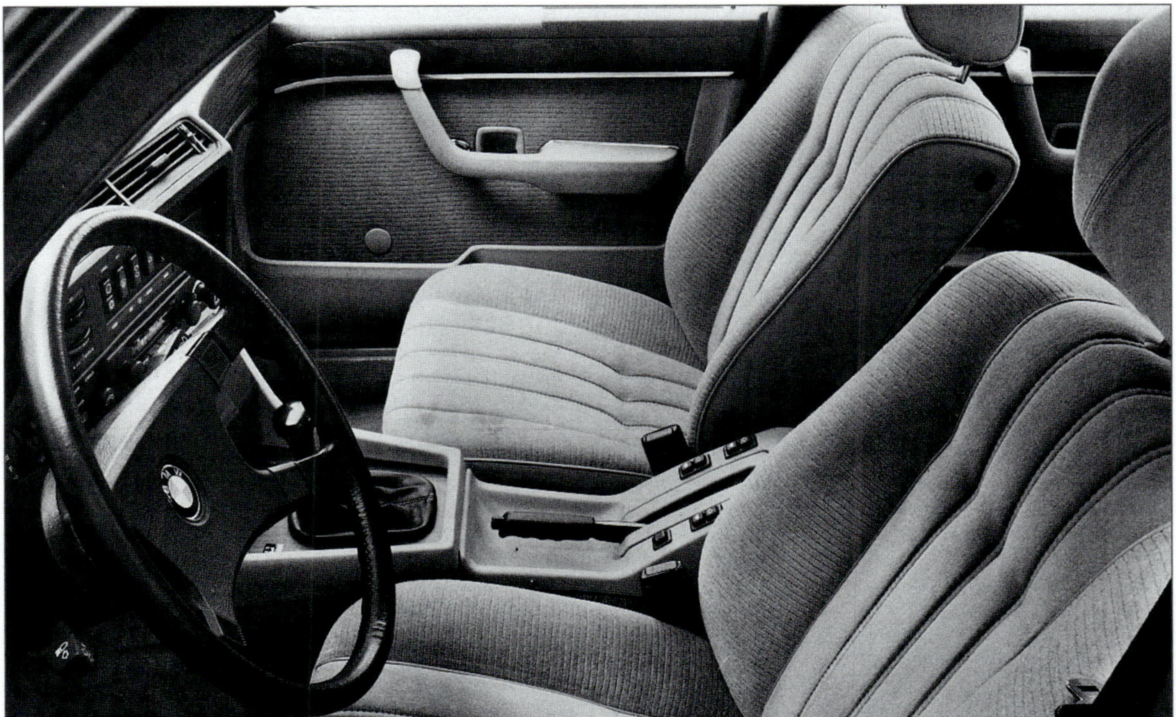

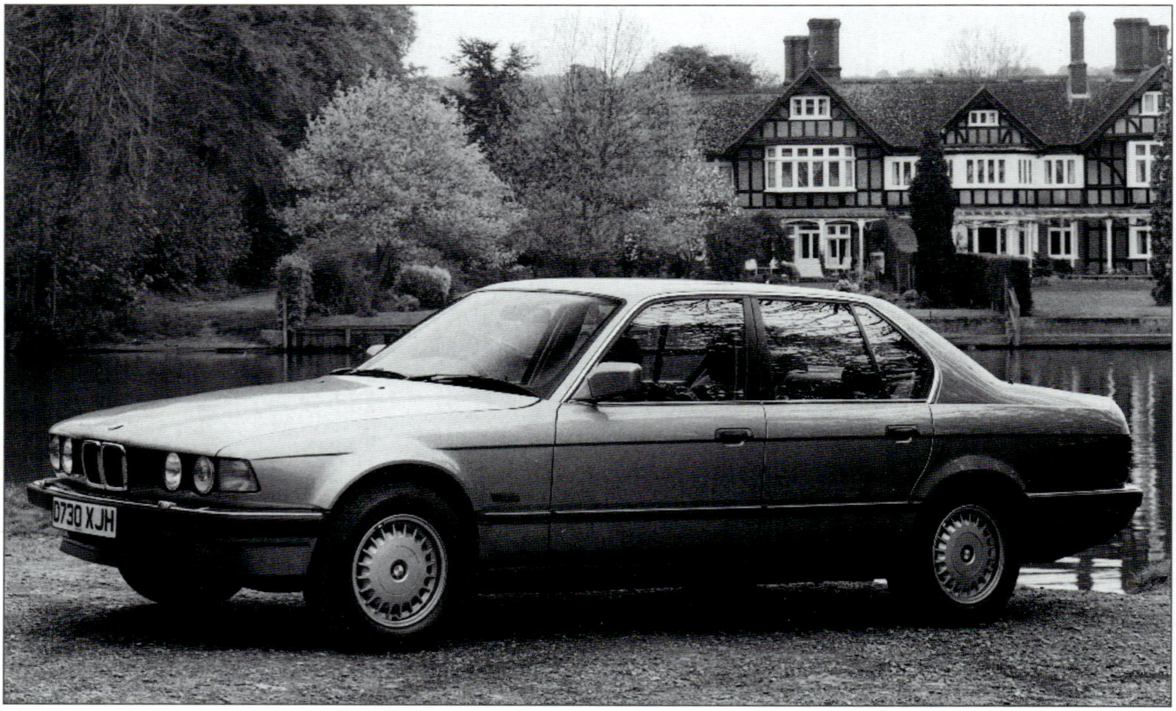

The smallest of the E32 models in 1986, a 730i. The new body looks substantially larger all round, but in fact the real gains were in the cabin. In detail, the nose is finer with no echoes of the earlier prow – the kidneys have lost some depth and gained in width. The 2986 cc 197 bhp engine gave this car a top speed of 130 mph (208 kmh). A drag coefficient of 0.32 helped.

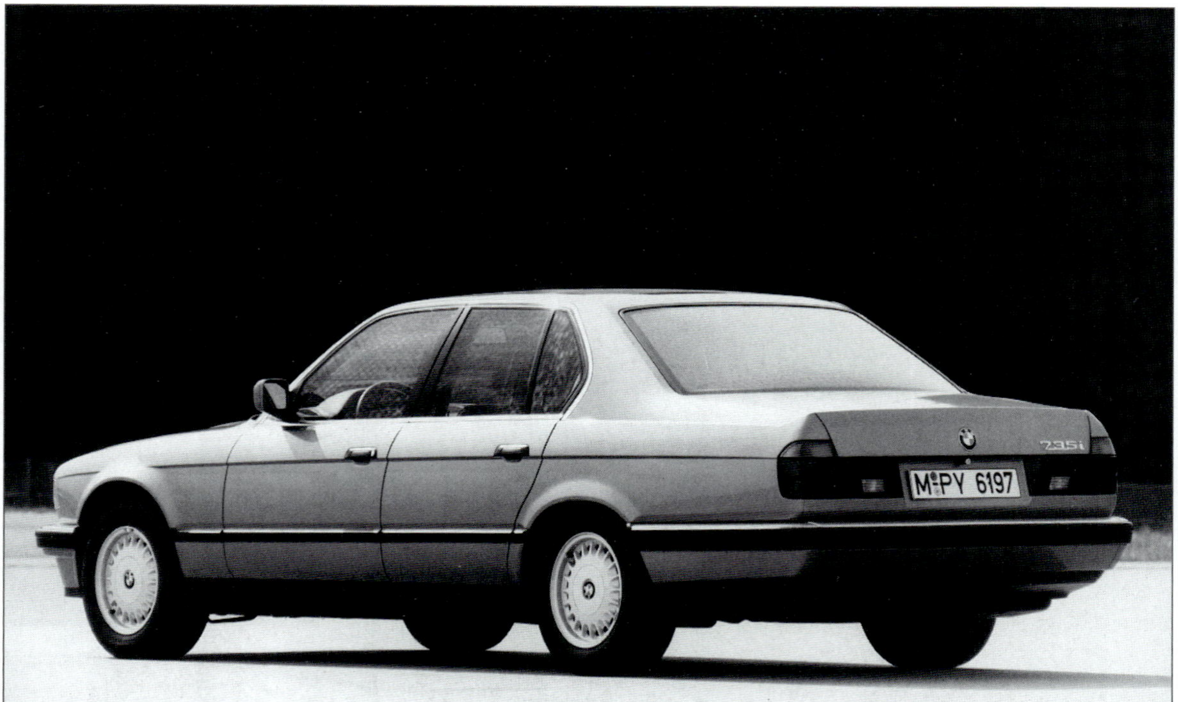

The rear quarters gave a stronger impression of size, and the tail treatment was outstandingly clean. This is a 1986 735i, with the larger of the two straight six engines listed, its 211 bhp giving a maximum speed of 143 mph (230 kmh).

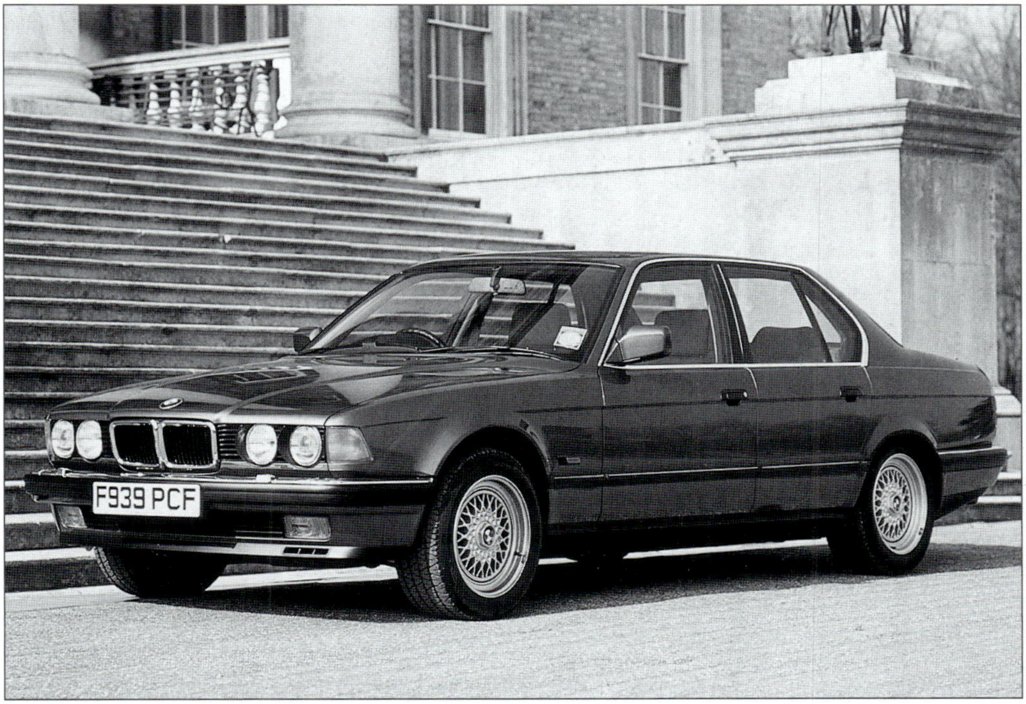

The 750s were landmark cars in the 7 Series, with a magnificent V12 in the existing chassis. The 750i was promoted as a 'sports model', but introduced with no alternative to the four-speed automatic gearbox. The engine was refined, smooth and quiet, giving a top speed at the 250 kmh limit. If there was criticism, it was aimed at handling and ride quality, perhaps because the only settings were 'Comfort' and 'Sport'.

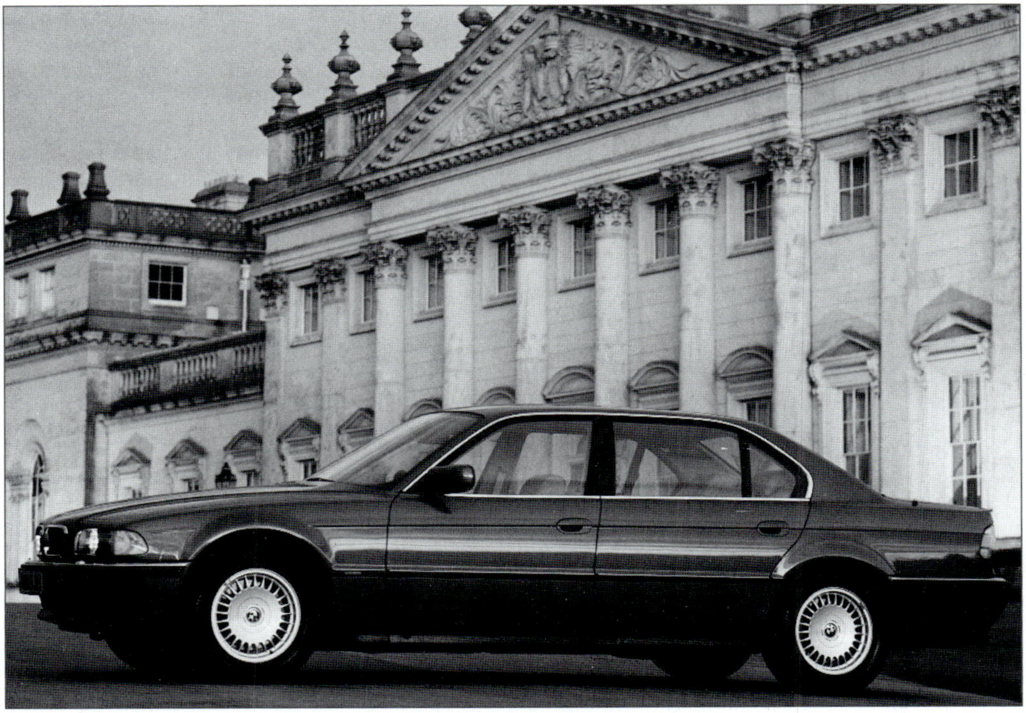

The 750iL was stretched by 5½ inches, and that made for real comfort for four; coupled with the high specification, it made the 750iL a worthy long-distance executive car.

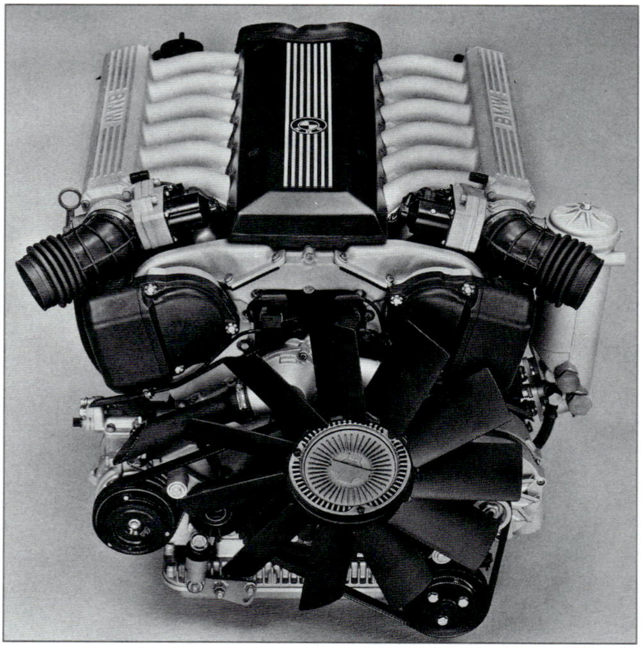

The 5-litre V12 was typically tidy and symmetrical, and it fitted into the existing E32 engine bay. When introduced, it was rated at 300 bhp at 5200 rpm.

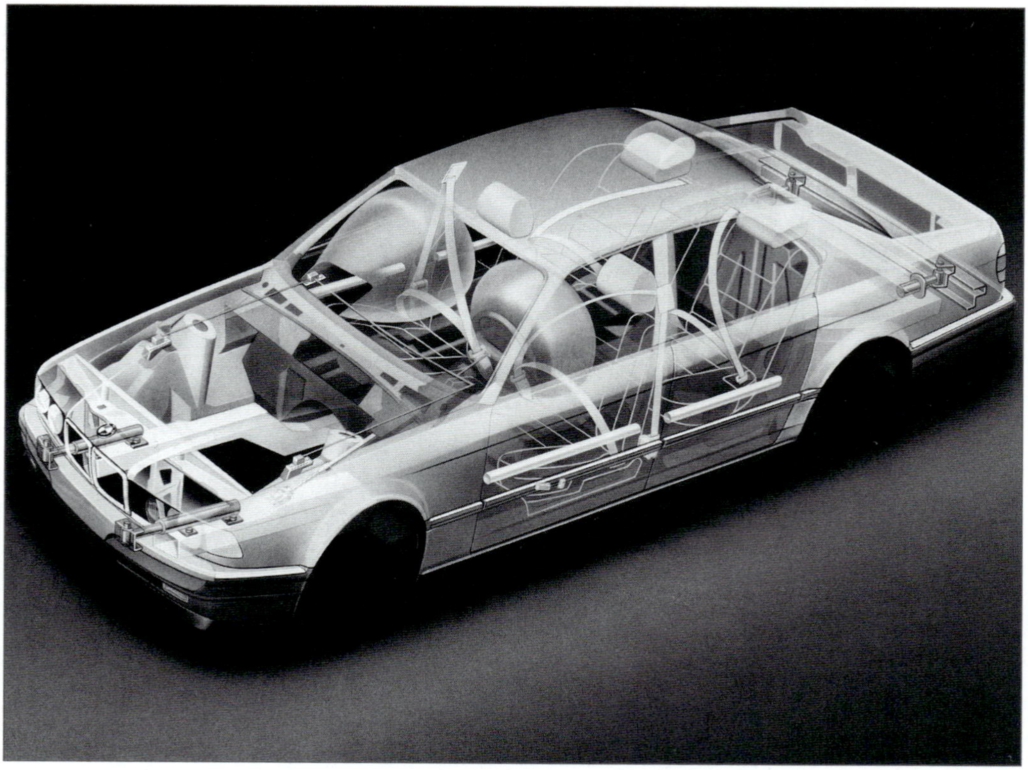

The passive safety features of the 7 Series cars as the 1990s opened naturally included driver and passenger air bags (side air bags were to come), belts, impact protection bars in the doors and crumple zones. This drawing also shows the impact-absorbing bumpers: at low speeds, up to 4 mph, hydraulic pistons absorb impact and return a bumper to its original position; at up to 10 mph the pistons are forced back into light boxes, intended to collapse and protect the engine mounts. (A dented bumper could be replaced in three minutes.) Then there was the primary safety inherent in the suspension and brakes, with the sophisticated stability and brake controls.

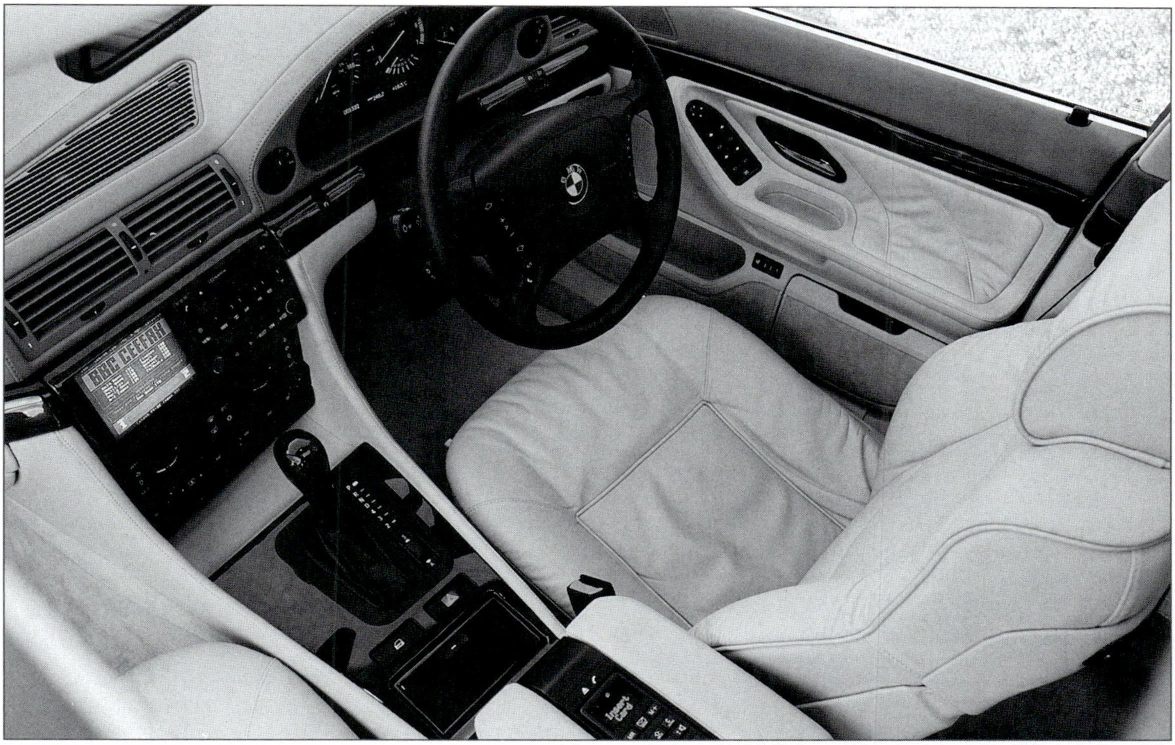

The luxurious interior of a mid-1990s 750iL, with leather prominent. The steering wheel has controls for radio, air circulation, cruise control, and a telephone. The console has an information display for the on-board computer and navigation system, as well as TV and videotext. There were no rear-seat leg room problems in this car: Eberhard von Kuenheim's brief to build a luxury car was executed to the full! The rear seats are electrically adjustable, in several planes, the windows are double-glazed, there are side window sun blinds, and so on.

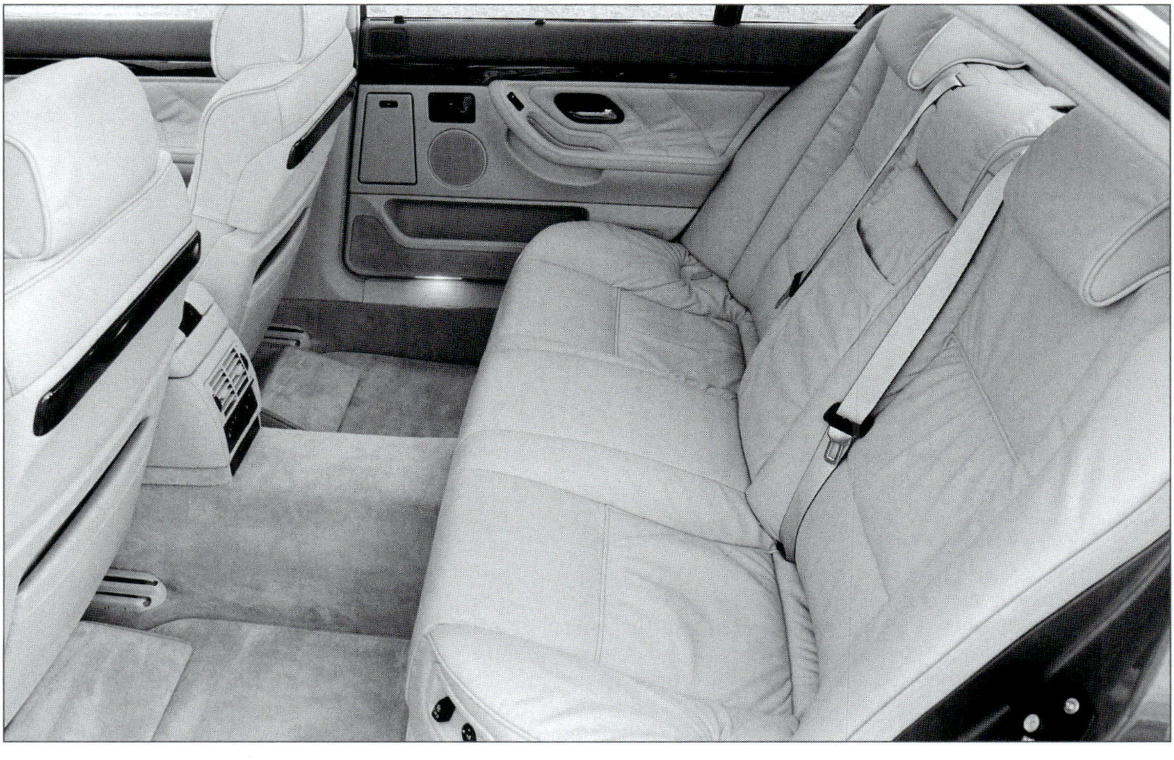

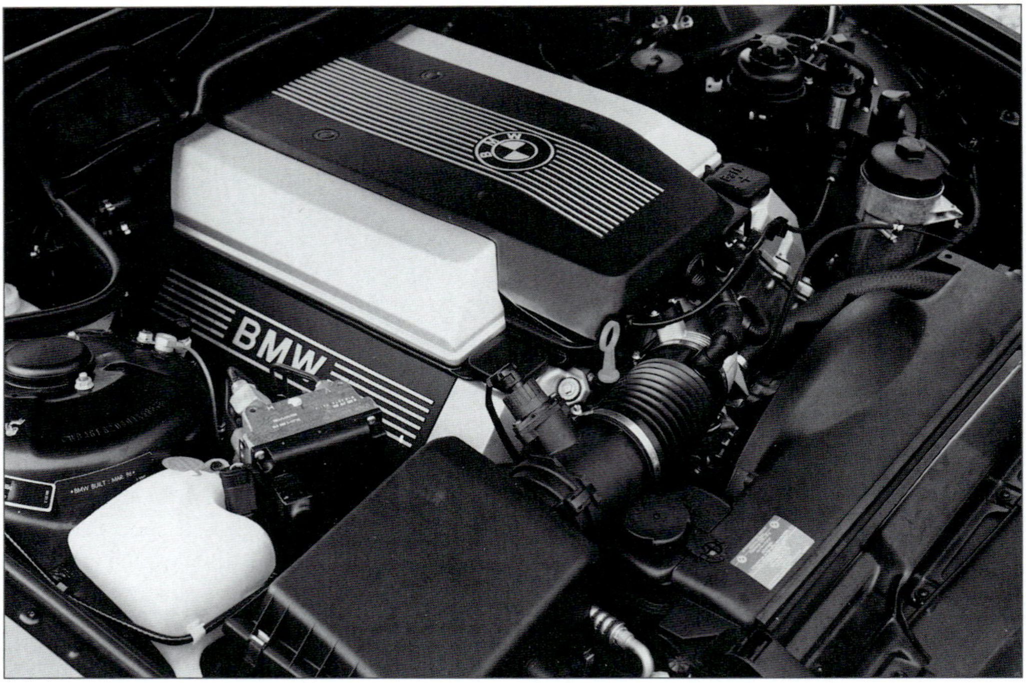

A V8 installation, which appeared identical in the 1996 models, and is typically tidy. The new 728 was introduced quietly for 1996, with a 2793 cc engine weighing only 348 lb (158 kg), with four valves per cylinder, variable camshaft control (VANOS) and Siemens digital management. Its 193 bhp gave it performance figures comparable to the 735i (top speed was 140 mph/228 kmh).

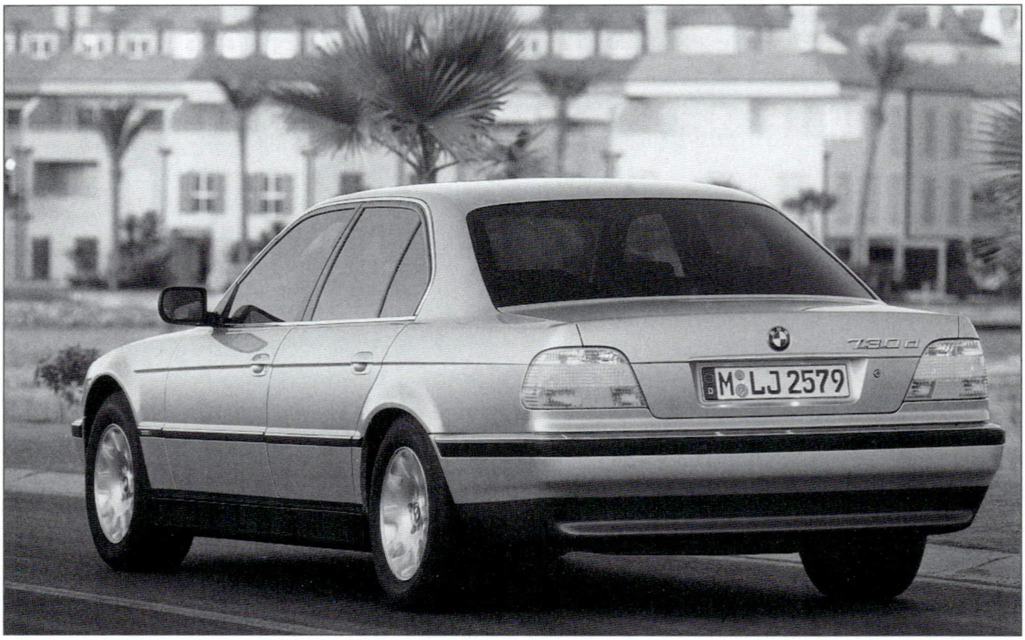

The chrome letters on the boot read 730d, and diesel luxury cars were acceptable in the 1990s, however odd they might have seemed in earlier decades. The 3-litre turbocharged straight six in this car delivered 184 bhp, for the 1999 model year, and it gave the 730d a top speed of 137 mph (220 kmh).

In 1999 BMW introduced the 740d, claimed to be a 'world first', as a luxury car with a direct injection V8 diesel engine. This 90-degree biturbo unit gave 238 bhp and drove through a five-speed gearbox, offering drivers fully automatic or sequential Steptronic gearchanges.

Chapter Eight

Modern Sports Cars

Most BMWs have had sporting characteristics – qualities such as above-average handling, acceleration and stopping powers – and in the company's first decade many of its models were sports cars by the traditional definition, open two-seaters. There were few in the period after the Second World War; if coupés with enormous sporting potential are excluded, there were then no sports cars until the odd little Z1 appeared in 1986. That was a success, in that the production run sold out, and at prices that seemed ridiculously high. Maybe BMW's real success with this car was in pitching it at wealthy poseurs?

The Z3 that followed early in 1996 was much more convincing, as a sports roadster and especially in its more powerful forms from 1997. In the first three years of production at Spartanburg more than 150,000 Z3s were built. That speaks volumes for the car, and for a market sector some pundits had predicted would continue to shrink to insignificance. In production terms, nothing came of earlier coupé proposals, but the Z3 was to have its hard top and coupé variants, and there was an M follow-up.

BMW surprised many industry watchers with the Z07 concept car unveiled late in 1997, for this was an unashamedly retrospective two-seater and cars on these lines supposedly did not find favour in Germany. But BMW likened it to the 507 of the 1950s, and pursued the theme in the production Z8 announced in 1999. Its styling might have looked to the 507 for inspiration, but it also emerged as a super sports car to challenge front-engined roadsters from Aston Martin, Jaguar and Mercedes-Benz.

Then, of course, there are the Motorsport cars covered in the next chapter, which in many respects could run on to this one, for the 'M' designation has been applied to some very sporting machinery.

Powered by BMW

Although it was by no means a sports car in the old sense, the McLaren F1 with its tailor-made BMW V12 belongs in this company, and it reflected enormous credit on both companies. It was the fastest road car of its time – of all time – and its successes on circuits included a victory in the Le Mans 24-hour Race, and many victories in private entrants' hands, always regarded as a great sports car attribute in years past.

Other specialist constructors who relied on BMW power, albeit 'off-the-shelf' engines rather than tailor-made power units, included Ascari and Weismann in the 1990s.

The X5

Even as an appendix, a four-wheel-drive vehicle which to all outward intents and purposes has off-road attributes seems out of place among sports cars. The pretext is BMW's claim that the X5 is a 'sports activity vehicle', for had Bernd Pischetsrieder not insisted in 1993 that the company would never build a four-wheel drive off-road vehicle?

X5 could have slipped into that class when it made its public debut in 1999. But it does look sleeker than most, with wedge lines running back from a nose with the kidney grille, and there was every indication that BMW had built a vehicle to fill yet another recreational market niche. . . .

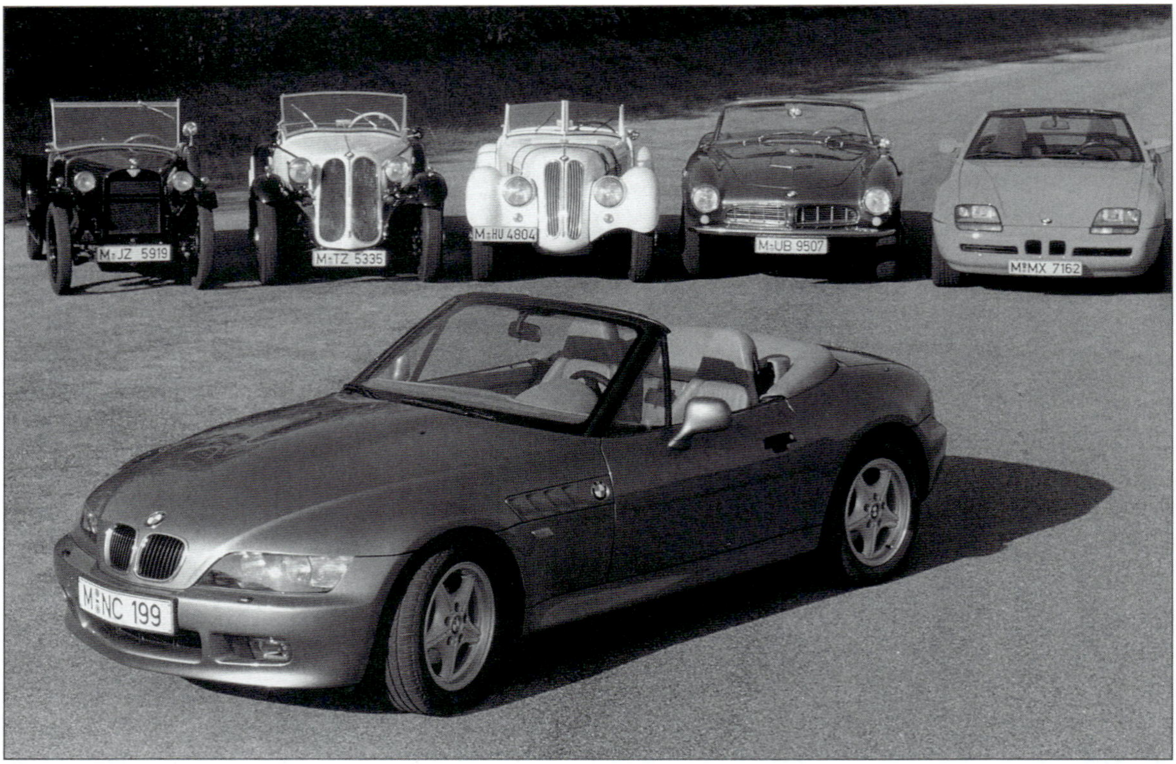

A Z3 roadster – an open two-seater meeting the traditional definition of a sports car – posed with five of its predecessors, a DA 3 Wartburg (née Dixi) of 1930–31, a 319 and a 328, a 507 and a Z1.

Opposite top: The Z1 looked every inch a sports car, and it had some outstanding driving characteristics, such as refined handling and a 136 mph (224 kmh) top speed from its 170 bhp 2494 cc straight six. The cockpit was cramped if the soft top was erected, and there was hardly any luggage space under any conditions.

Opposite bottom: The Z1 was based on 325i mechanical components, and numbered E36/7. Apart from sports-racing cars, it was the first open sports BMW two-seater since the 507, and this pair make for an interesting contrast in sizes. The electrically operated doors of the Z1 are dropped down in the open position, and the outline of the soft top stowage and tiny boot show.

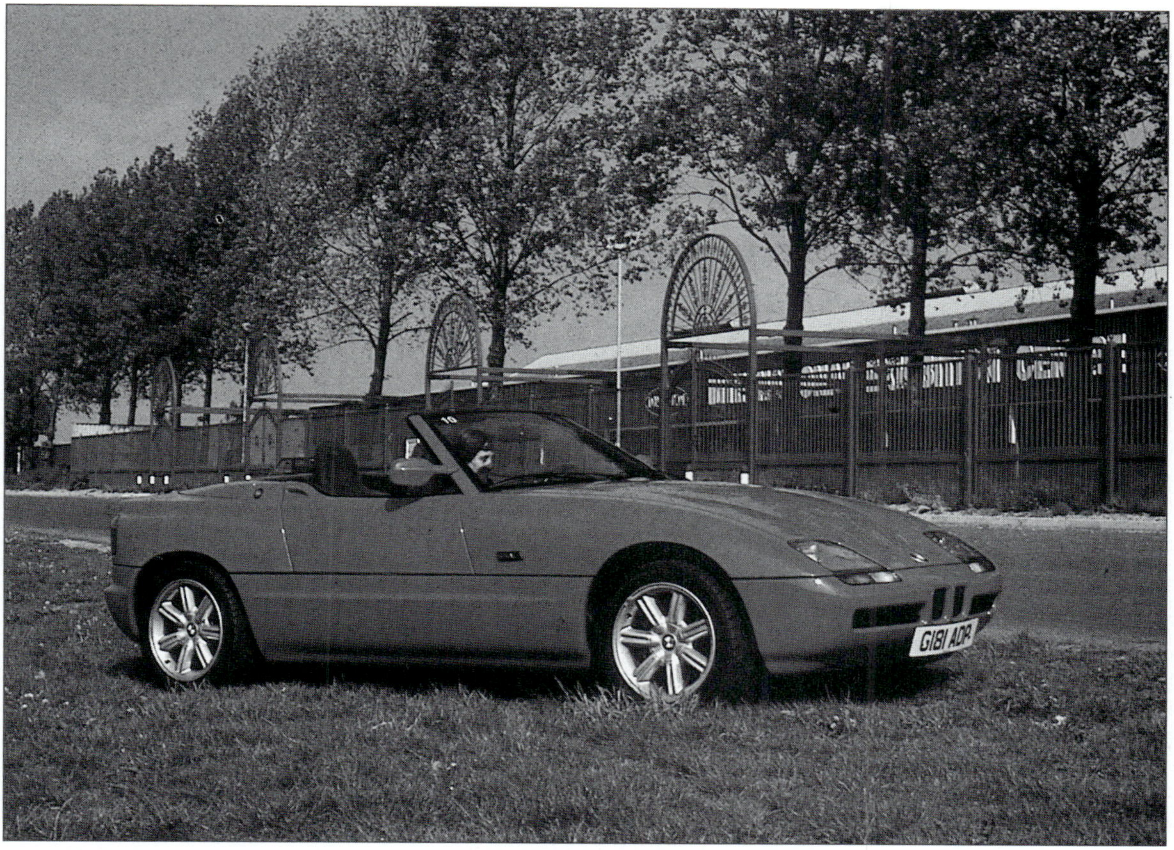
G181 ADR

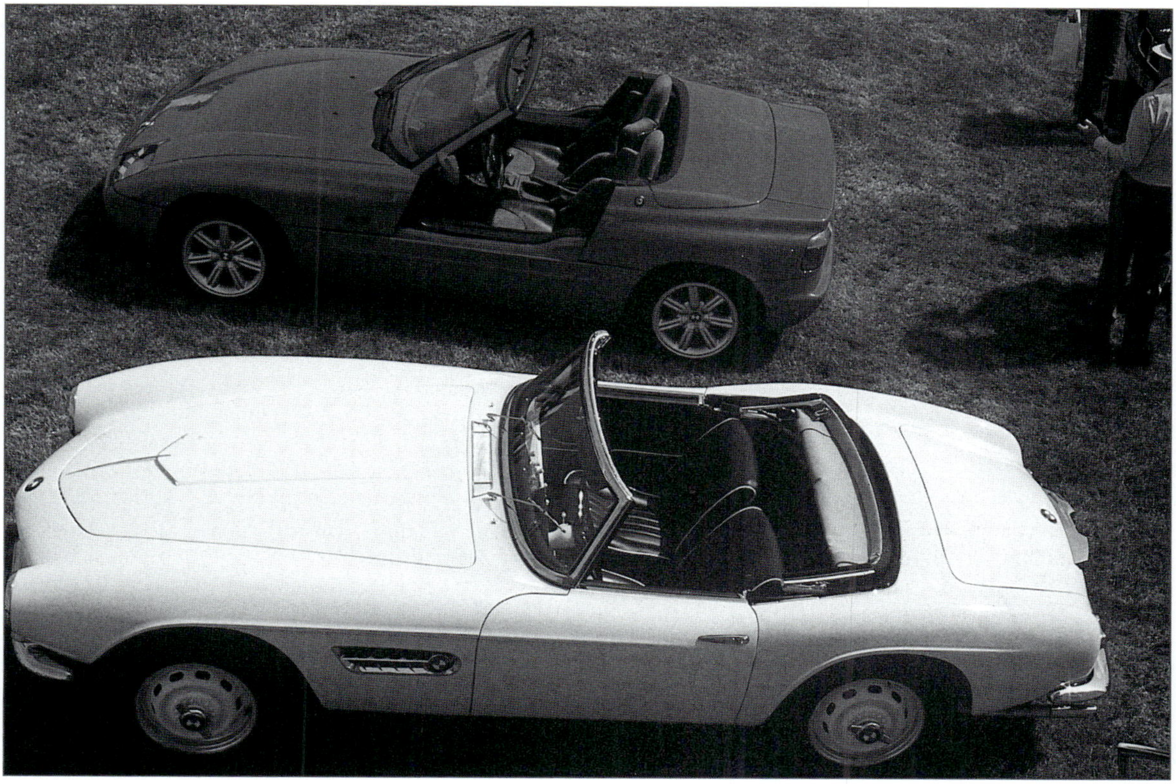

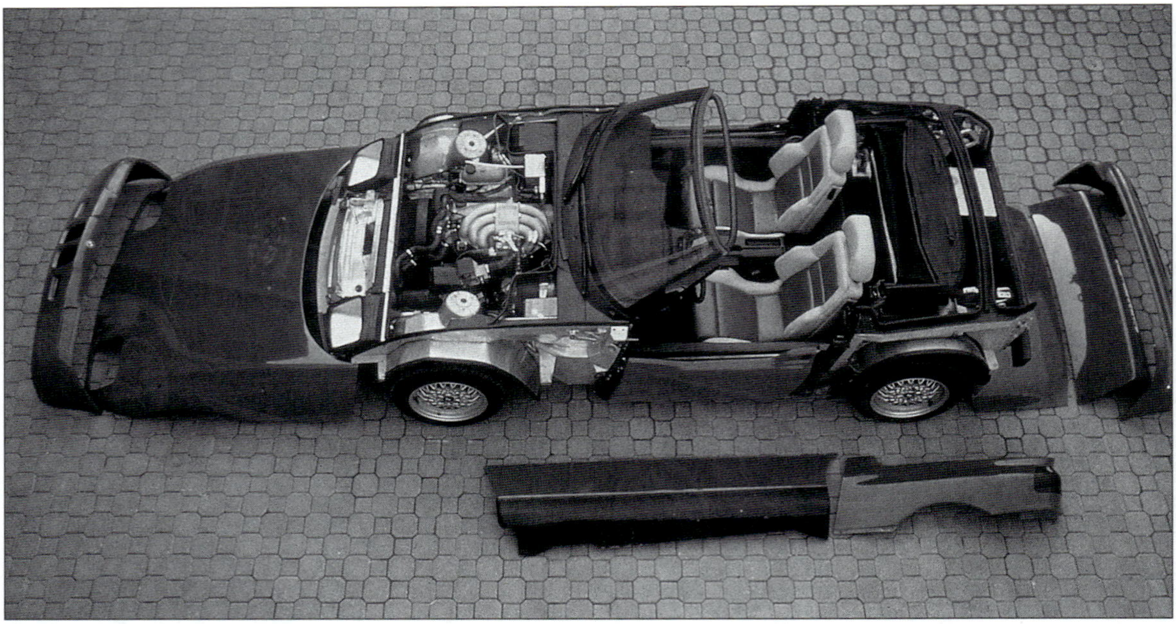

BMW obligingly displayed a Z1 with its bodywork removed. There was a composites sandwich underfloor, the plastics body panels were not stressed, and the 'Z axle' at the rear was intended for a future 3 Series line. Z1 production was 8,000, and any temptation to build more after 1991 was resisted.

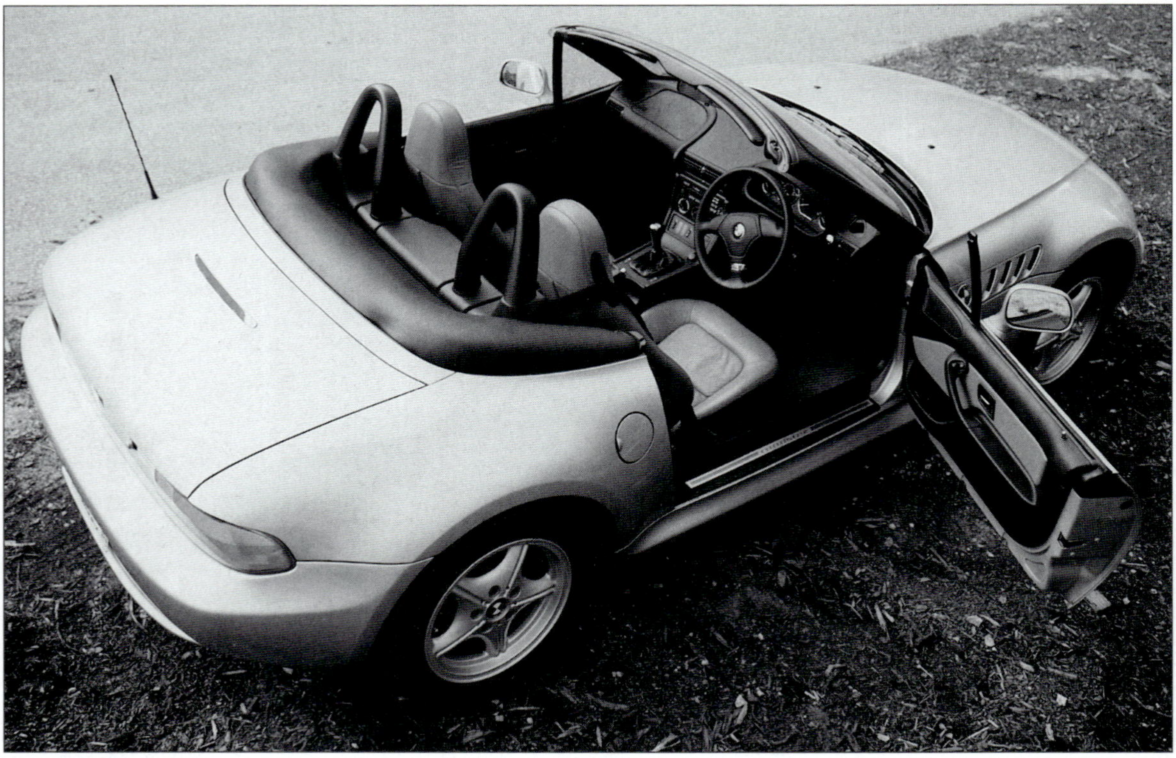

The Z3 two-seater on traditional front engine/rear drive lines was classified a roadster rather than a sports car. It was built on a 3 Series (E36) Compact platform, with the semi-trailing arm suspension and a 1.9-litre engine in 8- or 16-valve forms. Some critics suggested that it needed more power, and that came in the version with a 2.8-litre straight six that made its debut in autumn 1996, just nine months after the first Z3s went on sale, in the USA. The 1.9 was a 127 mph (205 kmh) car, the 2.8 reached 140 mph (225 kmh) and there was a 3.2-litre M version to follow.

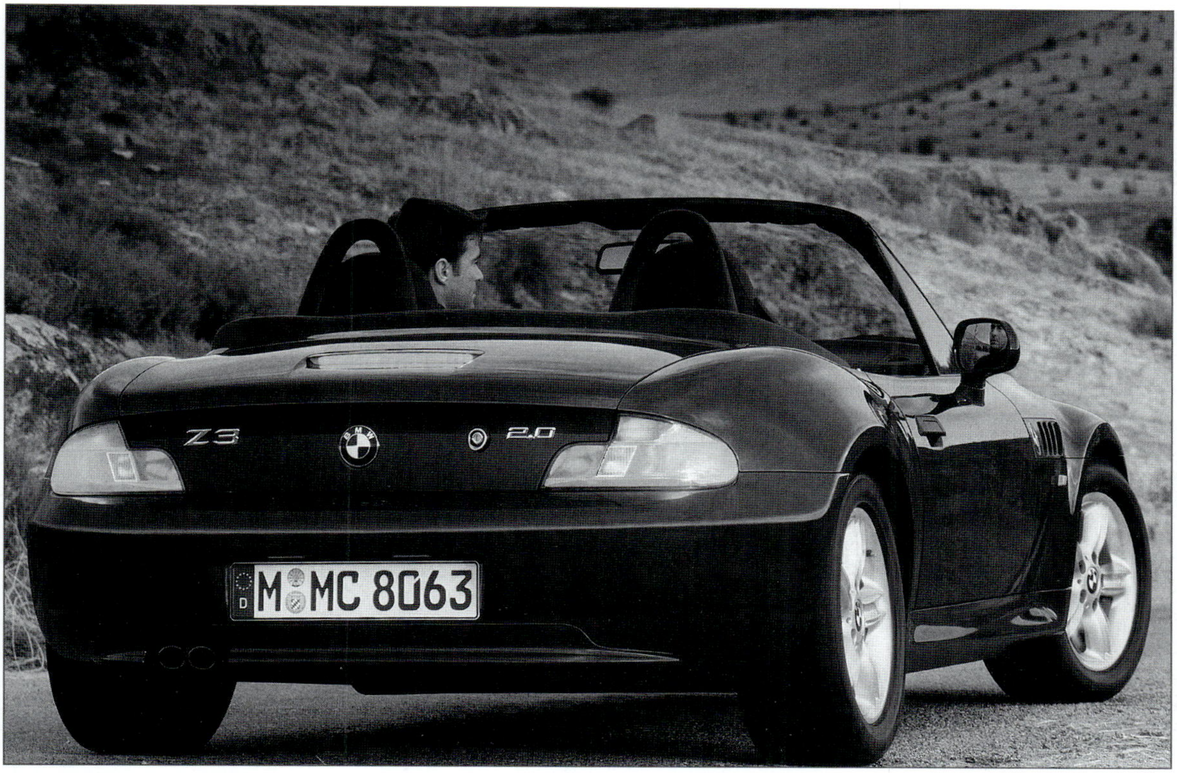

In 1999 the entry model was revised with the new 3 Series 1.9-litre four-cylinder engine, and a variant with a 2-litre straight six was introduced. This car also had minor styling changes, with a wider rear wheel arch flare, revised boot lid and rear light clusters.

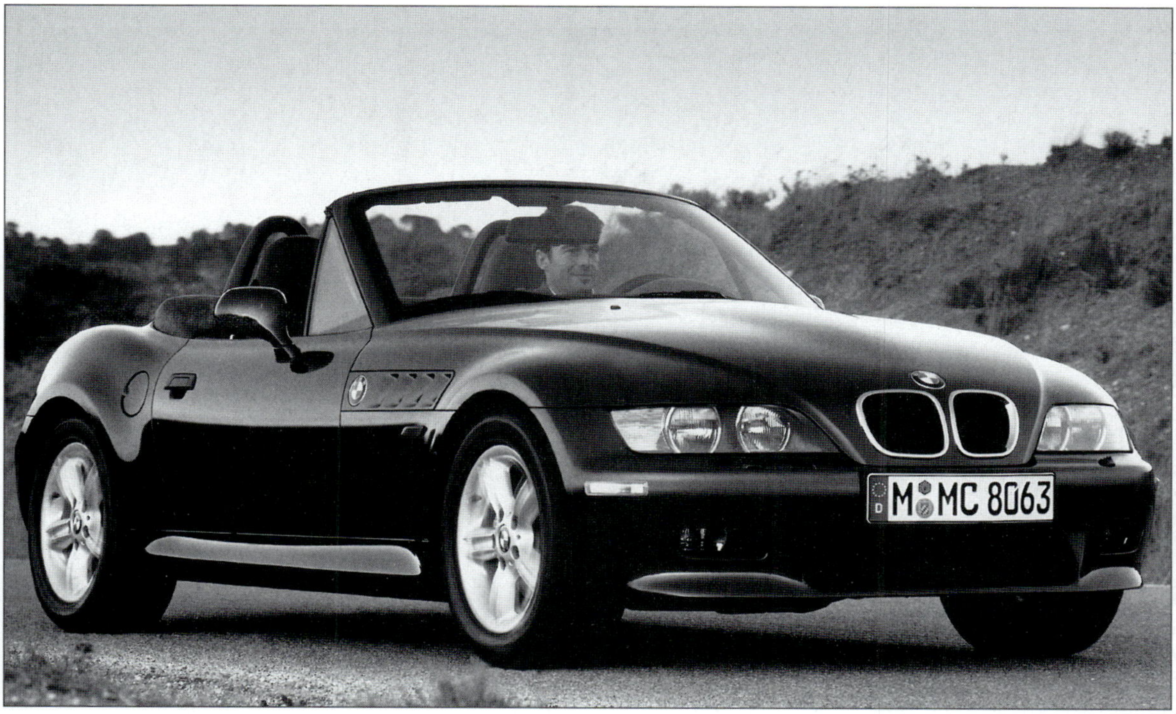

The 1999 Z3 face list was limited to minor revisions at the nose. This is a Z3 2.0 with a short-stroke, small-bore version of the straight six giving 150 bhp and a 130 mph top speed.

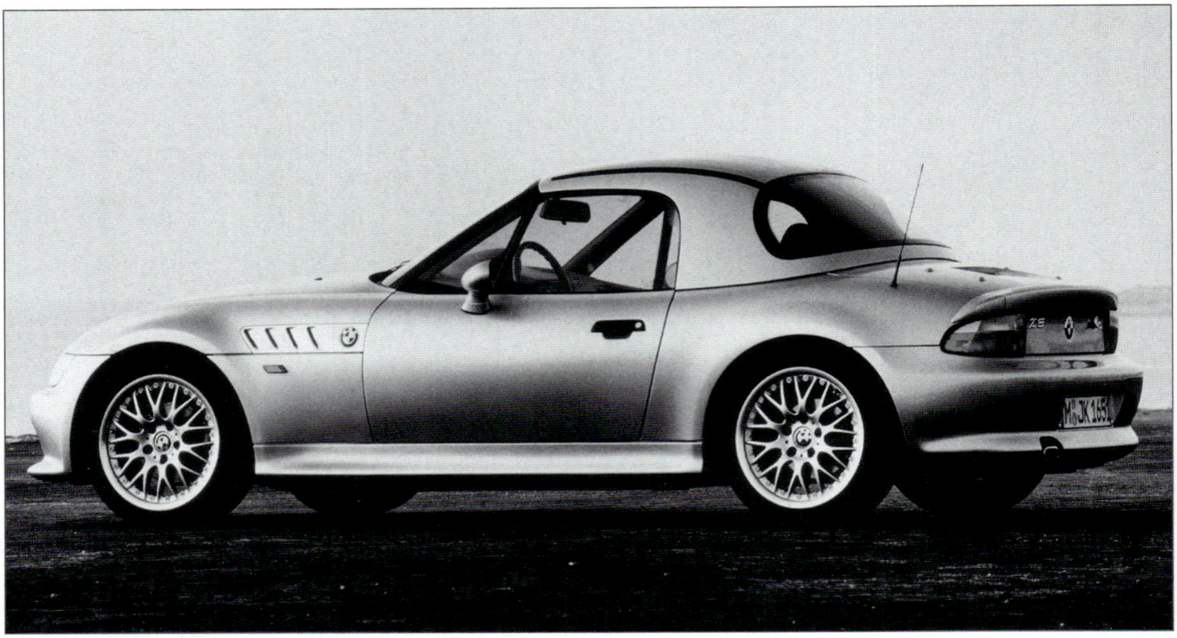

A smart removable hard top was added to the Z3 options. This angle shows the relatively long wheelbase of the car, which was coupled with wide track and minimal front and rear overhang, together with details such as the engine bay vents that recall the 507.

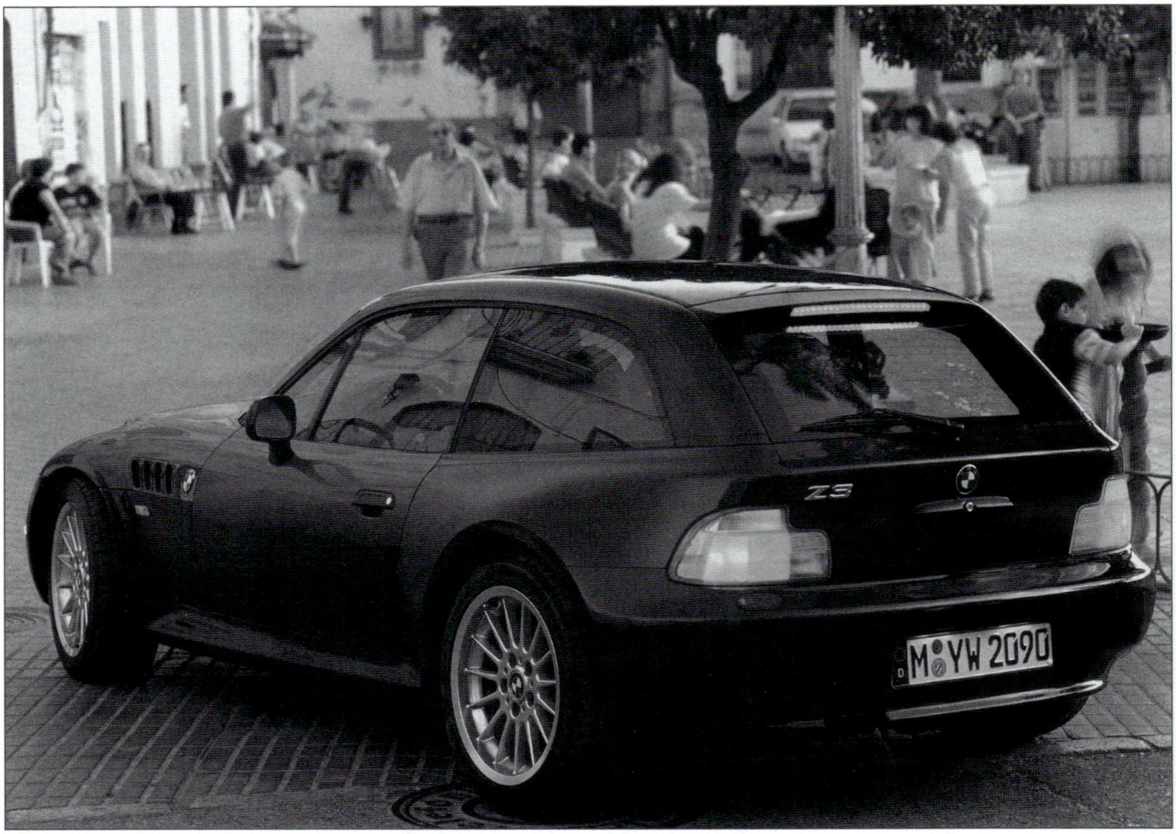

The rear-end styling of the coupé was not universally admired, but was efficient: with the 2.8-litre straight six in 193 bhp form, the coupé was a little quicker than the equivalent roadster, reaching 144 mph (231 kmh).

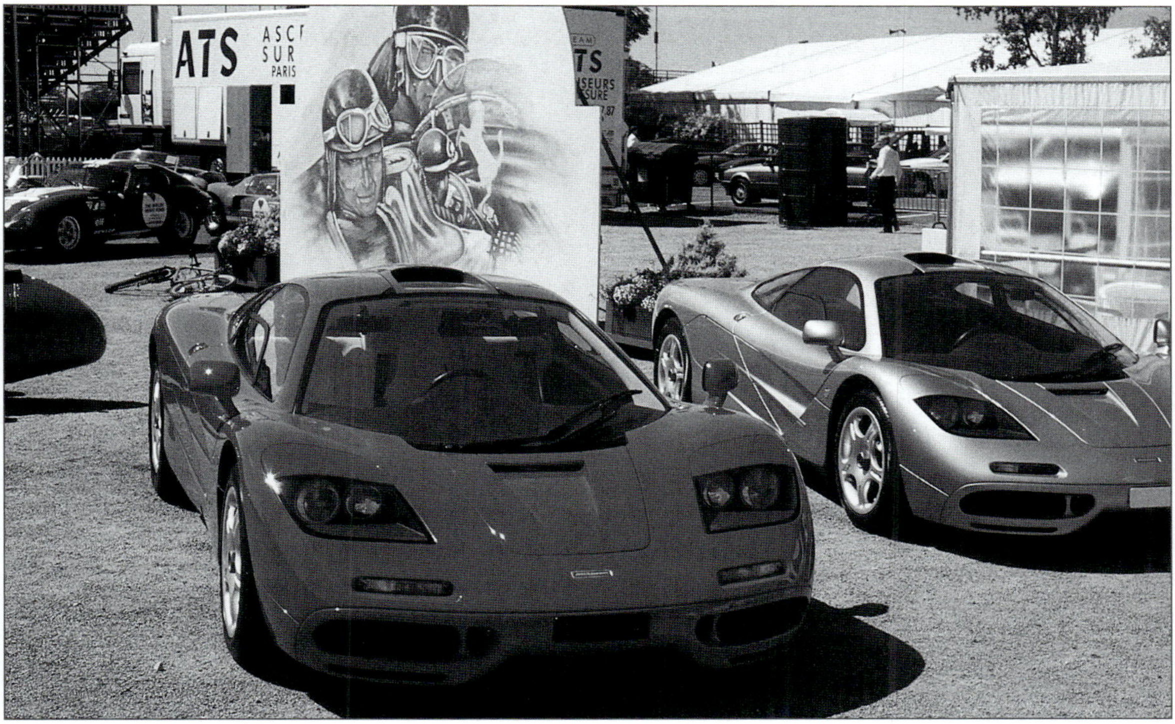

The McLaren F1 was the outstanding supercar of the 1990s, and very much the product of a collaboration between the British company and BMW, in particular as designers Gordon Murray and Paul Rosche picked up their relationship from the Brabham-BMW Formula 1 car days. Work on the V12 began early in 1991, and an S70/2 prototype was running by the end of that year. The 6.1-litre unit comfortably exceeded Murray's power requirements, producing some 627 bhp in the production F1.

The car was a sleek mid-engined three-seater coupé. The concept of a central driver's seat was important, and the arrangement proved practical. The maximum road speed was shown to be in excess of 230 mph (370 kmh), so most road users would normally see the rapidly disappearing rear of an F1, with its large exhausts – and presumably a number plate.

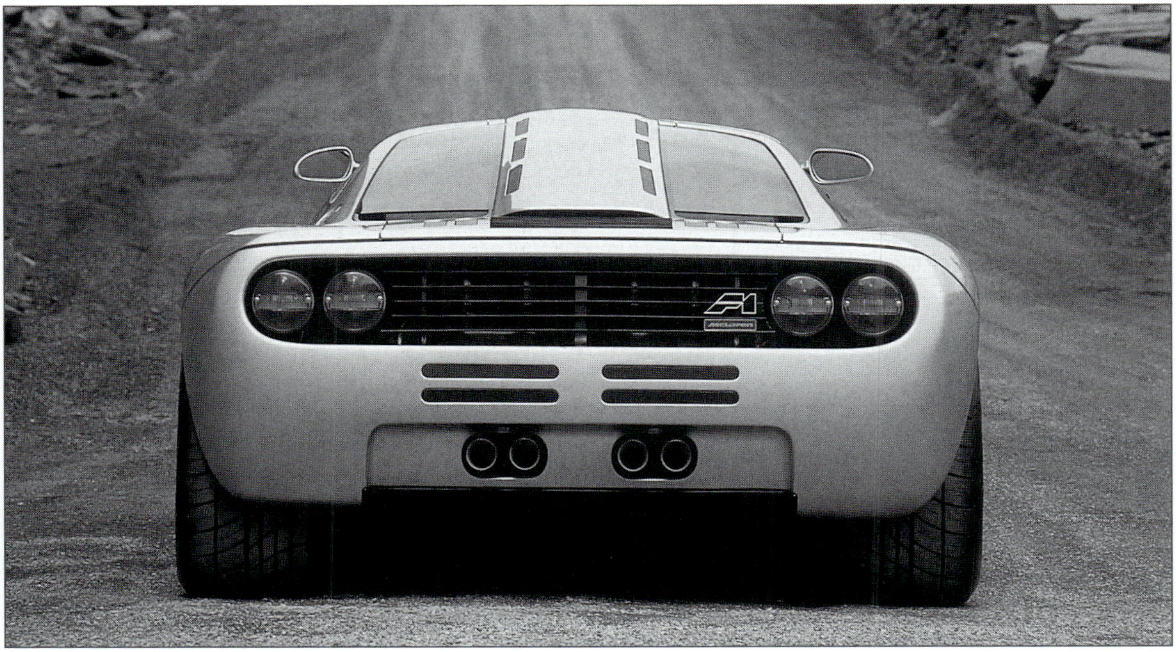

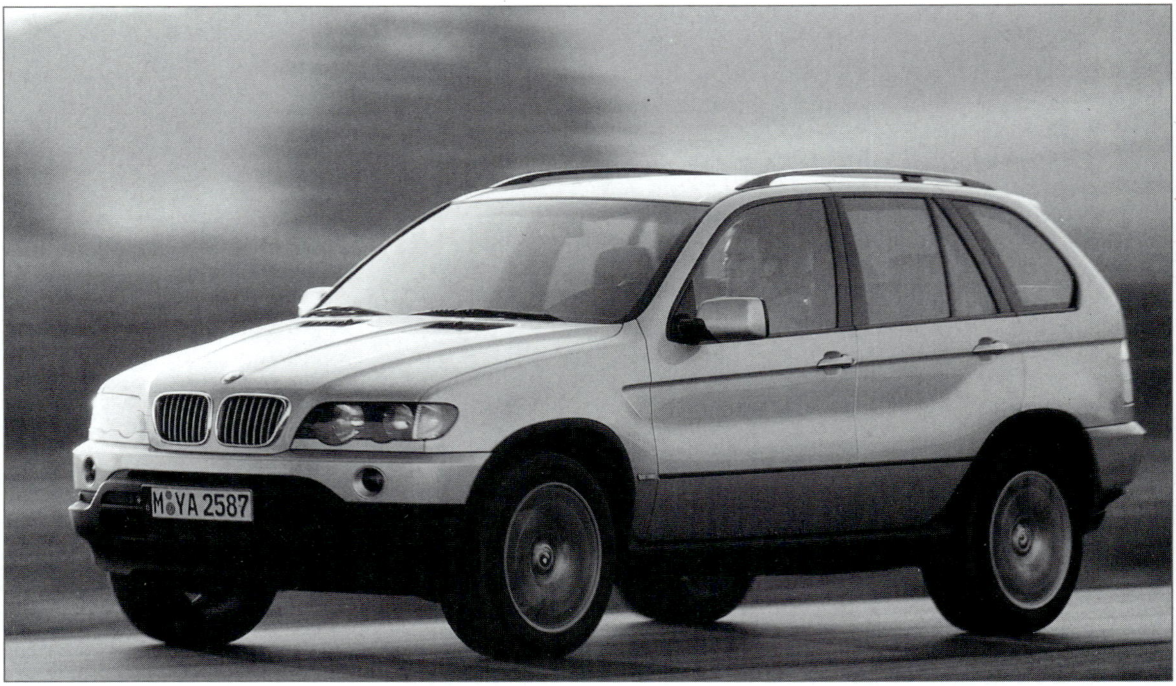

BMW insisted that the X5 (E53) is a Sports Activity Vehicle, not a sports utility vehicle, when it made its public debut early in 1999, appropriately at the Detroit Motor Show as it was to be built at Spartanburg. The engines were V8s, from the 5/7 Series lines, a straight six and a diesel. Although first-class off-road capabilities were claimed, the emphasis was on 'car-like road holding'. Four-wheel drive was combined with ABS, AS+T stability control and Dynamic Stability Control. The lines were deliberately stylish, when 4x4 vehicles tended to be chunky, and the grille and front lights followed contemporary 3 Series layouts. There was a slight wedge line to the flanks, and the rearmost windows showed the 'kink'. The roof pillars were necessarily strong, as unlike many off-road vehicles the X5 had a unitary body. Interiors were luxurious – almost over-equipped – and there could be as many as ten air bags.

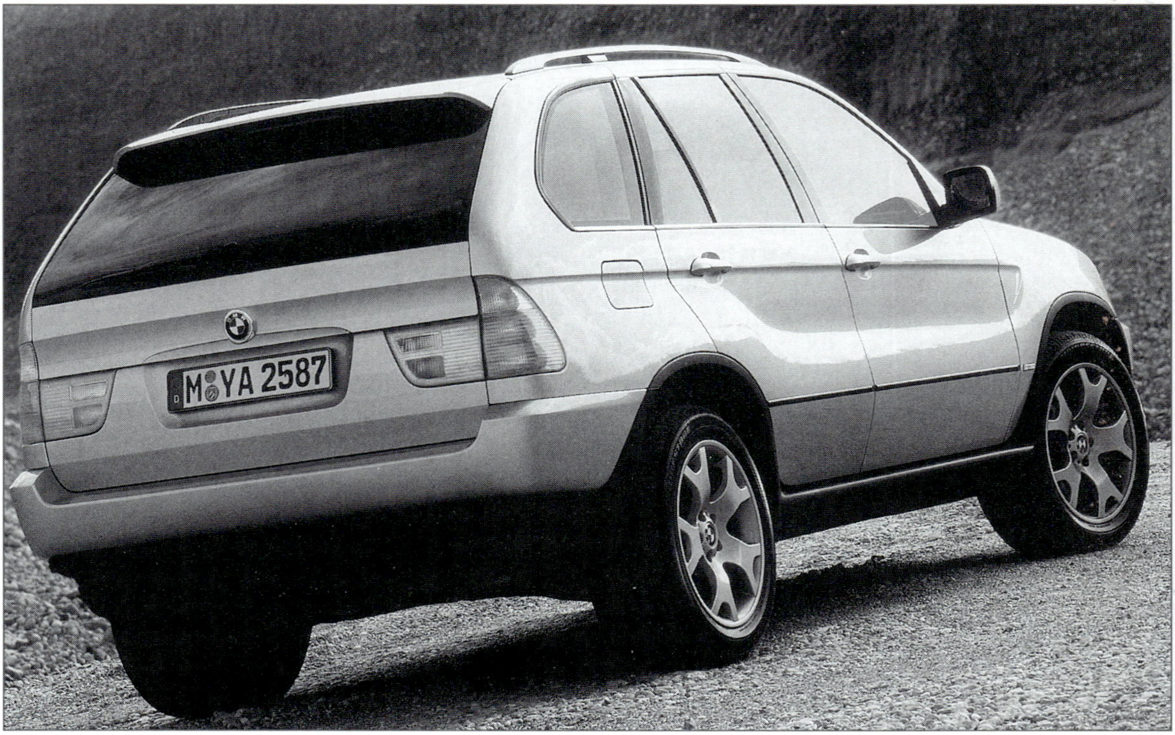

CHAPTER NINE

M SPECIALITIES

BMW Motorsport GmbH was a BMW subsidiary formed in 1972, effectively as the company's racing division. From its engine, chassis and suspension development work to further those competition activities, it was to move on to originate high-performance versions of road cars for sale, while after almost a quarter of a century it was to lose its direct motor sport responsibilities.

The first of its distinct high-performance models was an original supercar, while the versions of saloons that followed often had supercar performance potential. There were partnerships, usually but not invariably with strong racing intentions, and there were the circuit activities covered in the next chapter.

Jochen Neerspach set up this subsidiary, with Paul Rosche in charge of engineering, especially the engines that were to be at the heart of the Motorsport operation, and Martin Braungart responsible for chassis and aerodynamic work.

A strong identity was quickly established, and a dedicated factory in the Munich suburb of Garching came into use in the 1980s. Some M cars were still assembled in other plants, and the roadsters would be built at Spartanburg in the next decade. The success of the operation, and its continuity, was seriously threatened when there were enormous setbacks in getting the M1 into production, largely because of back-sliding outside contractors.

In the 1990s the competitive motor sport activities were separated from the commercial side. M Cars produced the vehicles, M Technic marketed components from a specialized suspension units to steering wheels, there was even M Style to market clothing, and there were driver training courses. It sometimes seemed that there could be conflict with staunch BMW allies, especially Alpina and Schnitzer, but there was actually room for all.

By 1972 the first racing engine had been developed, and this Formula 2 unit was to be used to win the European Championship in 1973, and four more titles before the series lapsed in 1984. (In 1975 Laffite won the title with Schnitzer-developed BMW engines.) Meanwhile, Motorsport's impressive CSL coupés won the European Touring Car Championship in 1973, and cars carrying the BMW badge were to win that title another thirteen times in the sixteen seasons it was run. Beyond that, there was to be a championship-winning Formula 1 engine. Racing – and a brush with rallying – are recalled in the next chapter, and are mentioned here simply because the competition record proved the Motorsport cars beyond dispute, before 'M' lost its responsibility for the company's sports activities. (A new company, BMW Motorsport Limited, was set up in 1995 and operated from a base in Britain until 1999.)

There have been ten M cars, some with dual road and track roles. In two decades from 1978 Motorsport made more than 100,000 cars. Some Motorsport-developed cars were sold earlier in the 1970s, when performance parts were also sold to enable owners to uprate their cars.

The first production car, M1 (E26), did not come until 1978. It had a troubled gestation, for an association with Lamborghini proved frustrating – indeed, it must have been infuriating – and it was abortive, for the Italian company's short-termism as it floundered

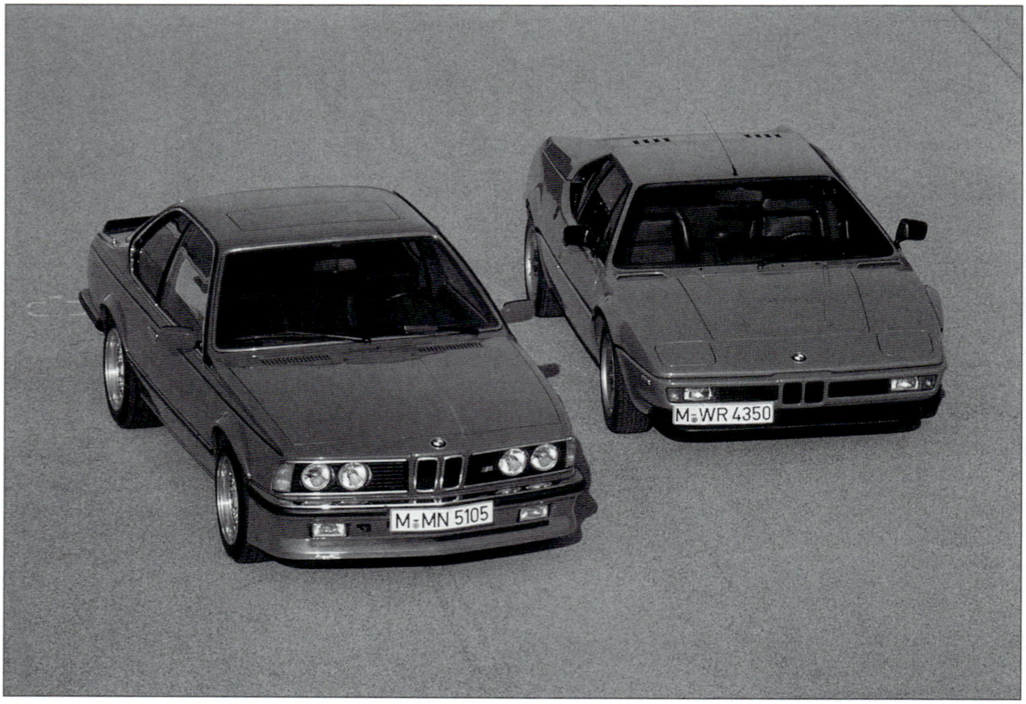

Potent pair. An M635CSi from the mid-1980s gives scale to an M1. The M635CSi was run by Nelson Piquet, when he was reigning World Champion driver. Both cars were most effectively painted blood red.

in a financial morasse meant that it lost a contract that could have been its salvation. Eventually M1 was built, as a road supercar and a circuit car, with modest success in open racing but gleaning much publicity for BMW in a one-model ProCar series.

The M cars that followed were derivatives of production models, and M3s were built on normal production lines, although early examples had been assembled at Garching where the later M5s were to be made. M3s were built in considerable numbers, passing the 50,000 mark with the E36 type late in 1997.

Some rumoured variants, such as an M5 convertible, did not appear. Nor was there an M8, although the M-developed 5.6-litre V12 was fitted in the 850CSi – 1,510 cars had this 380 bhp engine, which in 1992 was the most powerful road-car unit built by BMW: it could have been used in a lightweight version of the 850i, which might have been an M8, but was dropped.

Engines have been fundamental to the Motorsport operation, and almost unbelievably its first boss, Jochen Neerspach, essayed to take the first Formula 1 engine to Talbot when he left BMW in 1980! The vacillating BMW management of the 1950s might have agreed to that sort of proposal, but by 1980 there was the confidence to back Motorsport. 'BMW M Power' on cam covers was to command worldwide respect.

Paul Rosche was an intuitive engine designer who had joined von Falkenhausen's engine department in 1957, and was largely responsible for power units from the '*Neue Klasse*' onwards, to a Formula 1 engine at the end of the century. Notably, BMW racing engines were usually developed from production units, and none were bought in.

As road cars and motor sport were quite logically separated, the M responsibility became high-class, high-performance, road car development and production – a field where it has excelled, with, for example, more than 20,000 M cars sold in 1998 alone.

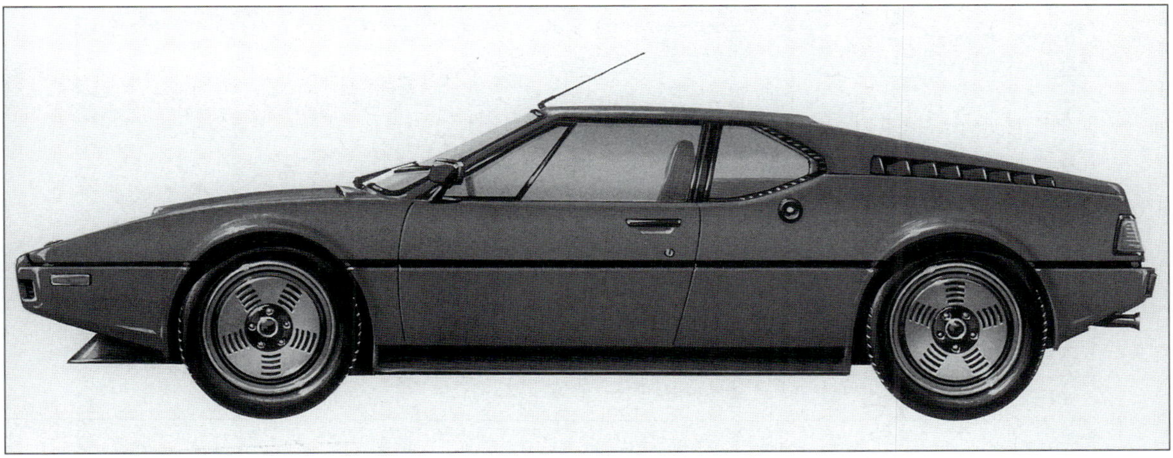

M1, the first M car, did not quite match some Italian supercars in superficial eye-catching qualities, but it was a rarity – a sensible and usable supercar. Giugiaro designed the body and Lamborghini's Dallara had a hand in chassis development, but that Italian constructor contributed little else but delays. Eventually Marchesi built up the square-section tubular chassis, Ital Design made the body parts, and final production was taken on by Baur in Stuttgart.

The M1 went on sale a year later than intended, with a production target of 400 to qualify it for Group 4 racing, where it would have taken on Porsches. Eventually 450 were completed by 1981, 397 of them road cars.

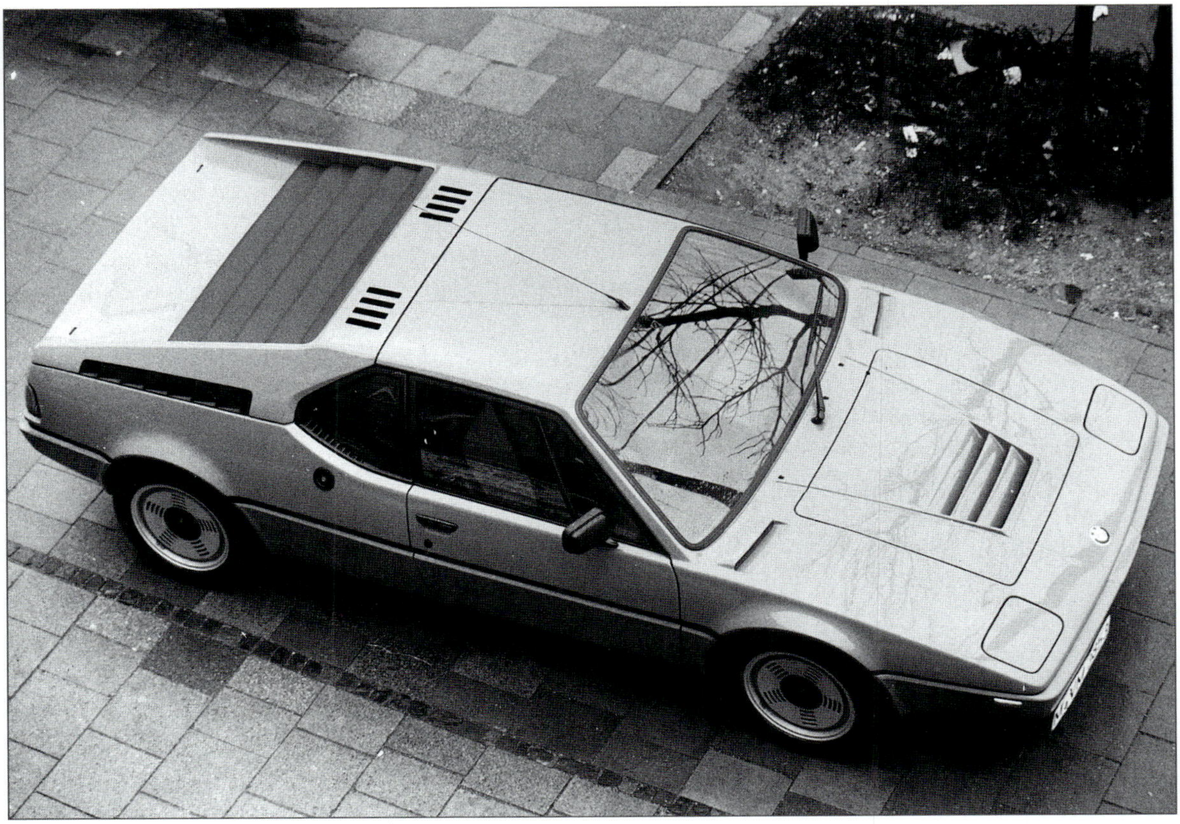

This angle shows the M1 as a no-nonsense two-seater coupé, from pop-up headlights to engine air outlets in the tiny buttresses at the rear. Flamboyant aerofoils were seen only on circuit M1s. The glass fibre body panels were not stressed, and were bonded and riveted to the substantial steel frame.

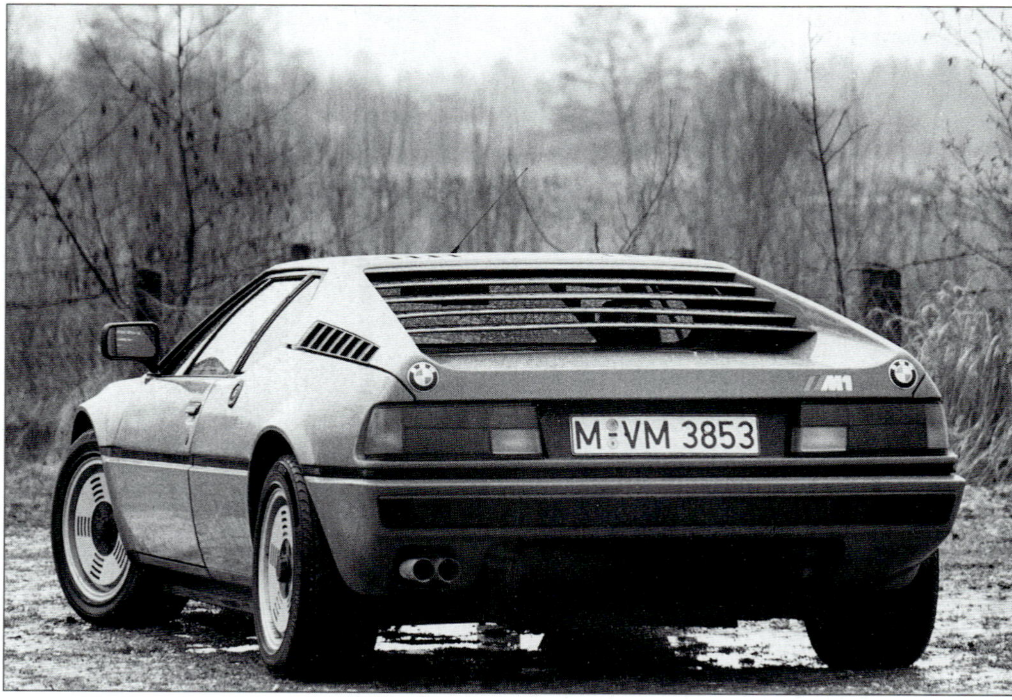

Twin BMW badges, plus M1, left other drivers in little doubt about this car when it passed them. Louvres behind the cockpit echo Italian supercars. Low-profile tyres were mounted in 7-inch light alloy front wheels and 8-inch rears (where it was legal, the spare was a space-saver), and there were ventilated disc brakes all round.

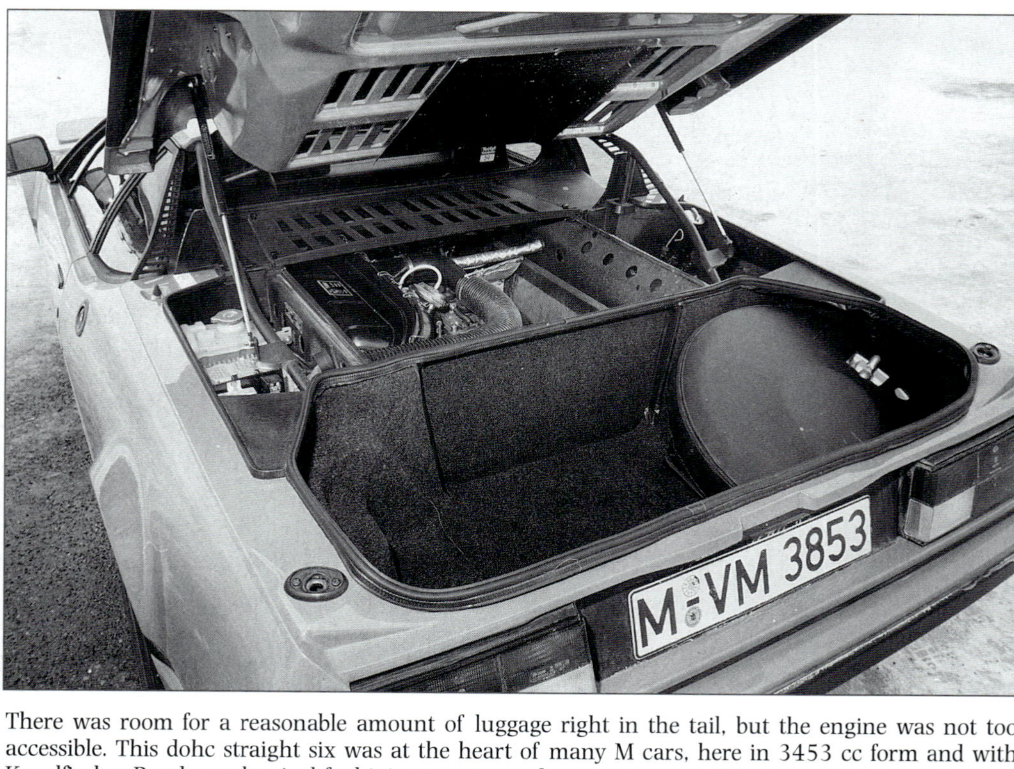

There was room for a reasonable amount of luggage right in the tail, but the engine was not too accessible. This dohc straight six was at the heart of many M cars, here in 3453 cc form and with Kugelfischer-Bosch mechanical fuel injection. Its road car rating was 277 bhp, and it drove through a ZF five-speed transaxle. The maximum speed with standard gearing topped 160 mph (258 kmh).

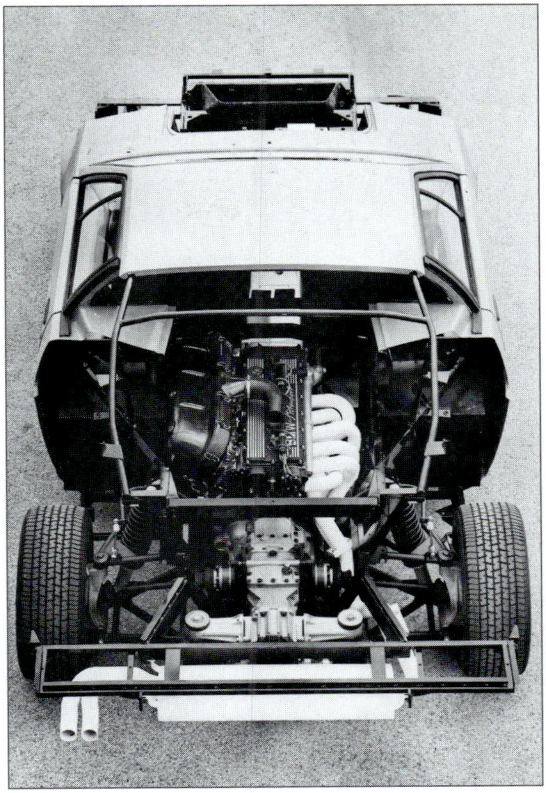

Rear-engined BMWs have been rare, but the position of the straight six between engine and transaxle is clear in this shot. The suspension members were substantial, but relatively light tubes supported the bodywork above the engine.

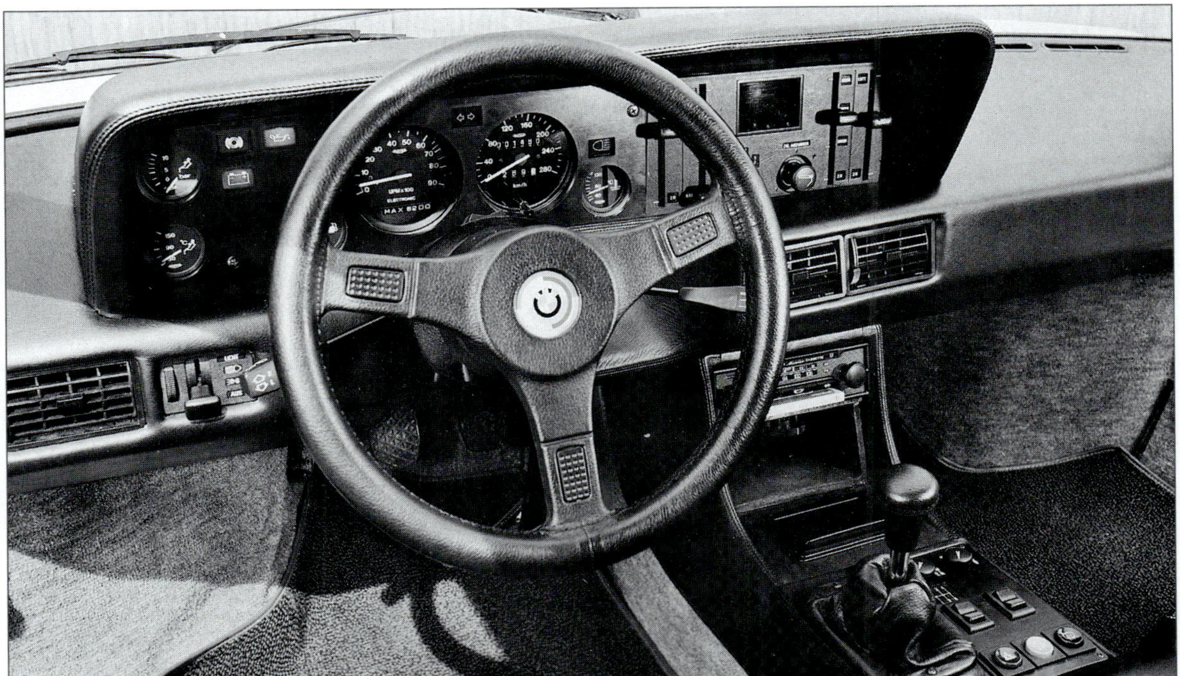

The M1 cockpit was roomy – this was a wide car – and the seats were supportive and comfortable, but the fascia hardly lived up to BMW's high standards. The main instruments were a rev counter on the left and a speedometer on the right (reading to 280 kmh); secondary gauges on the left could be obscured by the wheel. The pedals were well offset towards the centre, because of that obtrusive wheel arch.

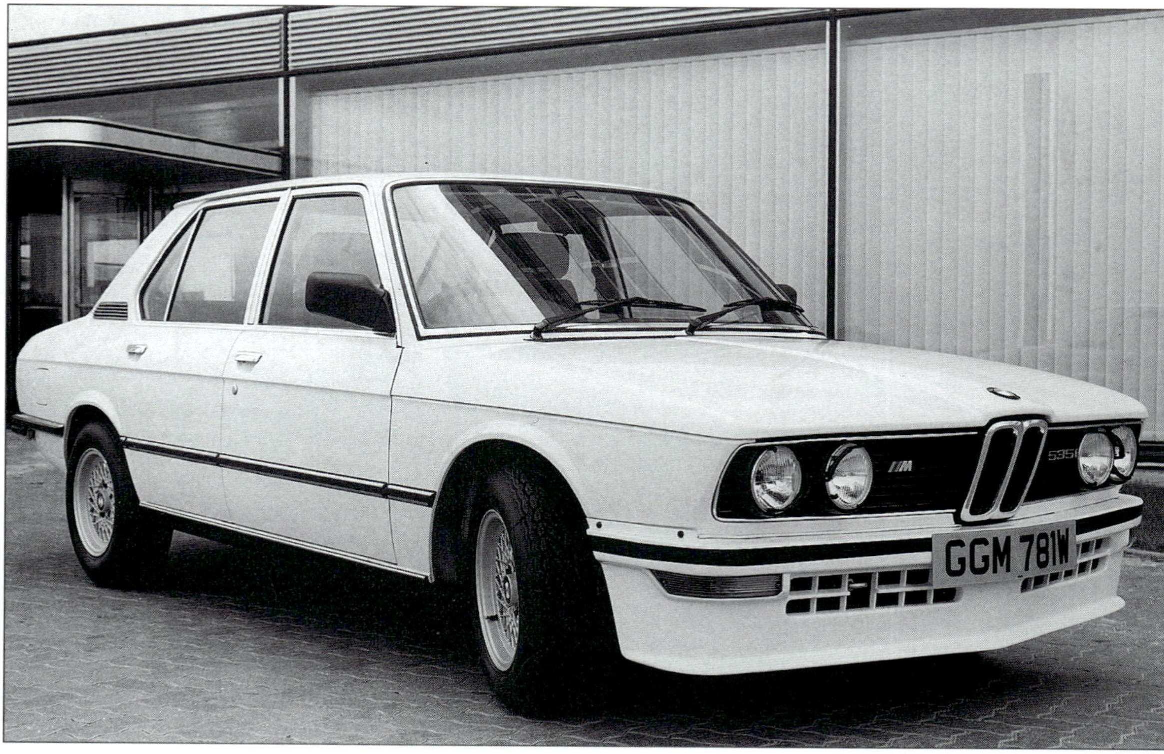

First of the saloons in the M line, the M535i combined the E12 body shell with a 3.5-litre straight six, effectively the 735i engine. Some 1,650 were completed in 1980–81, for sale in Europe: this was only one of a number of models not exported to North America down the years. The air dam and wide wheels, as well as the badges, distinguished this 140 mph (225 kmh) saloon from other 5 Series models.

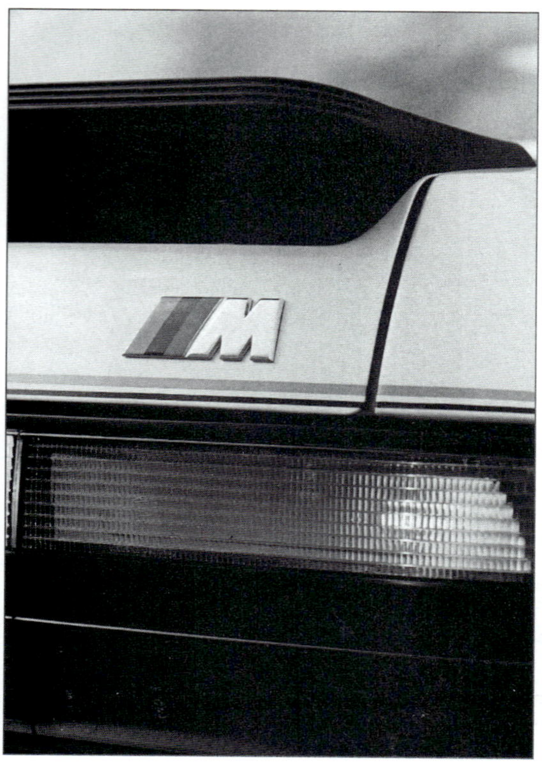

The familiar M badge carried on the nose as well as the tail of M cars from 1980 combines the letter 'M' with diagonal red, blue and turquoise bars (here on an M535i). In 1973, incidentally, the blue and white BMW badge had been part-encircled by those colours in a Motorsport house emblem.

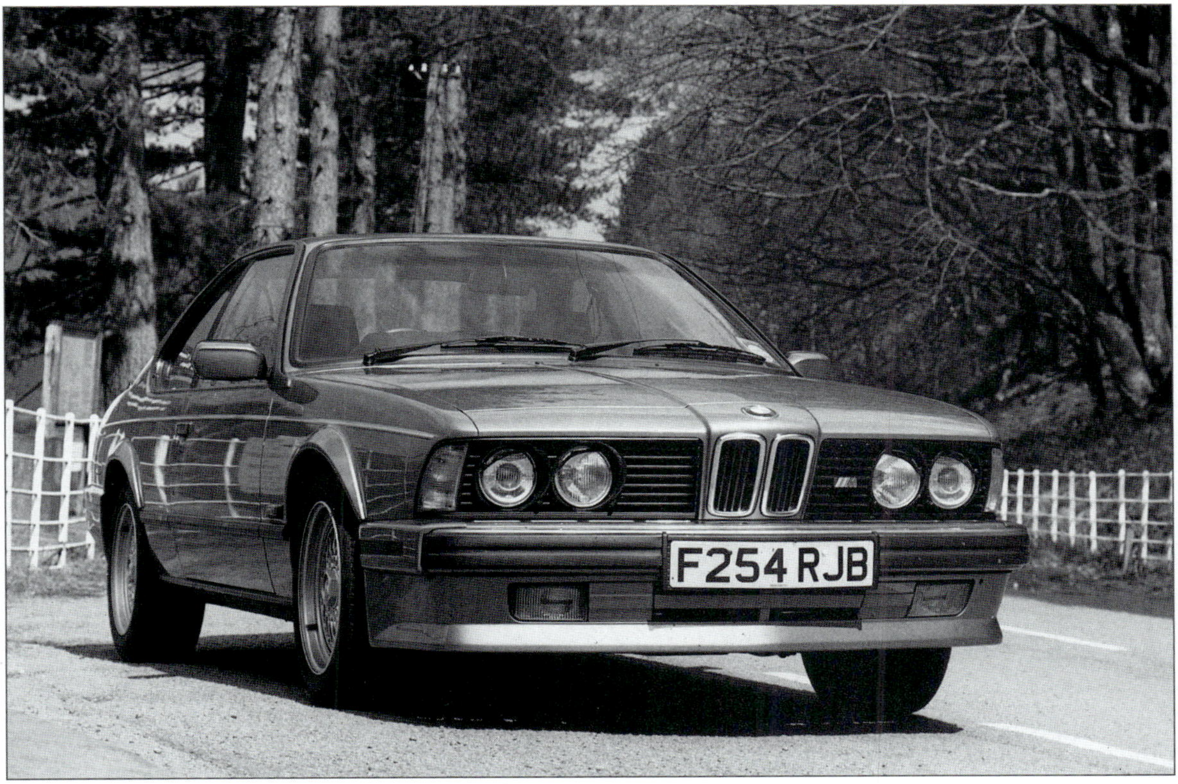

The M635CSi of 1983–9 was a handsome coupé – by comparison, the M535i of only a couple of years earlier looked almost homely in a three-quarter front view – and 5,855 were built. The engine was the straight six, in a twin-cam 24-valve high state of tune version giving up to 286 bhp. There was a ZF close-ratio four-speed gearbox and limited slip differential. Firmer suspension made for excellent handling, but the ride was hard by luxury car standards. Few full four-seaters could match this car's performance. (Maximum speed was of the order of 150 mph/240 kmh.)

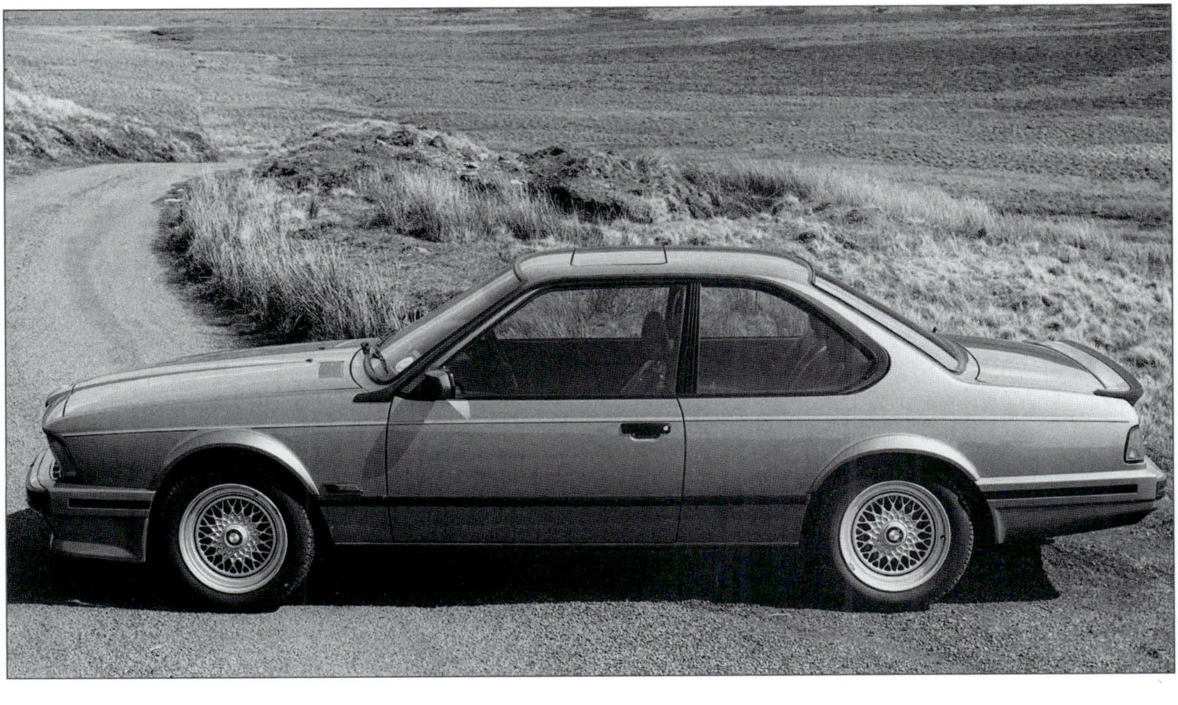

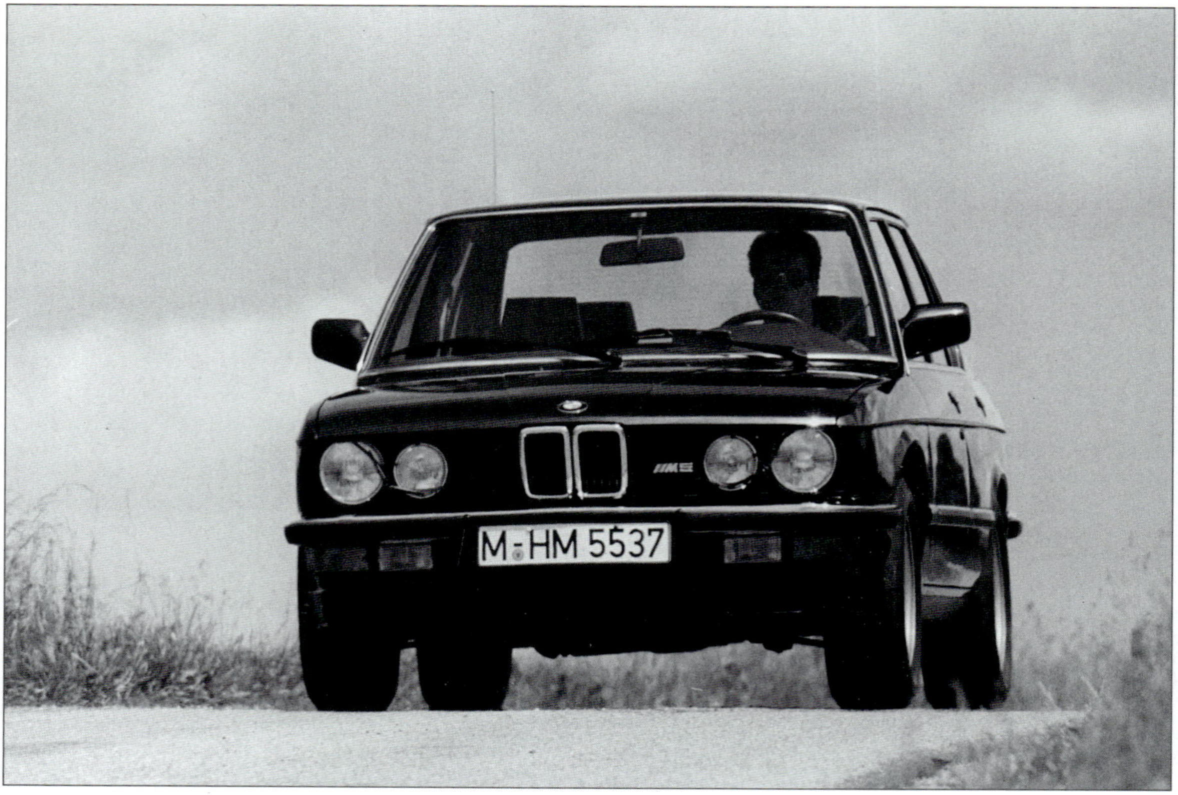

The first M5 (E28) in 1984–87 was another high-performance road car, the 5 Series four-door saloon with the 24-valve 3.5-litre engine developed for the M1 and M635CSi. Its 286 bhp gave a top speed similar to the M635CSi. BMW sold 5,855 of these M5s.

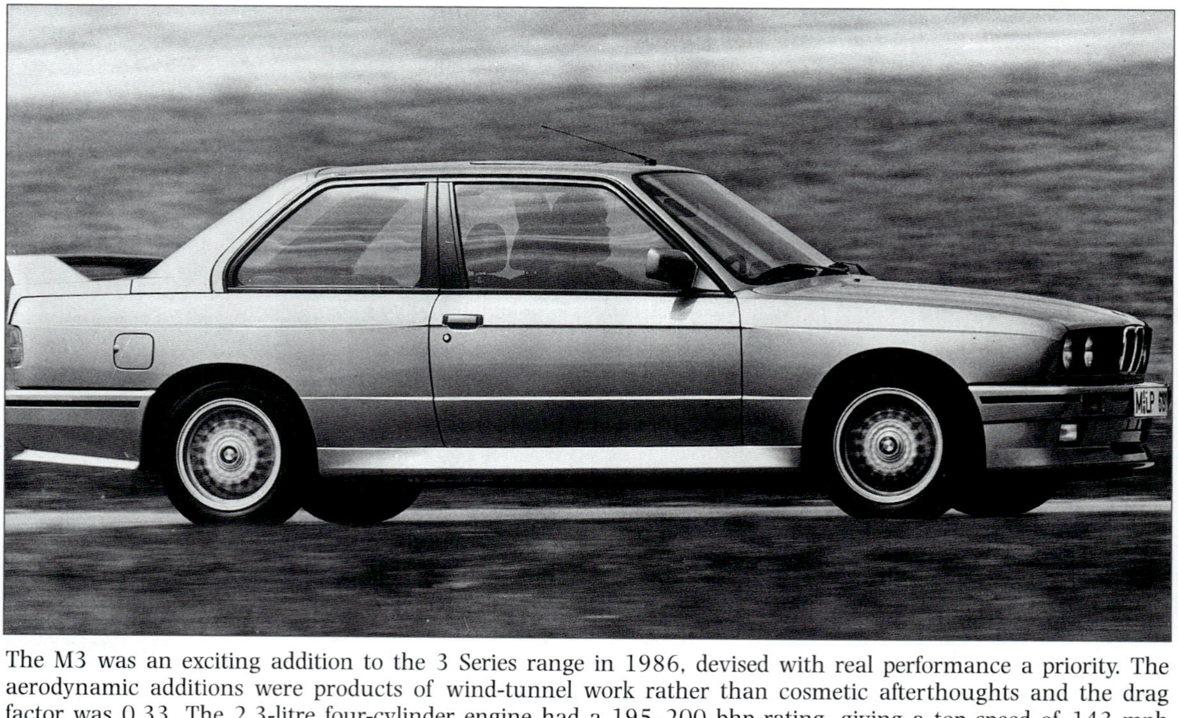

The M3 was an exciting addition to the 3 Series range in 1986, devised with real performance a priority. The aerodynamic additions were products of wind-tunnel work rather than cosmetic afterthoughts and the drag factor was 0.33. The 2.3-litre four-cylinder engine had a 195–200 bhp rating, giving a top speed of 143 mph (229 kmh), and the power/weight ratio meant sparkling acceleration.

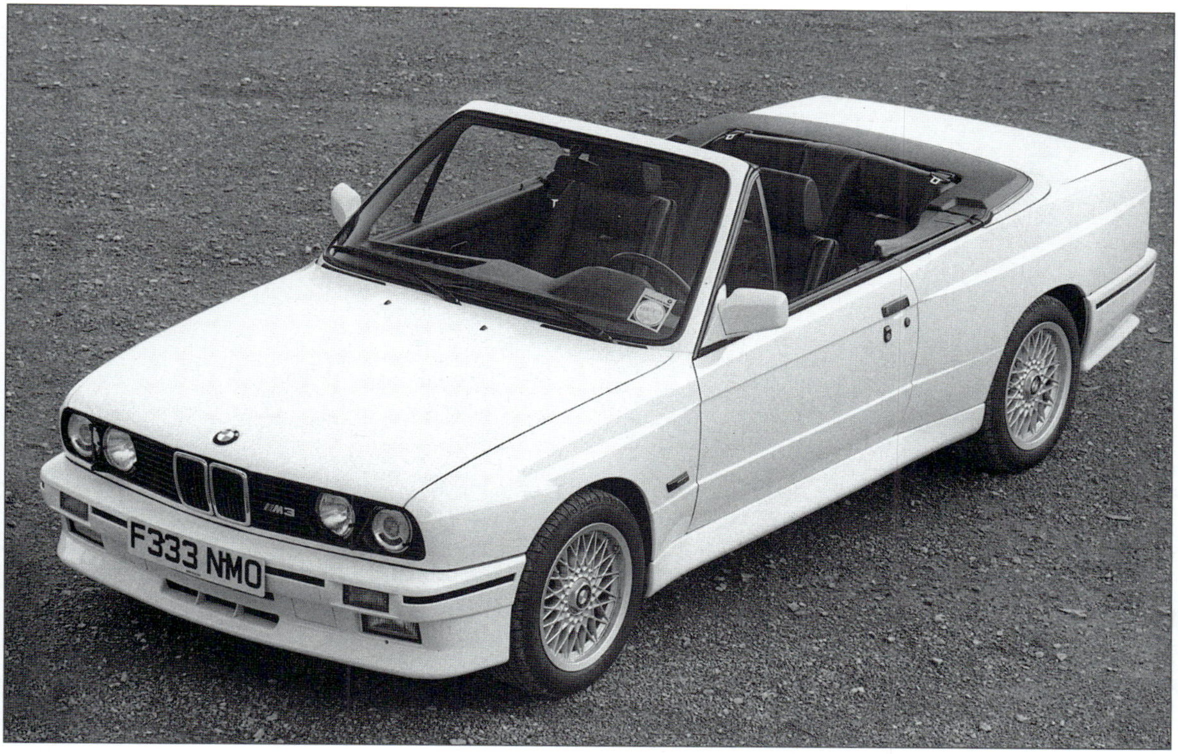

The second generation 3 Series convertibles were deservedly successful in sales terms. This mid-1980s M version compares with Alpina's near-contemporary soft-top (page 155). The flank extensions to accommodate wider wheels are more subtle with a nice flowing line.

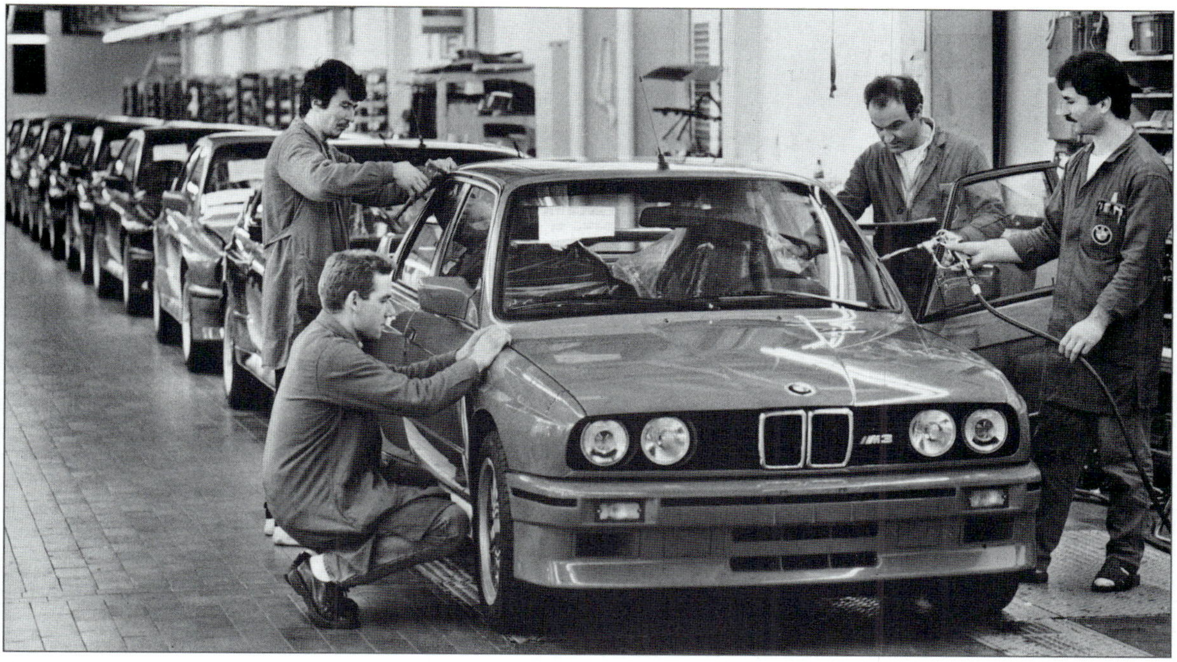

There was an M3 Evolution model in 1988, with engine modified to give 220 bhp, revised aerodynamics, and increased wheel-arch flare to accommodate 16-inch wheels (and allow for 18-inch wheels on competition cars). The next M3 Evolution type came at the end of 1989, named Sport Evolution, with a 2.5-litre engine giving 238 bhp – it was the first M3 to have a claimed top speed over 150 mph (240 kmh). This posed quartet do not look convincingly busy around an M3 in the Motorsport plant early in 1990. . . .

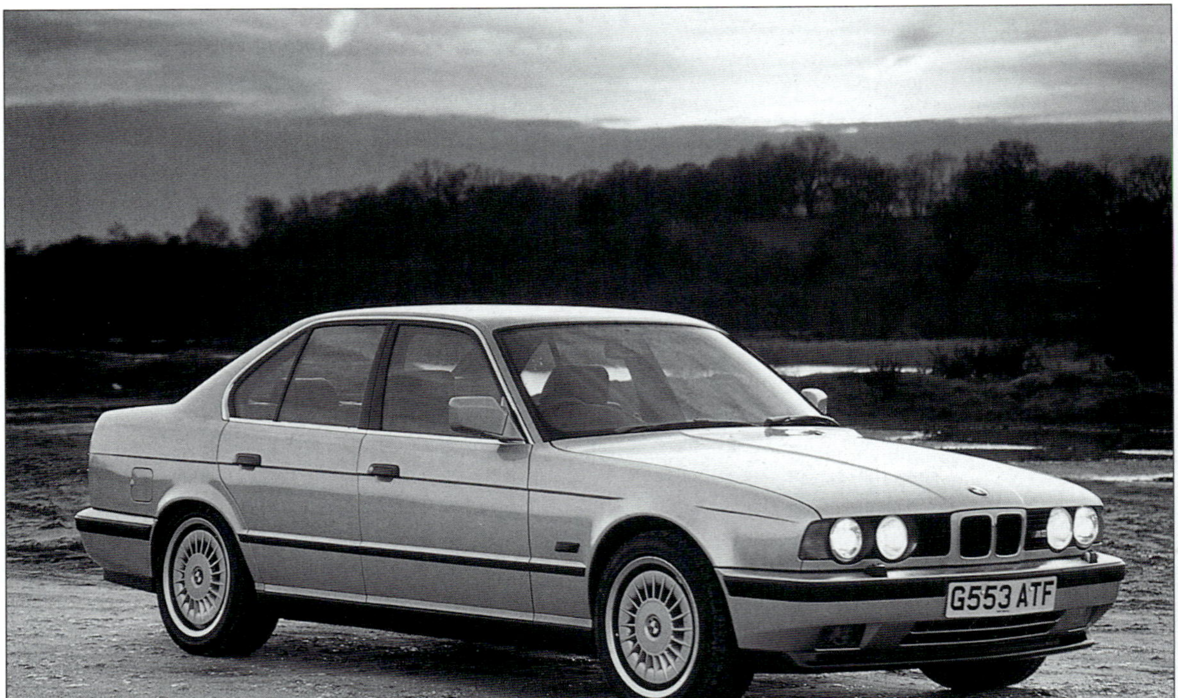

The M5 introduced in 1986 was an outstanding sporting saloon, with the E34 bodyshell giving few outward clues to its 150+ mph potential. There was a 3.5-litre 24-valve straight six, a beefed-up Getrag five-speed gearbox and a limited-slip differential, sports suspension and ABS, and a well-equipped interior.

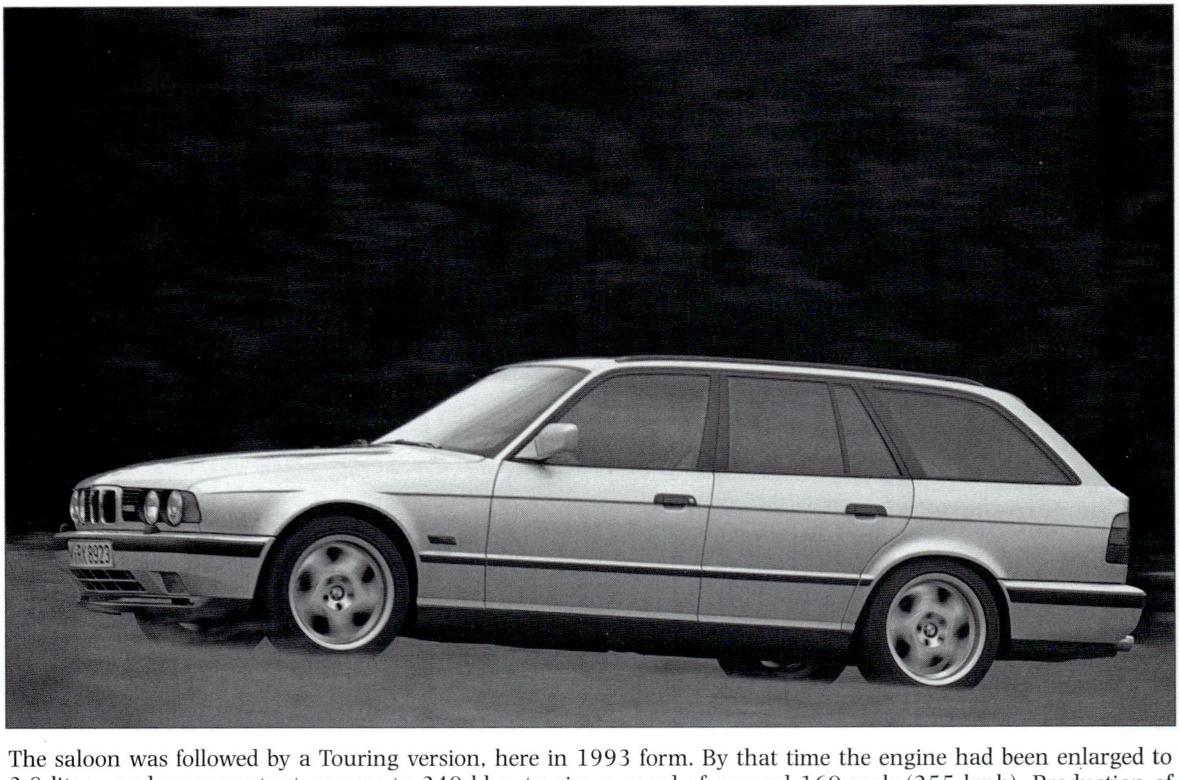

The saloon was followed by a Touring version, here in 1993 form. By that time the engine had been enlarged to 3.8 litres, and power output was up to 340 bhp, to give a speed of around 160 mph (255 kmh). Production of these first-generation M5s reached 12,254 units.

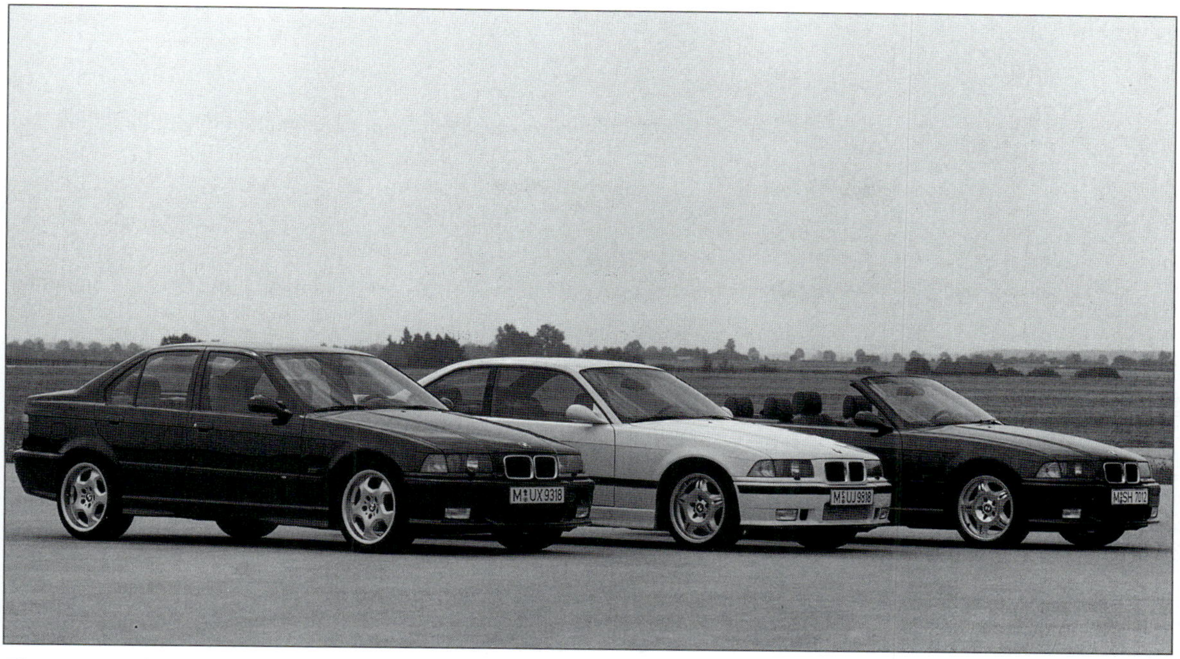

The E36 M3 line-up of the mid-1990s – saloon, coupé and convertible. Once again the saloon was understated, the others among Europe's most attractive cars, in visual terms. The saloon was introduced in 1992 with a 3-litre 286 bhp engine, and the trio was completed two years later; by the end of the decade only the coupé and convertible were listed.

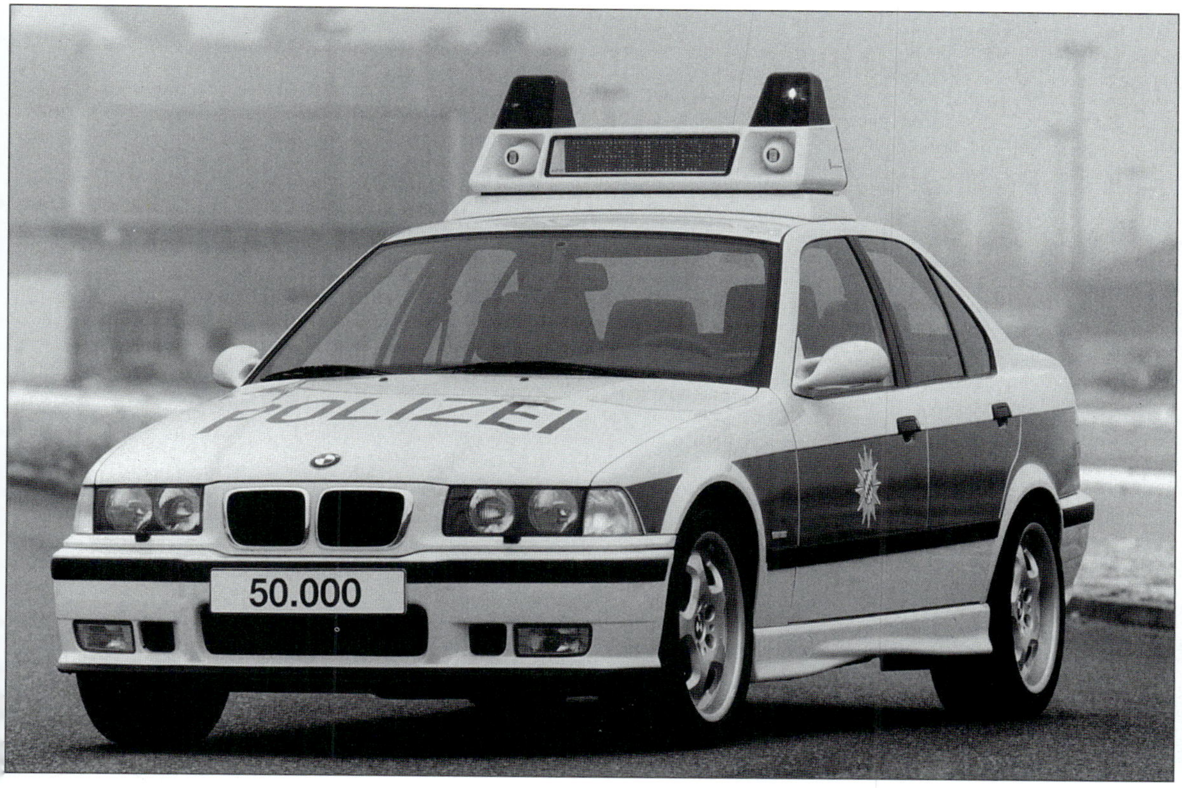

Late in 1997 the 50,000th M3 was completed, in the German police livery for the Bavarian force. It was fully equipped, and there were assurances that the roof-top lights were tested at the car's maximum speed. That was electronically limited to 155 mph (250 kmh).

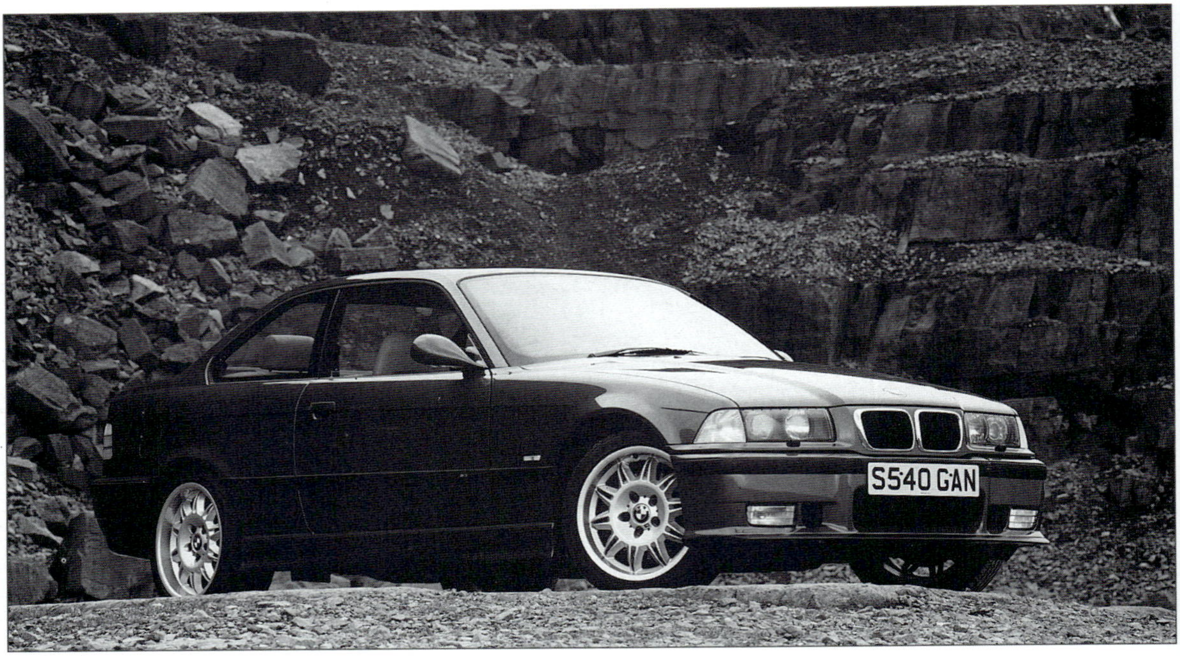

The M3 coupé had the sleek overall lines of the normal two-door E36, with details such as the front and rear aprons, blending into a spoiler at the front, that actually made for a slightly worse drag coefficient. The 3-litre straight six incorporated refinements such as variable camshaft timing (VANOS) and there was an advanced management system. It was rated at 286 bhp – that's 96.8 bhp/litre – and torque figures were excellent.

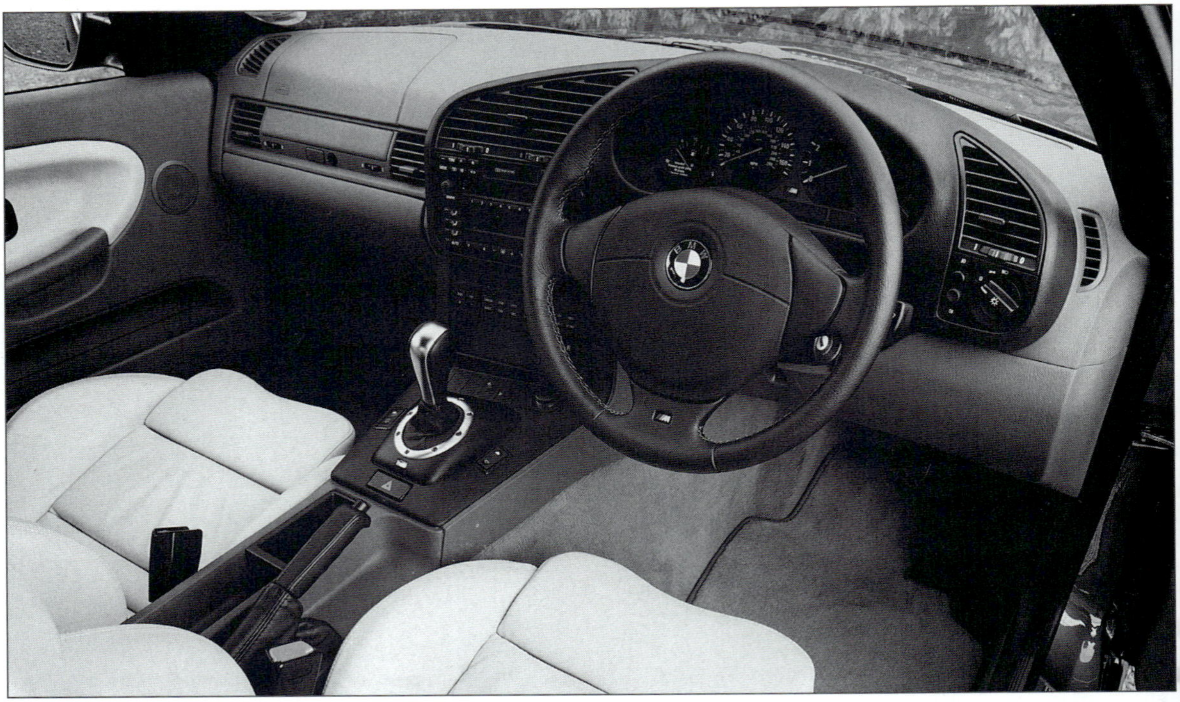

The M3 cockpit had a typical clean BMW layout. This car has the Sequential M Gearbox (SMG) introduced early in 1997, which combines the ease of operation of a normal automatic with the instantaneous control of a manual system, without the need to operate the clutch when making changes. The automatic (E) mode was initially provided with one pre-determined programme, but experience showed that this was not in harmony with driving conditions in all countries, so for example a separate UK programme was devised. There were also 'Sports' and 'Winter' programmes.

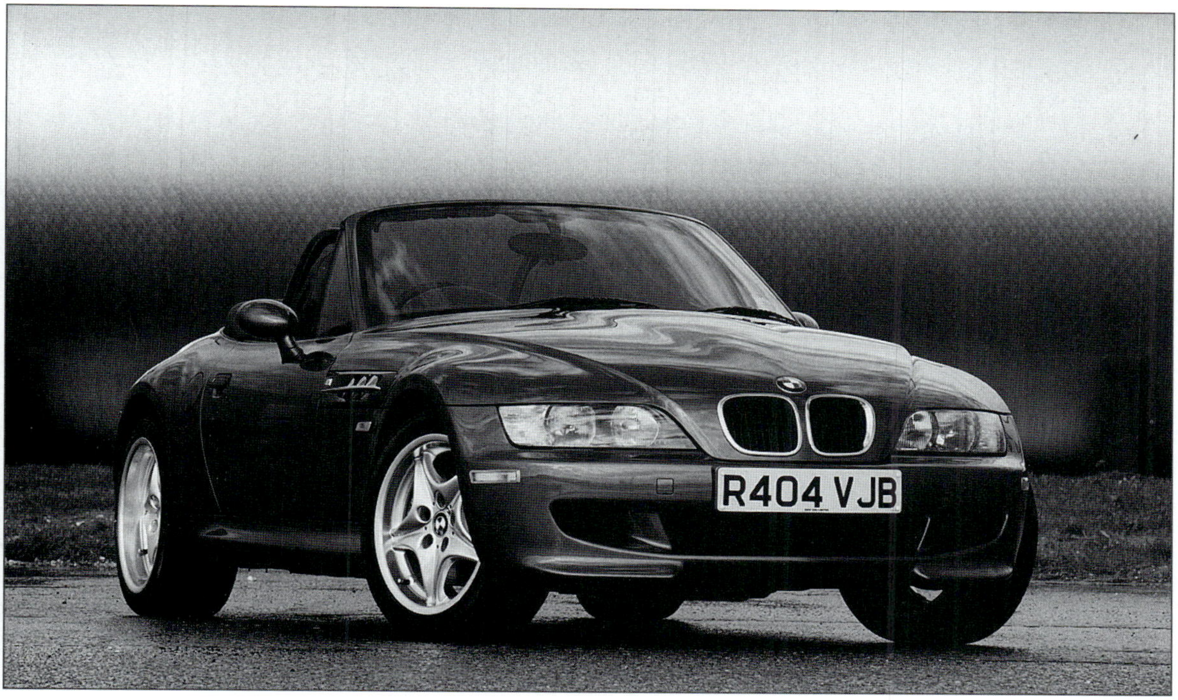

In many respects, the M roadster came as the top model in the Z3 range, but it was set apart by its M designation. It shared features such as the electro-hydraulic soft top operation with the 2.8-litre Z3, but under the bonnet (below) there was a prominent 'BMW M Power' notification: this meant it had the 3.2-litre six-cylinder engine, giving 321 bhp and that familiar governed 155 mph top speed. Visual details such as the front air dam and flared arches helped distinguish it from other Z3s.

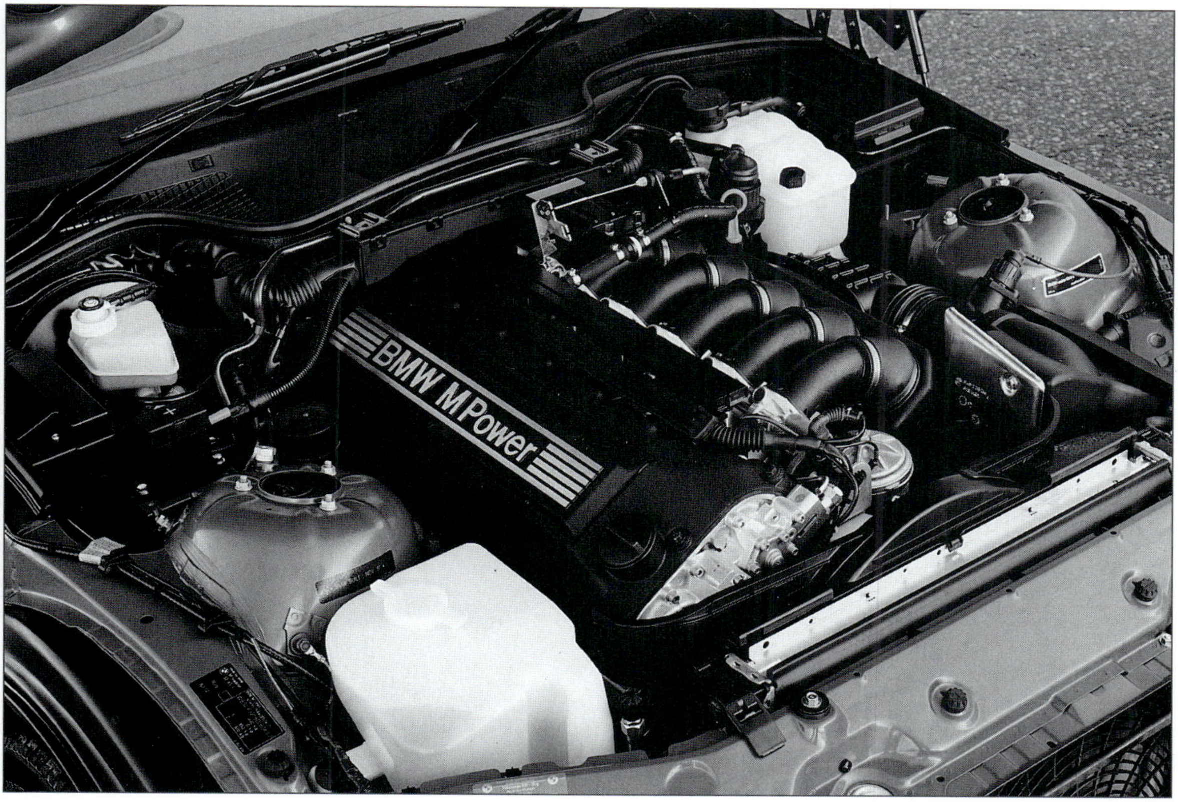

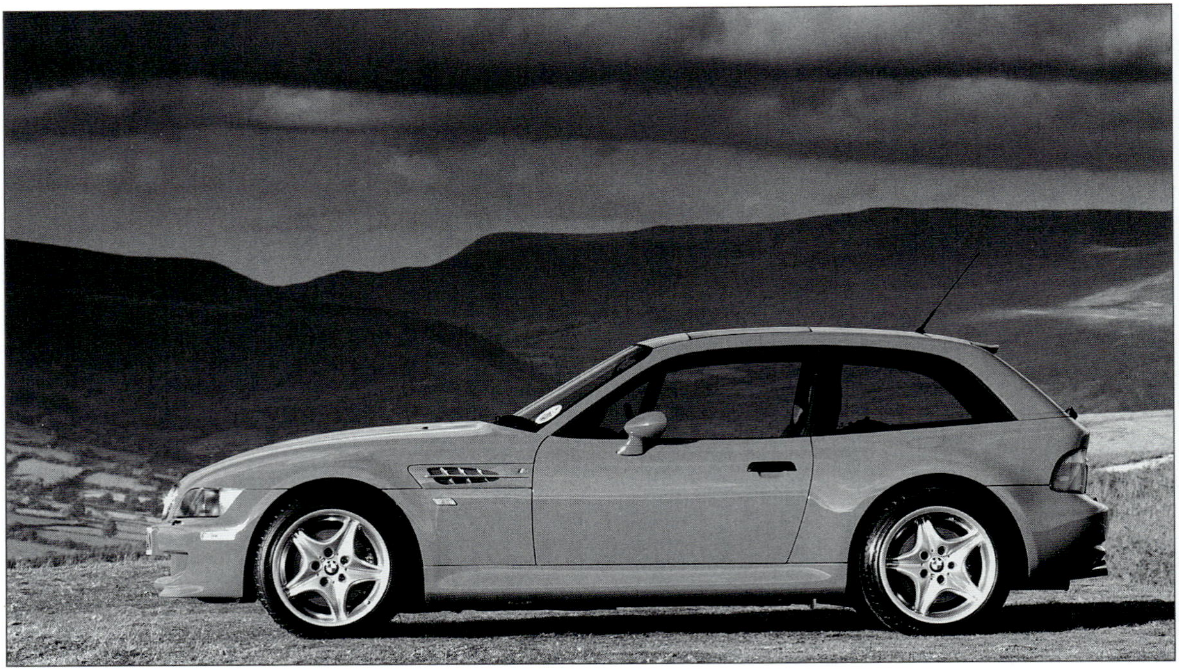

The M coupé in side elevation shows a possibly more flattering view of that controversial rear. It shared most of the roadster components, from the engine to those deeply dished alloy wheels. Cockpit space was not generous, and while the hatchback was sensible there was not much room for luggage in the tail.

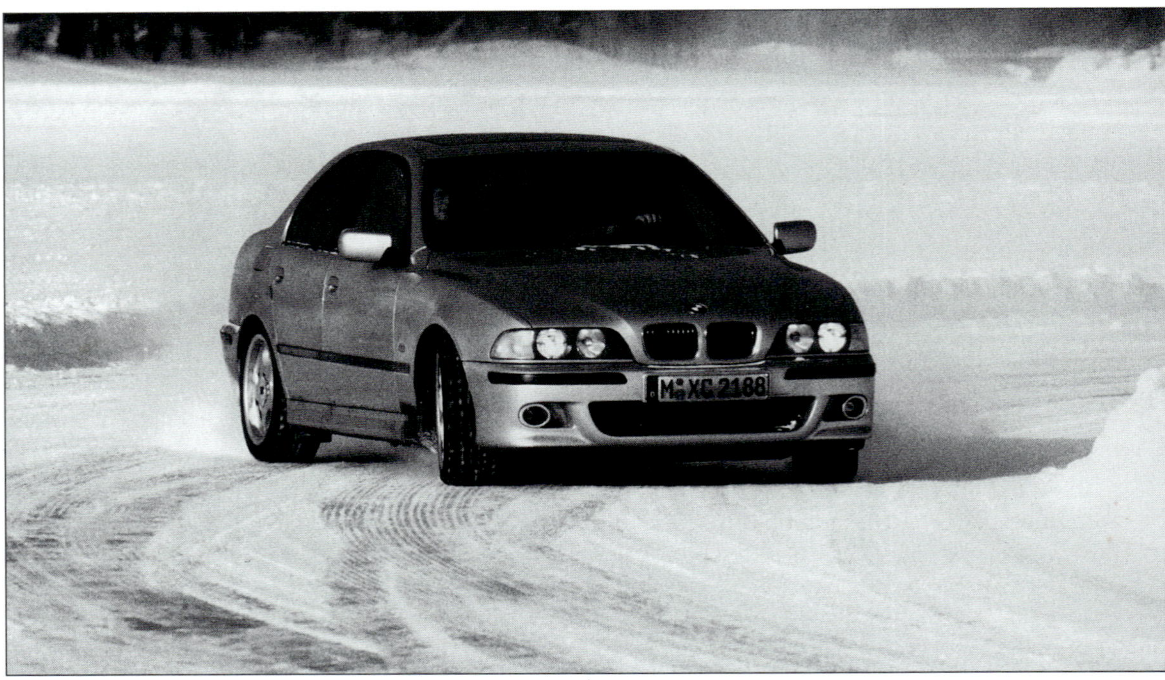

On paper the E34 M5 that went out of production in 1996 was replaced by the 540i in the E39 series that same year. In fact, there was a market gap for other manufacturers to fill, so BMW had to bring in a third M5. It came in 1998, with a 5-litre V8 producing 408 bhp. Basically, this was the engine used in the 5 and 7 Series, tuned and with clever features such as an oil pumps system to ensure that oil was returned to the sump during the most extreme cornering. A six-speed manual gearbox was fitted. The suspension was lowered and tuned, and wide low-profile tyres fitted – those 18-inch tyres were one of the few signs that this was a four-door luxury saloon with supercar performance.

CHAPTER TEN

A QUARTER OF A CENTURY OF MOTOR SPORT

BMW was prominent in motor sport throughout the last quarter of the twentieth century, and BMW engines powered notable cars in the premier categories. An era in touring car racing ended with the CSL's career at the end of the 1970s, so in some respects this snapshot review of motor sport does not dovetail with the production periods or types, or the chapters in this book.

Some big BMWs were still raced as the 1980s opened, the 635CSi in the then-new Group A category in 1981, for example, or the 528i which Helmut Kelleners and Umberto Grano drove to win another European Championship in 1982, followed by Dieter Quester, champion with the 635CSi in 1983. From 1976 Motorsport had been developing 3 Series cars, in its engine work drawing on Formula 2 development, and in 1977 the Junior Team used the 320i, spectacularly if not always successfully! In the next decade the M3 took the place of the big coupés. It had been laid down with motor sport very much in mind, and there is probably no record of the vast number of races that fell to it in the 1980s and 1990s. It was also driven to win hill climbs and, rarely for a BMW, it was successful in rallies; it combined the qualities of speed and reliability, perhaps best shown in a string of victories in the Spa 24-hour race (three falling to Gabriele Rafanelli's Bigazzi team, another of the associate teams that contributed so much to the BMW success story through this period).

That story continued until the end of 1998, when BMW officially withdrew from touring car racing, although independents still used M3s, and the South African subsidiary went ahead with a circuit version of the E46 Series 3 car.

In single-seater racing BMW made its mark as an engine manufacturer, first with an extraordinary record in Formula 2 through its 2-litre years, 1972–84. The engines were based on production units, and the first 2-litre M12 F2 type was quietly developed by Paul Rosche. It was therefore in hand when BMW Motorsport was formed, the programme could pick up impetus quickly, and a fruitful association with March was set up. While the March-BMW works team gained enviable records, engines were also sold to other teams, and eventually more than 500 F2 engines were produced. 'Motorsport' engines powered the champions' cars in 1973–4, 1978–9 and in 1982 (Schnitzer's BMW-based engine was used by the 1975 champion) sales made profits, and racing engine experience built up.

Rosche bravely stayed with the production-based four-cylinder format for the company's first Formula 1 engine. There had been rumours of a BMW F1 engine in the 1970s – inevitably, perhaps, in view of Formula 2 success – but this had not taken shape in the metal. Turbocharged versions of the four-cylinder engine had been run in saloons, and this line of development was continued in the iron-block unit for Formula 1, which in effect was a 16-valve adaptation of the engine used in cars such as the 316.

Brabham was a partner, and designer Gordon Murray had to overcome problems with the engine, as it had to be installed upright and could not be used as a stressed chassis member, but the team's early misgivings came after test failures suggested that unreliability could be a factor. These were never fully overcome, but eventually the engine produced more than 900 bhp in race trim, and a phenomenal 1200 bhp in 'qualifying trim' – at a cost in 'short-life' engines! The first of eight Grand Prix victories for Brabham-BMW fell to Nelson Piquet in Canada in 1982, and the Brazilian was World Champion in 1983. The 1986 Mexican GP fell to a Benetton-BMW driven by Gerhard Berger, later to become BMW's Motorsport Director. BMW withdrew at the end of the 1987 season, although the engines were still used in GP cars, as Megatrons. The teams which used these customer engines, Arrows and ATS, had little success. BMW confirmed its return to Formula 1, with Williams, in 1997 at the start of a long-term programme initially aimed at a race debut in 2000. A V10 engine took shape under the direction of Paul Rosche, soon to retire, and ran tests in an adapted Williams chassis in 1999.

Incidentally, in 1991 BMW and the ADAC launched an *ab initio* single-seater class in Germany, using chassis designed by Johann Knapp, with adapted motorcycle engines. In 1997 this had two classes: cars with 1.1-litre 130 bhp engines, and a junior type using 987 cc 110 bhp engines.

Sports car racing – the term is used loosely, as so often – has brought mixed results for BMW. The original main-line aims for the M1 were thwarted, but the all-M1 Pro Car races at Grand Prix meetings at least reaped some publicity. Ventures with March in the mid-1980s were hardly rewarding, but then came the McLaren-BMW, conceived as a road supercar but at once the world's fastest road car and a GT race winner at the highest level.

Unlike some road/racing car types, the McLaren-BMWs for both roles were closely related – the racing GTR was lighter, components such as gearbox and brakes were specialized, a slightly smaller S70 V12 was built to exploit a regulation, and a longer version was devised. But the independent F1 GTR that won at Le Mans in 1995 was not far removed from its road counterpart.

BMW returned to sports car racing in its own right in 1998 with the open V12 LM – developed and built by BMW Motorsport Limited in Britain and run by Rafanelli's team. The cars retired at Le Mans, but came out again in 1999 when they were joined by a refined V12 LMR (Le Mans Roadster), run by Schnitzer as BMW signalled its intentions to concentrate in-house racing activities on the Formula 1 partnership with Williams. The V12 LMR was one of those rare cars that enjoyed a debut race victory, at Sebring early in 1999. The company bowed out of sports car racing with a magnificent victory in the 1999 Le Mans 24-hour Race.

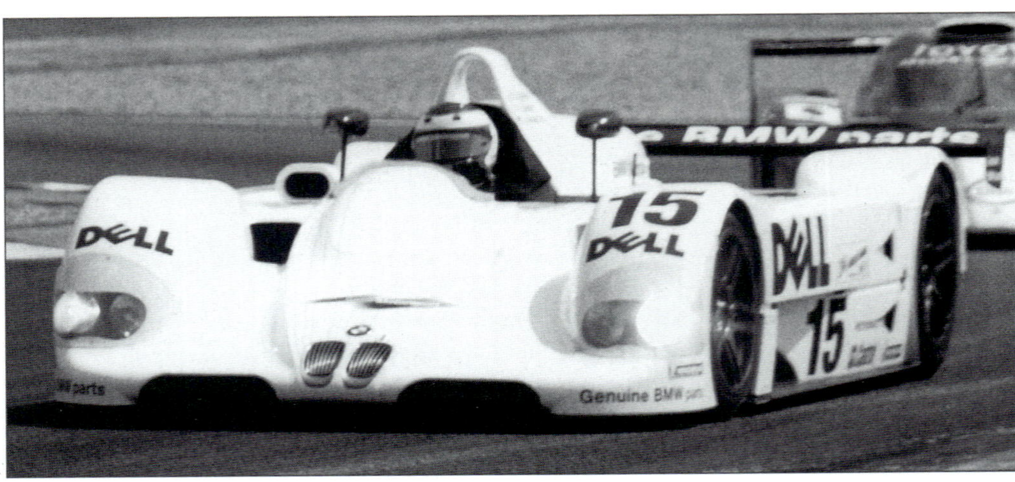

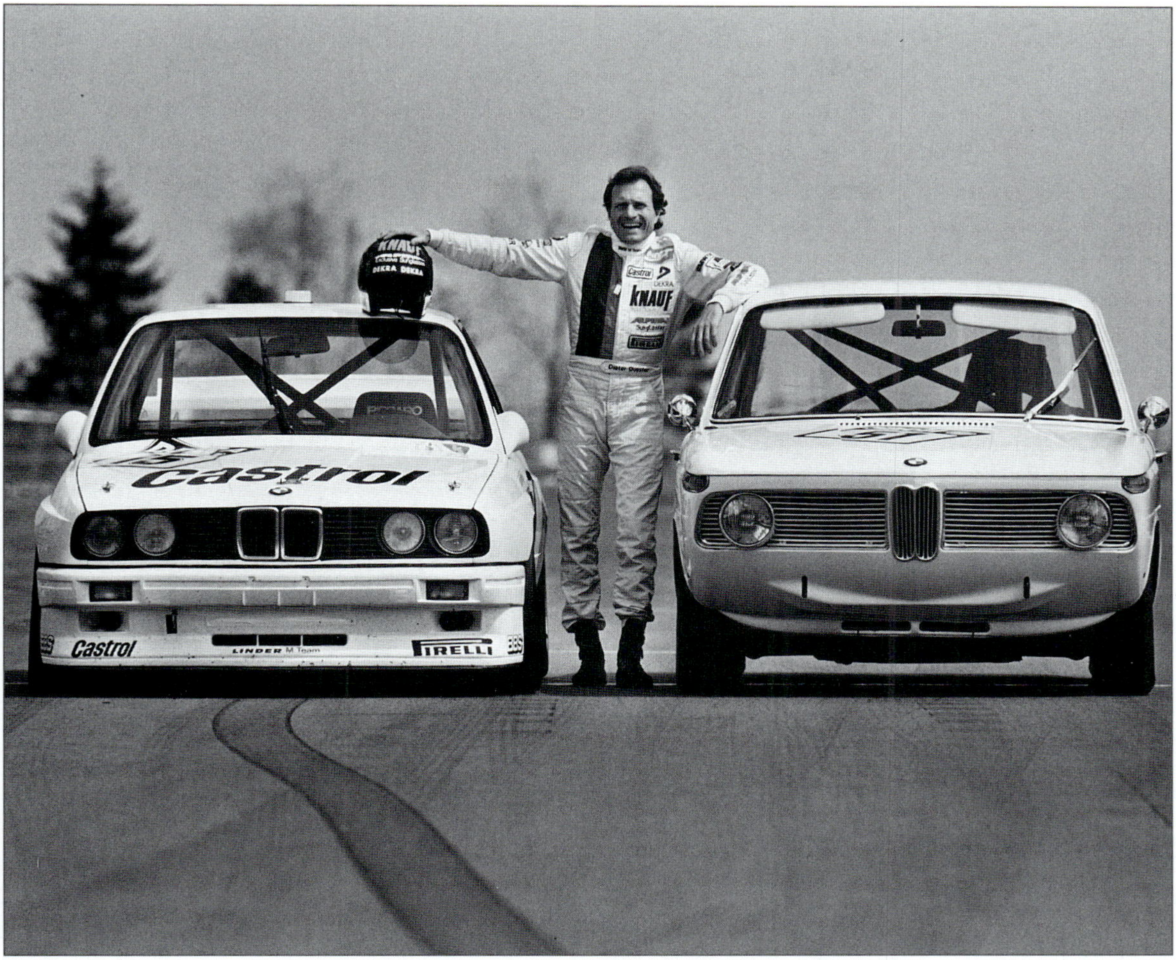

BMW stalwart Dieter Quester between a 1991 Linder team M3 and the 1800 TISA from the mid-1960s that he drove in the European Historic Touring Car Challenge series three decades later.

Opposite: BMW's greatest sports car racing triumph came in the high summer of 1999 when Yannick Dalmas, Joachim Winkelhock and Pierluigi Martini drove this V12 LMR to victory in the Le Mans 24-hour Race. A sister car had led the event for fourteen hours; this car led for the last four.

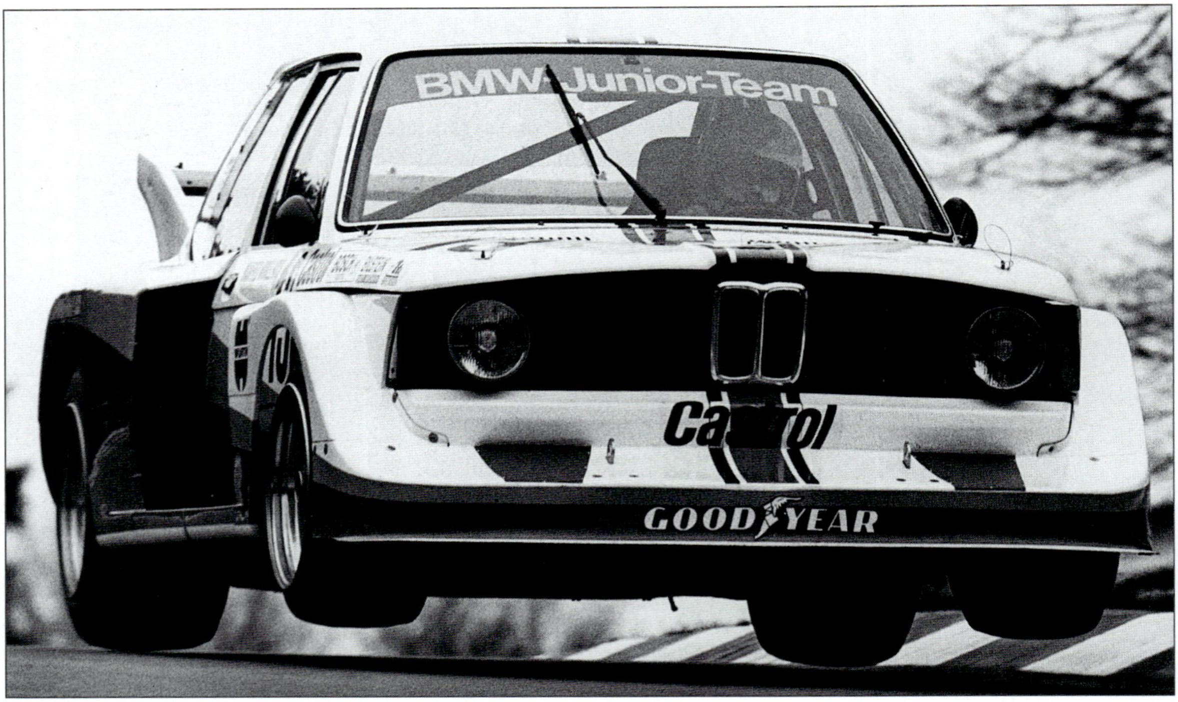

3 Series racing machines. Highly developed 320s were raced by the BMW Junior Team in 1977, with 2-litre engines developing more than 300 bhp in Group 5 form. There is abundant evidence of development in the aerodynamic appendages on this car – notably the big rear wing – glass fibre body panels were used, and wide wheels. 'Junior' misleads, for experienced drivers were used: this jumping 320 is driven by Manfred Winkelhock, and the team won the German championship. The cars were sold on to independent entrants in 1978, when the Junior Team turned to Formula 2.

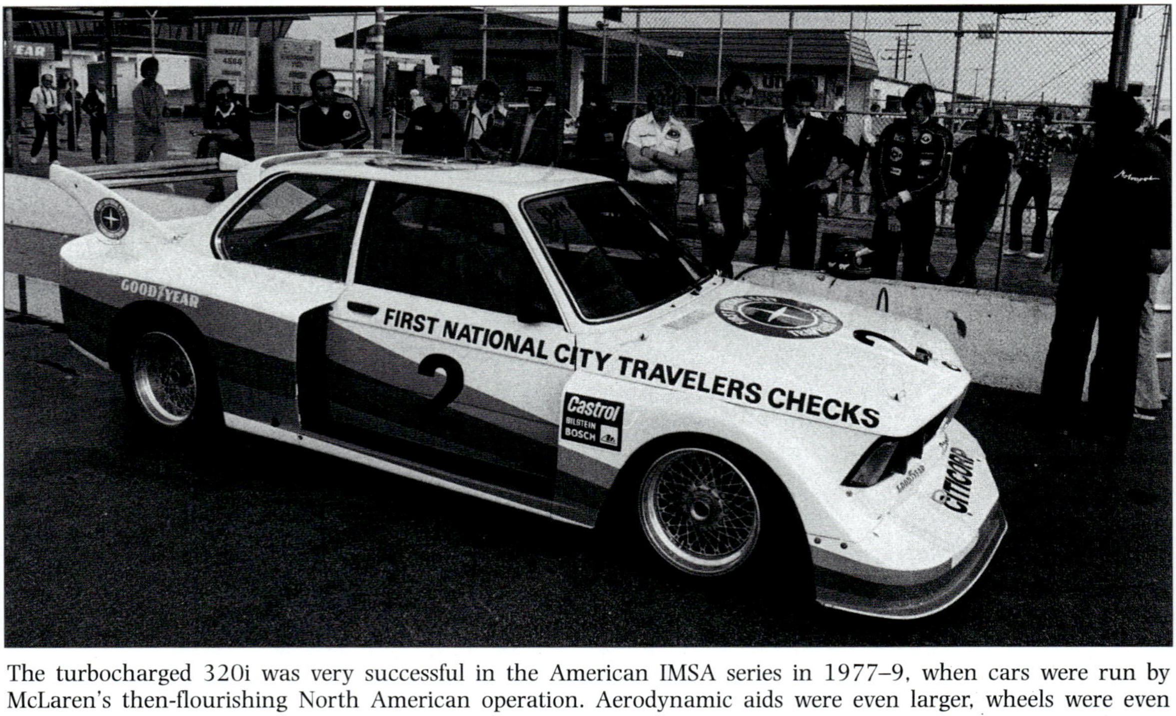

The turbocharged 320i was very successful in the American IMSA series in 1977–9, when cars were run by McLaren's then-flourishing North American operation. Aerodynamic aids were even larger, wheels were even wider, and engines gave up to 650 bhp. Here Ronnie Peterson and two Motorsport mechanics seem lost in thought during a test session.

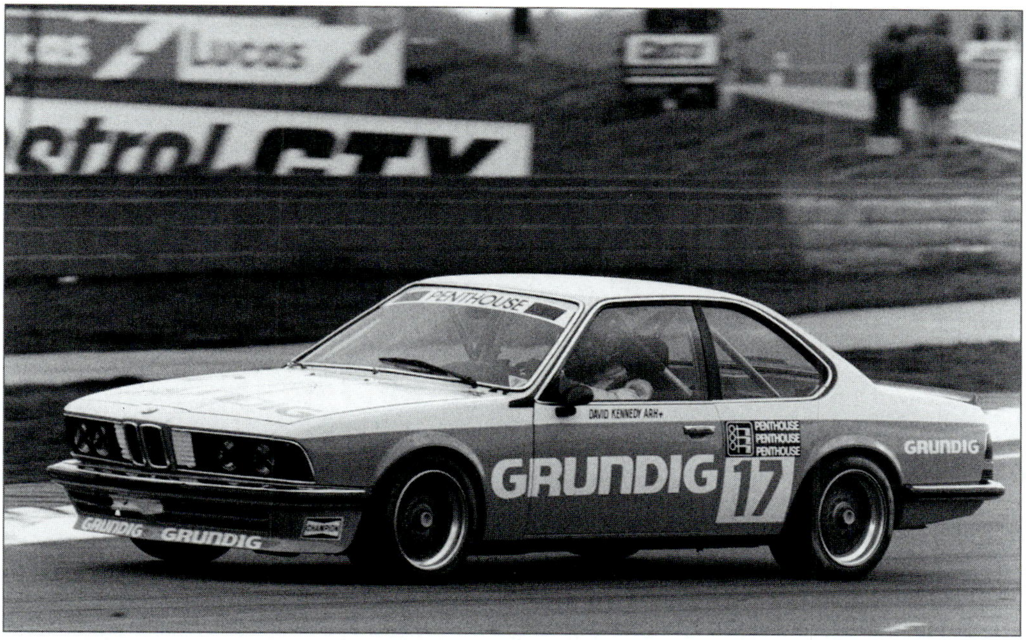

The British Saloon Car Championship series have usually been demanding, and in 1984 the battles for overall honours were between Rovers and BMW 635CSi coupés. That year's cars had to follow production specifications fairly closely, so while some suspension components were modified and safety equipment added, engines were near-standard and there were no lightweight body panels. This BS Automotive car was driven in an early-season Silverstone race by David Kennedy.

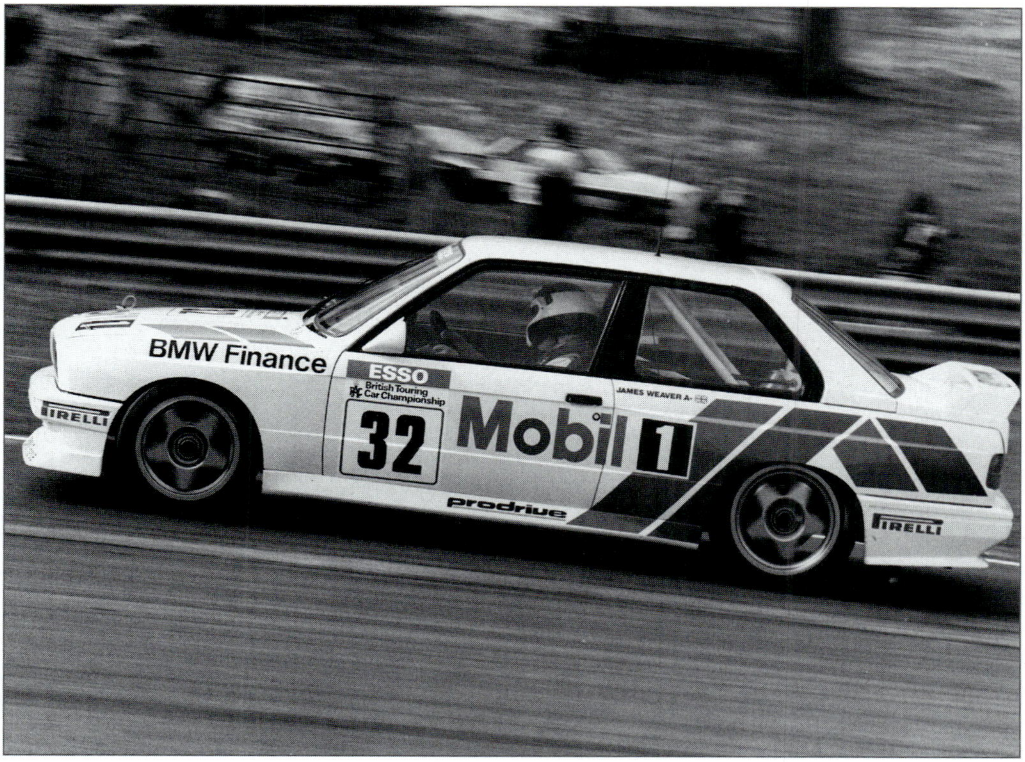

By 1989 'Touring' had replaced 'Saloon' in the series title, and that year M3s totally dominated the 2–3-litre class. Prodrive prepared the pair of BMW Finance cars, in effect the works cars, that were driven by Frank Sytner and James Weaver.

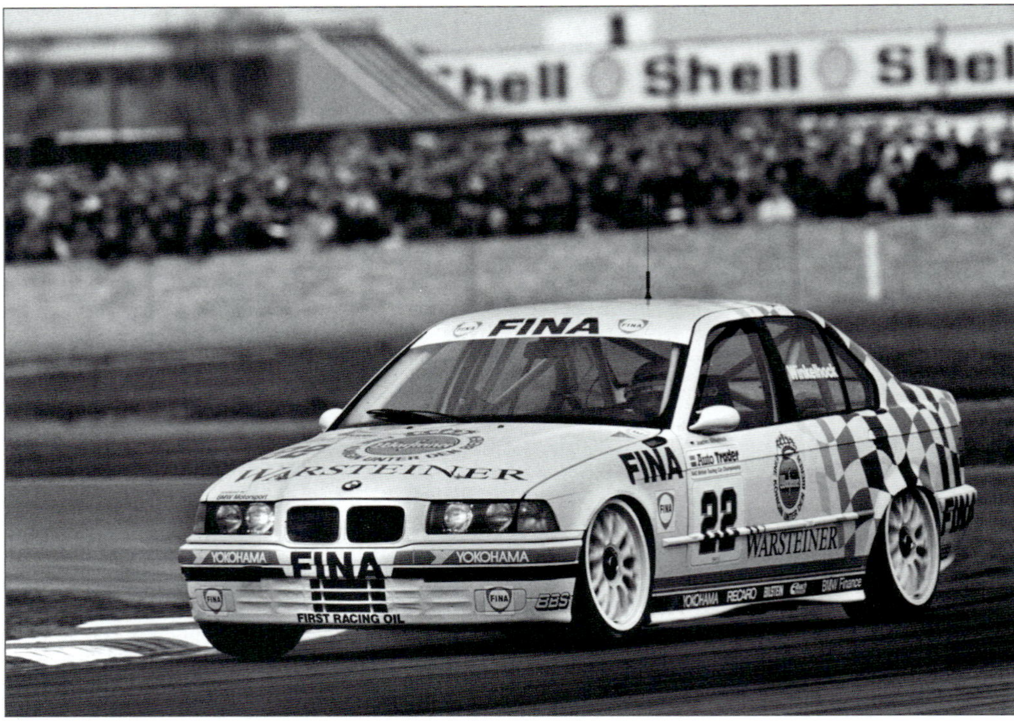

The works, or quasi-works, teams in the 1990s ran cars in very distinctive colours, with a chequer pattern strengthening towards the tails. This Schnitzer 318I also sits lower on its wide (18-inch) wheels, but otherwise looks close to standard. The driver is Winkelhock, who won the British championship with it in 1993.

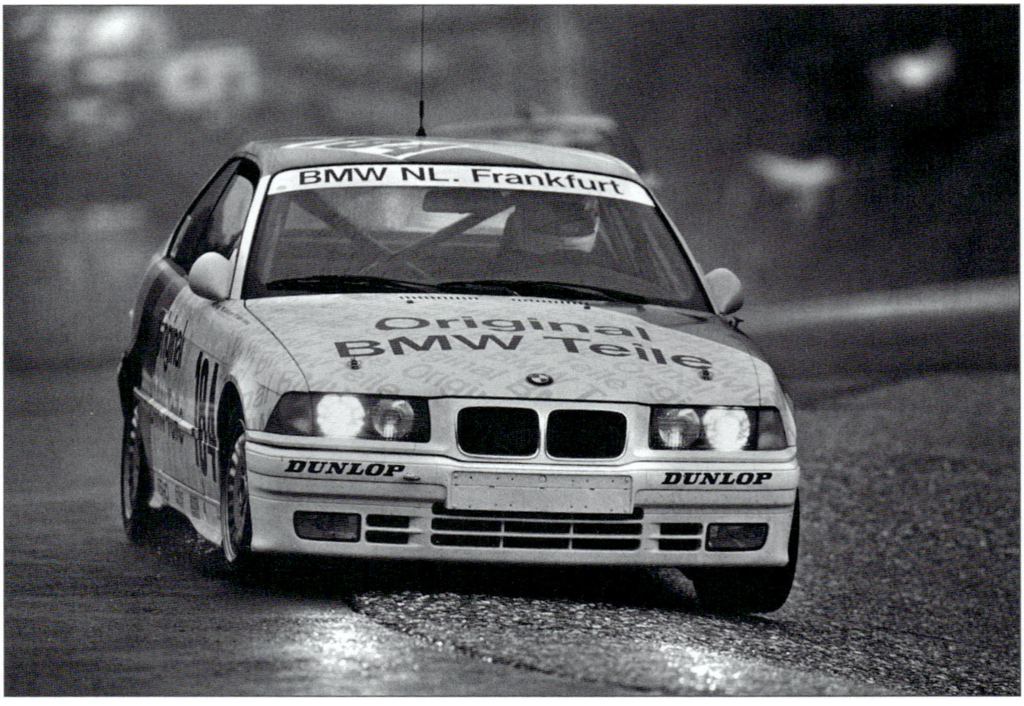

In the 1990s 'super-touring' races captured most media attention, but series for near-stock cars were still run, and these became much more popular late in the decade. This 325i coupé is a Group N ('showroom') car, in a Dutch championship race.

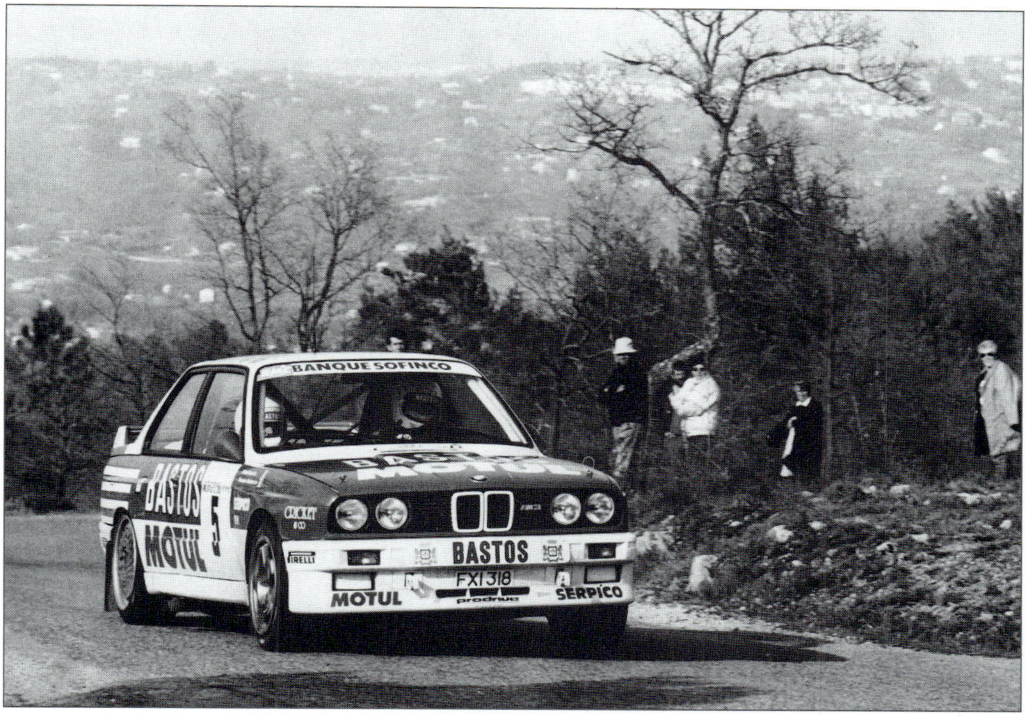

The Group A 'homologation special' M3s made an impact on European rallies in the second half of the 1980s. The cars prepared by Prodrive in Britain were notably successful in the French series, and Bernard Beguin drove one to a World Championship rally win in Corsica in 1987. This 1988 Group A M3 is driven by François Chatriot.

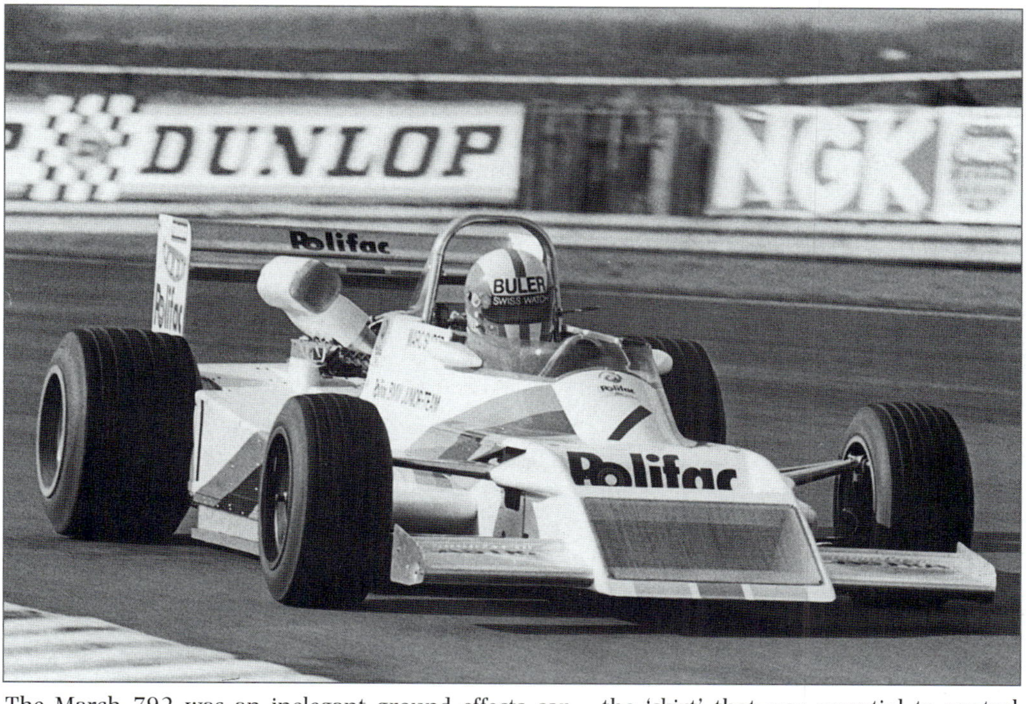

The March 792 was an inelegant ground effects car – the 'skirt' that was essential to control under-car airflow can be seen ahead of the rear wheels – but it was most effective, especially in the hands of Marc Surer. He won the 1979 European Championship with these cars, here in BMW Junior Team colours at Silverstone.

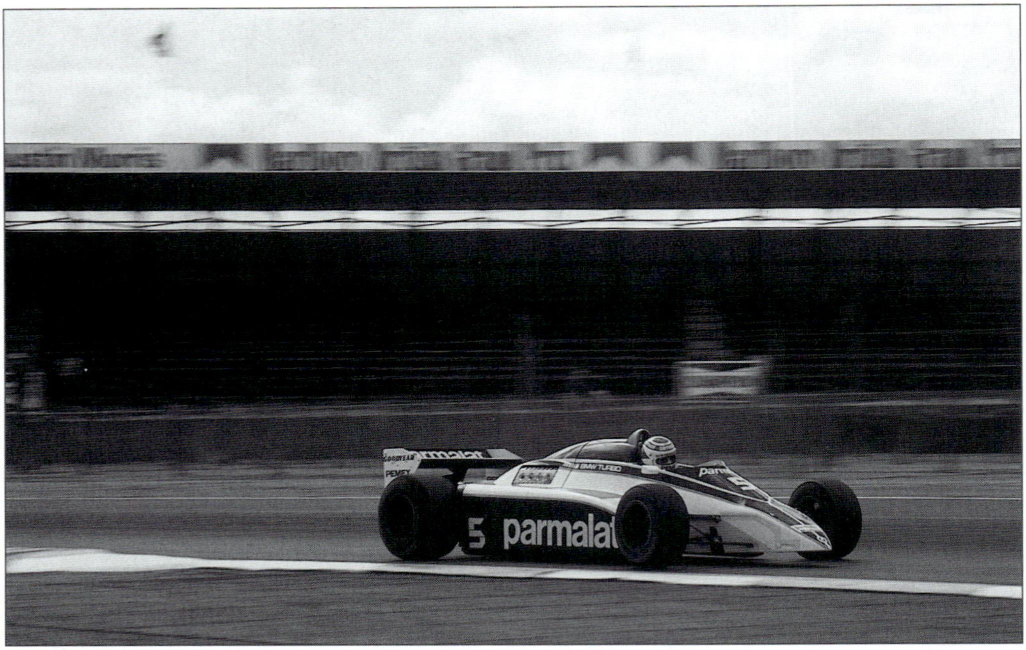

The first public appearance of a Brabham-BMW F1 car came during practice for the 1981 British Grand Prix, when Nelson Piquet lapped quickly but chose not to race the BT50. That type won just one GP, in Canada in 1982.

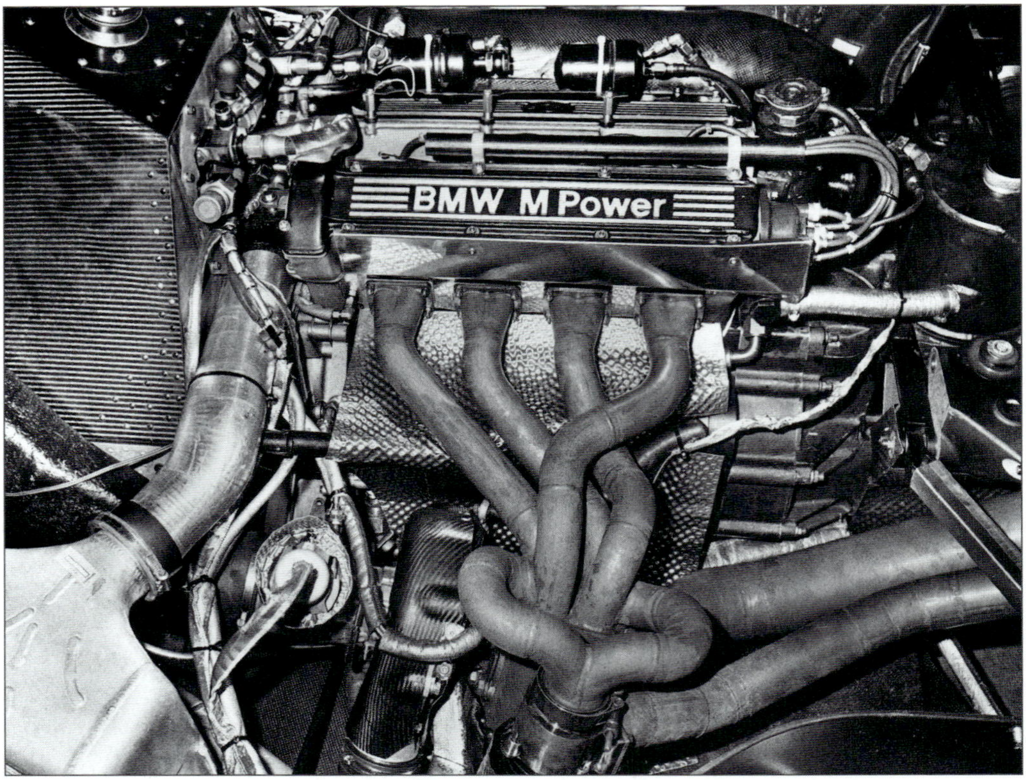

An M12/13 Formula 1 engine in its 1985 form, when a Garrett turbocharger had taken the place of the original KKK equipment, with a gain in reliability to back up a racing output of around 900 bhp. The intercooler and turbocharger are at the bottom; part of one of the very large radiators is on the left.

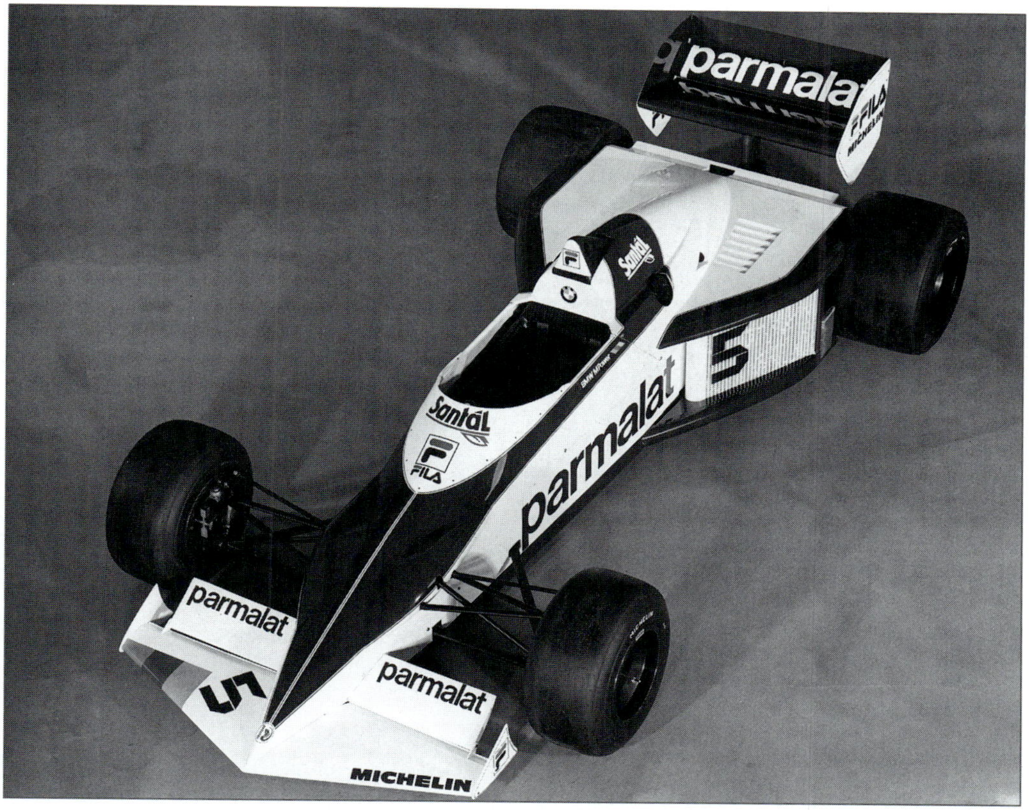

The BT52 had a very pronounced rearward weight bias, dubbed an arrow layout. Piquet drove this type to win the World Championship in 1983.

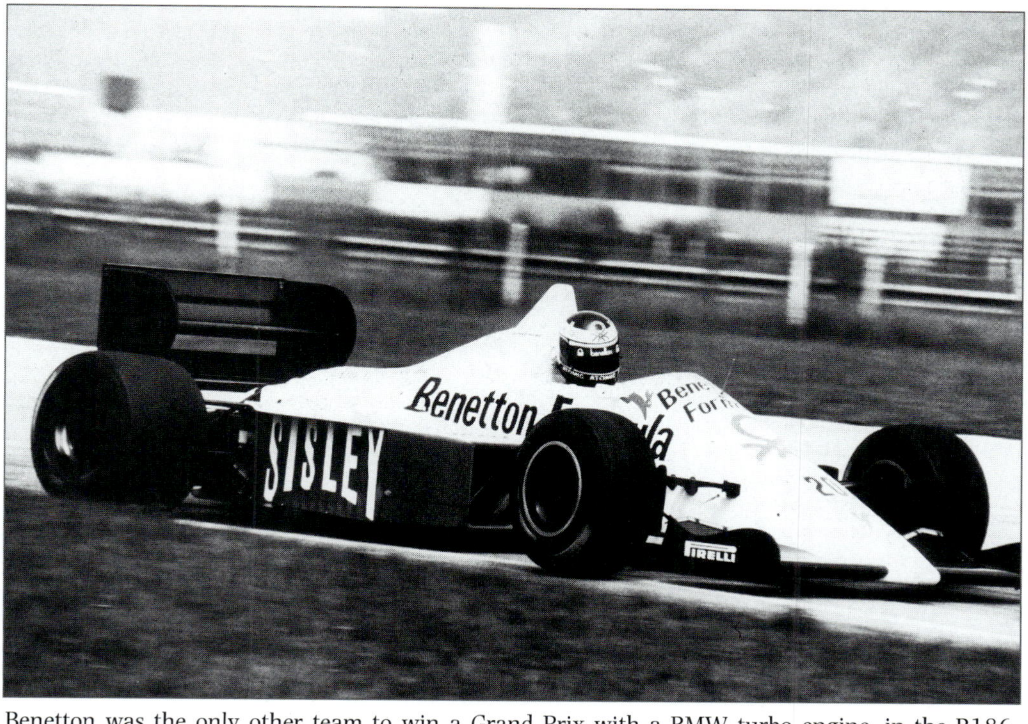

Benetton was the only other team to win a Grand Prix with a BMW turbo engine, in the B186. Gerhard Berger drove the winning B186 in the 1986 Mexican GP.

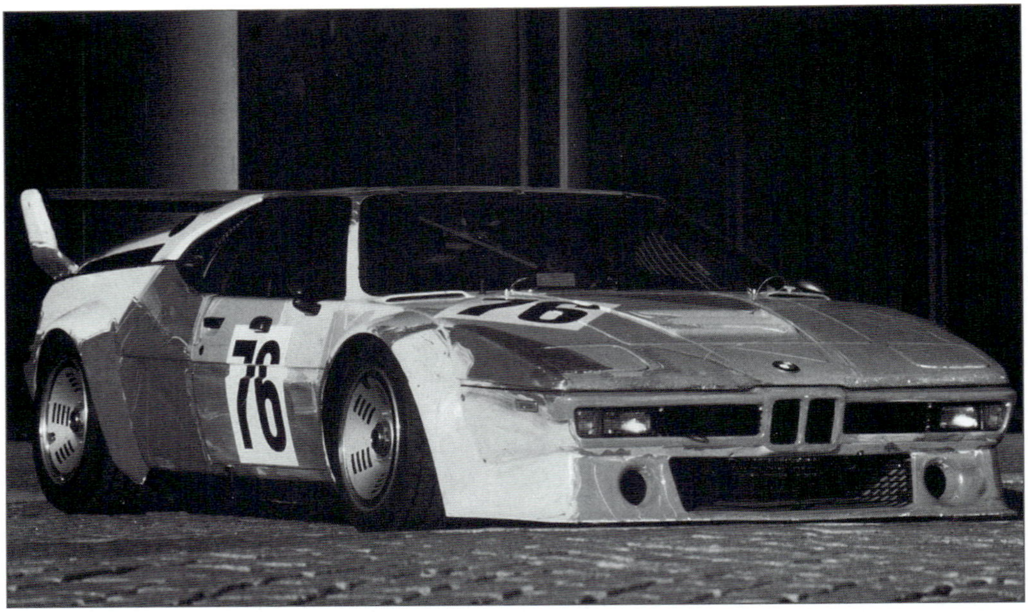

Delays undermined chances of BMW success in Group 4 or IMSA racing with the M1. The best place achieved at Le Mans was sixth in 1979, with this car driven by Mignot, Poulain and Winkelhock. It was daubed in curious colours by Andy Warhol, who claimed that his carefree mess 'portrayed speed pictorially'.

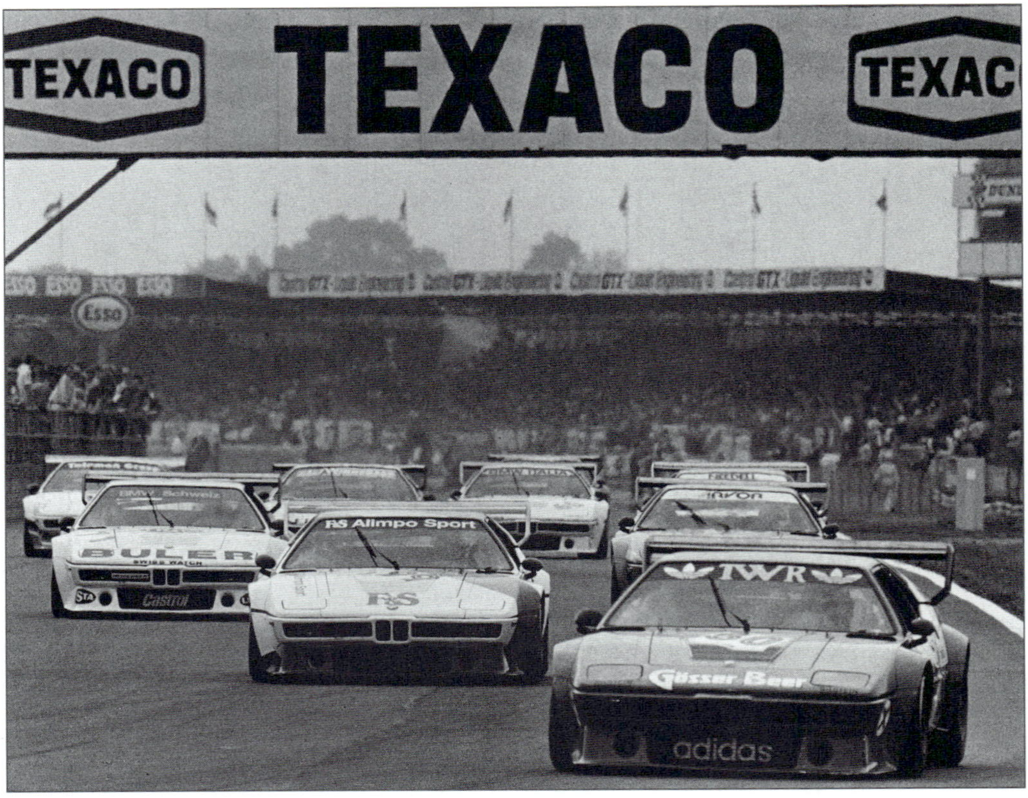

Pro Car racing brought publicity in 1979, when M1s were raced by leading drivers in supporting events at Grand Prix meetings. This field is jostling into the first corner at Silverstone at that year's British event, with a very familiar set of initials on the windscreen of the leading car: Tom Walkinshaw seems to have been involved with any and every type of racing!

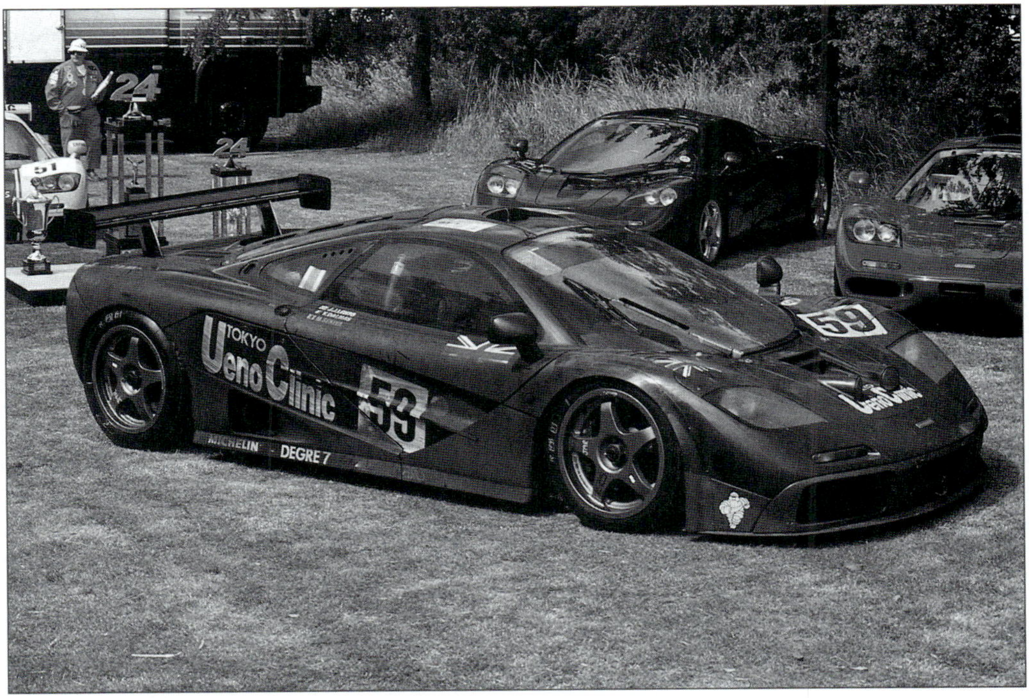

The McLaren F1 GTR which won the 1995 Le Mans 24-hour Race, displayed at the Goodwood Festival of Speed a few days later.

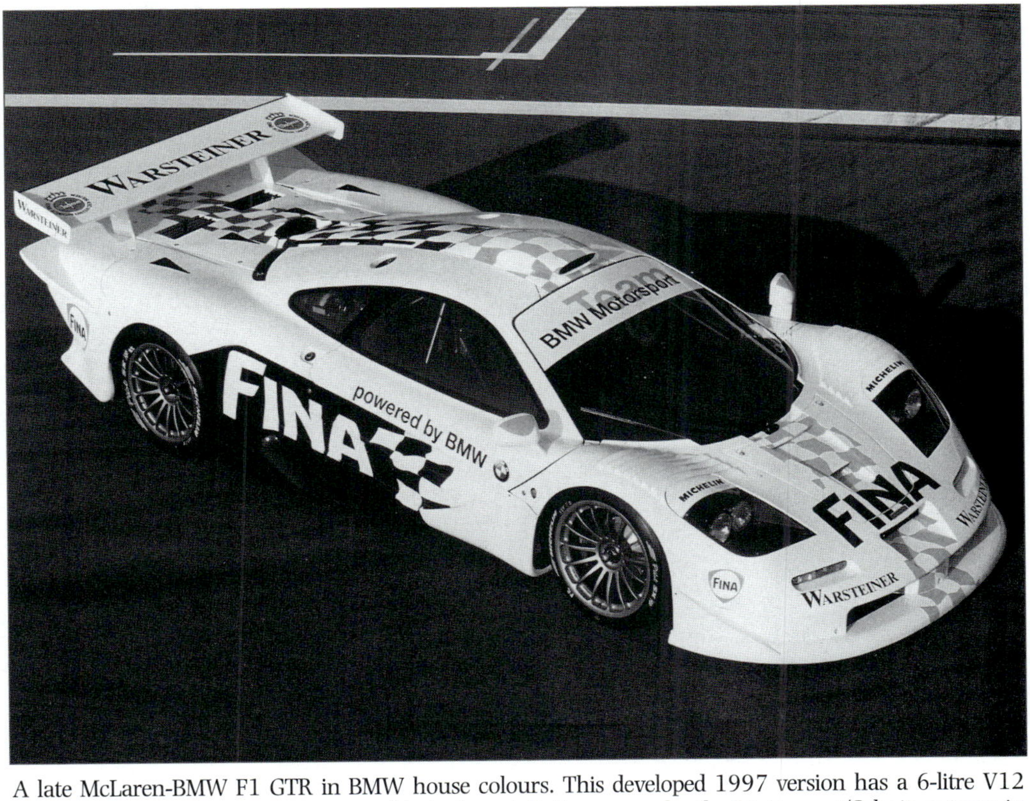

A late McLaren-BMW F1 GTR in BMW house colours. This developed 1997 version has a 6-litre V12 and modified aerodynamics, most notably its long tail. It was run by the Motorsport/Schnitzer team in the European GT Championship, winning five races. However, the team failed to repeat the Le Mans victory scored by an earlier F1 loaned to a Japanese outfit in 1995.

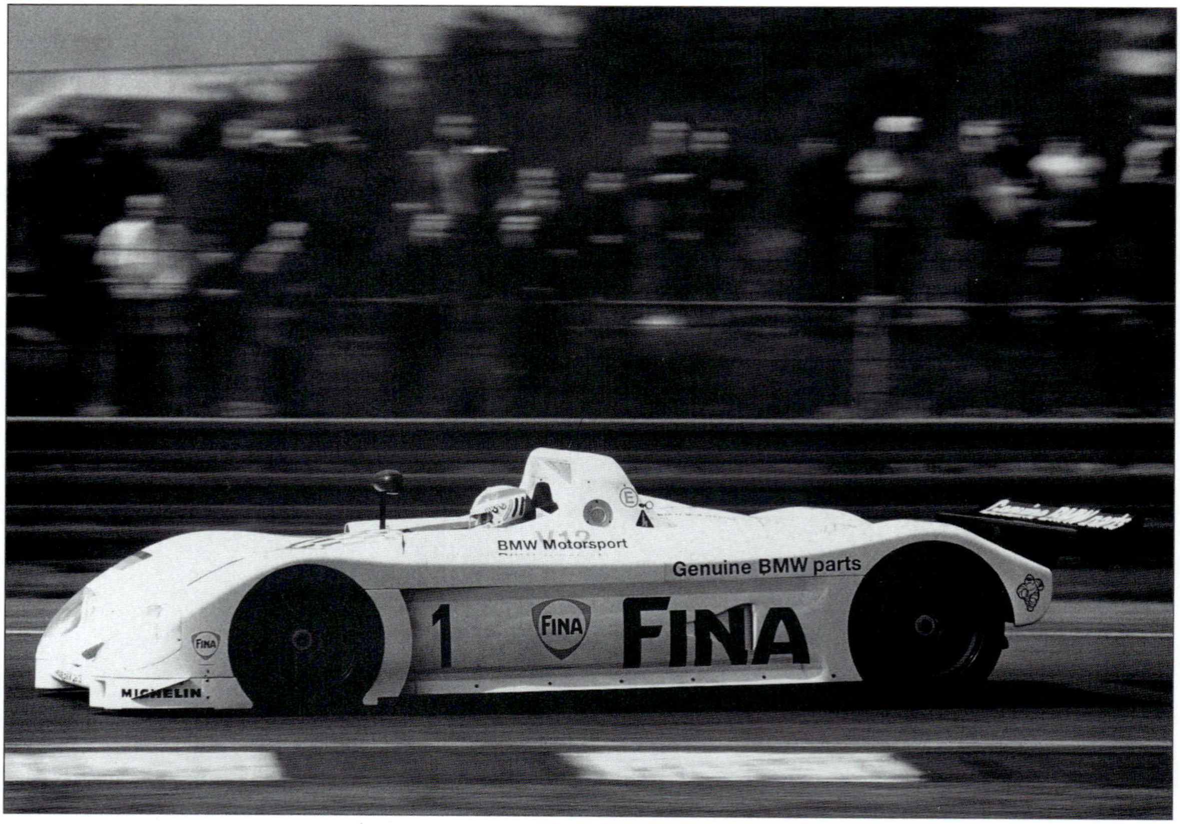

BMW's 1998 open sports-prototype, V12 LM, was built at the Motorsport plant adjacent to the Williams headquarters in Britain, and run at Le Mans by the Rafanelli team. It 'inherited' the V12 originally laid down for the McLaren F1. Both cars which started at Le Mans in 1998 retired with wheel bearing failures.

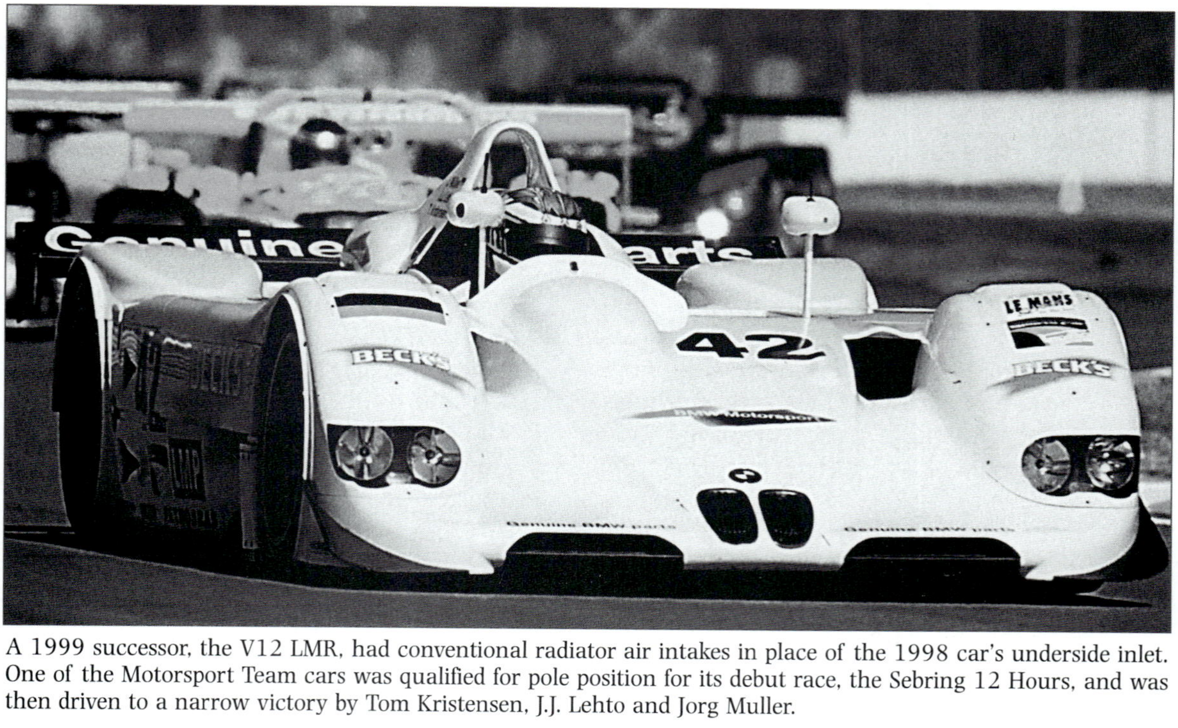

A 1999 successor, the V12 LMR, had conventional radiator air intakes in place of the 1998 car's underside inlet. One of the Motorsport Team cars was qualified for pole position for its debut race, the Sebring 12 Hours, and was then driven to a narrow victory by Tom Kristensen, J.J. Lehto and Jorg Muller.

CHAPTER ELEVEN

PROTOTYPES AND CONCEPT CARS

BMW has seldom indulged in attention-distracting motor show concept cars – just 4/2 for the 1995 Tokyo Motor Show was an exception – but it has sometimes shown realistic prototypes or encouraged coachbuilder allies to build cars that could be seen as quasi-official prototypes. Weismann's MF25 roadster which anticipated the Z1 was an example and there were Baur's cabriolets and coupés such as the TC3 in 1987. ('Limited production' of this very slippery car was mooted, but did not get the nod from BMW.)

BMW produced little motorcycle-engined off-road military prototypes as late as the 1980s, and earlier had developed farm runabouts on similar lines, for production by another firm, but these were not really fields for the Munich company.

It was looking to the future in the Turbo. This 1972 venture did not lead directly to a production car (although it was the first BMW to show an approach to the wedge shape), unlike the 1997 Z07 design study which did lead to the Z8. Z, incidentally, has been used for production sports cars, but is not exclusive to this type, and as with the E sequence there are gaps; Z designations have been applied to city car projects far removed from the Z sports cars. . . .

Various electric cars have been built during work at the FIZ research centre of BMW Technik, and these have carried E numbers which can confuse: E can indicate Electro. So E7 and E29 were electric cars with *Entwicklungs* numbers, while E1 was a later electric essay. Just to confuse a little more, a spider design study developed with Hewlett-Packard in the late 1980s was also designated E1. . . .

A long-term programme aimed at practical hydrogen cars has been undertaken. Although the first installations were cumbersome, largely because of storage problems, the concept has been shown to be feasible, and something of a contest developed between manufacturers to market the first hydrogen-powered car. As part of the BMW programme, a small batch of 7 Series saloons with fuel cells to provide current for the electrical system was laid down. Supplied by International Fuel Cells, these had sufficient capacity to supply other functions, such as an independent air-conditioning system.

Natural gas-fuelled cars were also marketed (see Chapter Four), but only in Germany where a fuel service infrastructure existed. Even there, they were regarded by the public as city cars, and the response was not strong.

There have been many bodies by coachbuilders, from full estate cars (for example, an adaptation of a 7 Series saloon offering much more load space than a Touring body) to sports cars. Then there was the Italdesign Nazca which some saw as a latter-day M1 when it appeared in 1991, inspiring speculation that Alpine might produce it in limited numbers. This was another project that interested BMW, and another that was not approved – and its production would have depended on a supply of BMW V12 engines.

Commercial considerations must mean that many projects remain confidential, so this chapter can do no more than illustrate a representative spread.

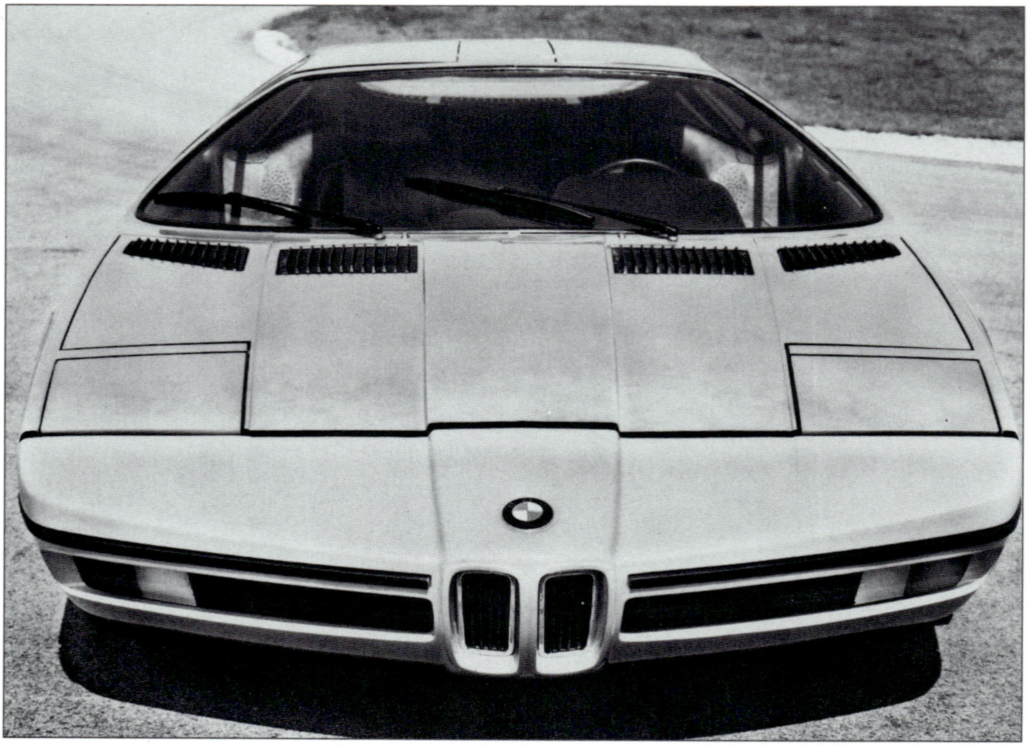

Two experimental Turbos were built in 1972, with avant-garde styling by Paul Bracq. The low coupé lines could have meant that it was not out of place among high-performance two-seaters of the period, but the cars were built as rolling laboratories rather than prototypes.

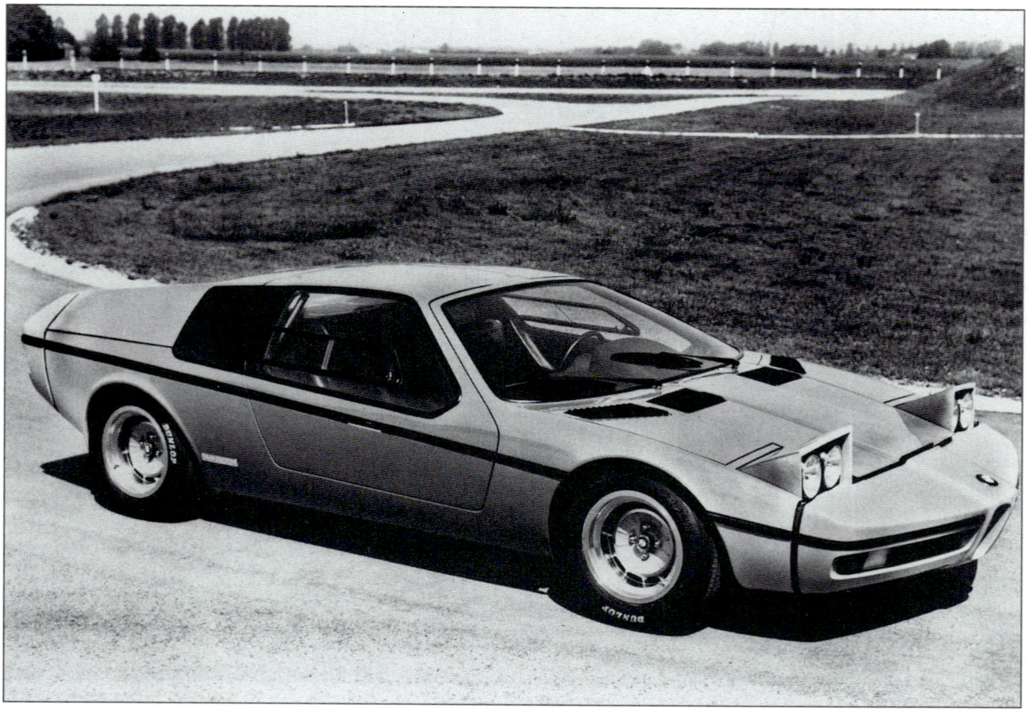

The Turbo looked exciting, and had serious purposes. The fibreglass body was colour-impregnated, with distinguishing sections. The engine was a 1990 cc turbocharged unit which anticipated the 2002 Turbo engine, mounted transversely behind the cockpit.

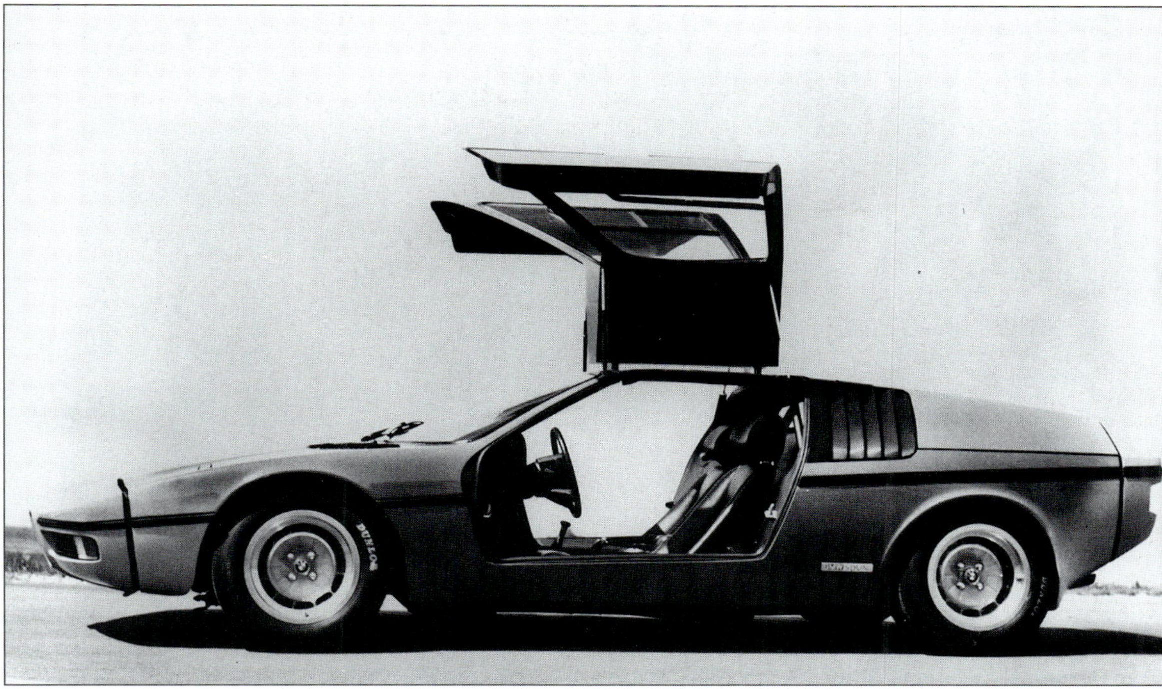

Safety was important. There were impact-absorbing nose and tail crumple zones, impact bars in the gullwing doors, even a radar device to warn a driver when he was too close to a vehicle ahead of him.

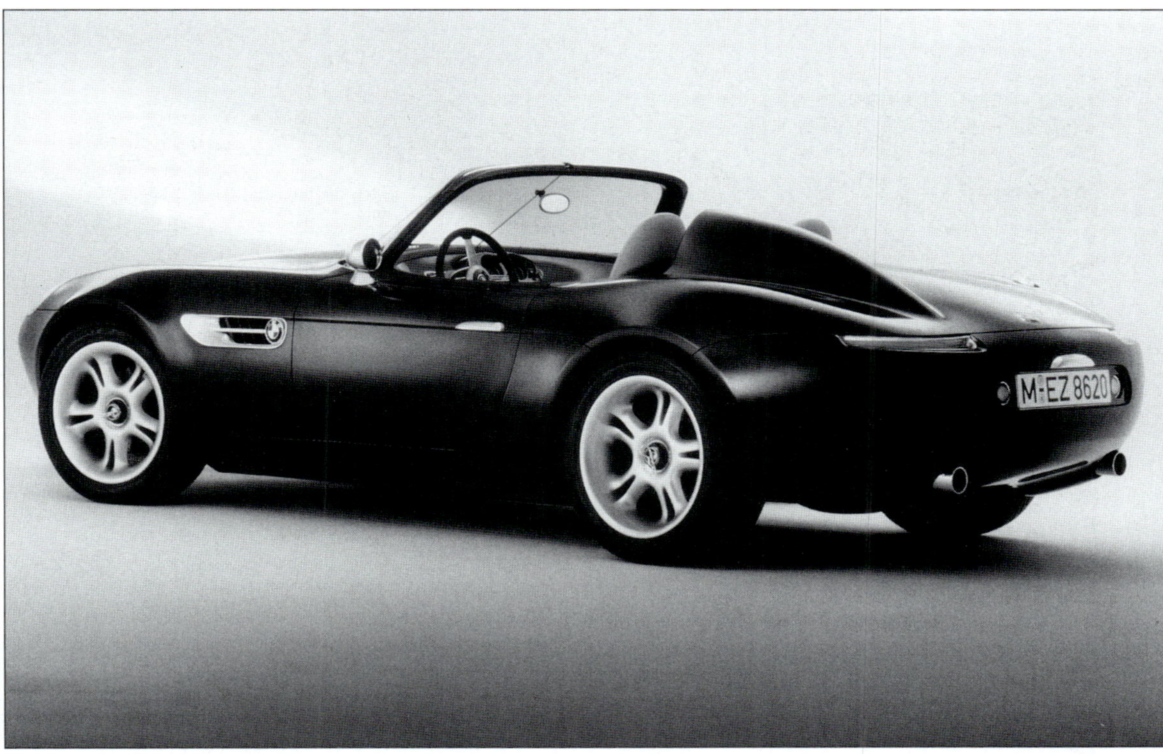

The design brief for the 1997 Z07 was to recapture the style and flair of the mid-1950s 507. It had elongated kidney grilles, round headlights behind covers at the front of the wings, engine air vents in the flanks, flared wheel arches and a fin behind the driver's headrest. Under the skin there was an aluminium space frame, front suspension derived from the 5 Series and rear suspension using 7 Series components, and a V8. Z07 led to Z8.

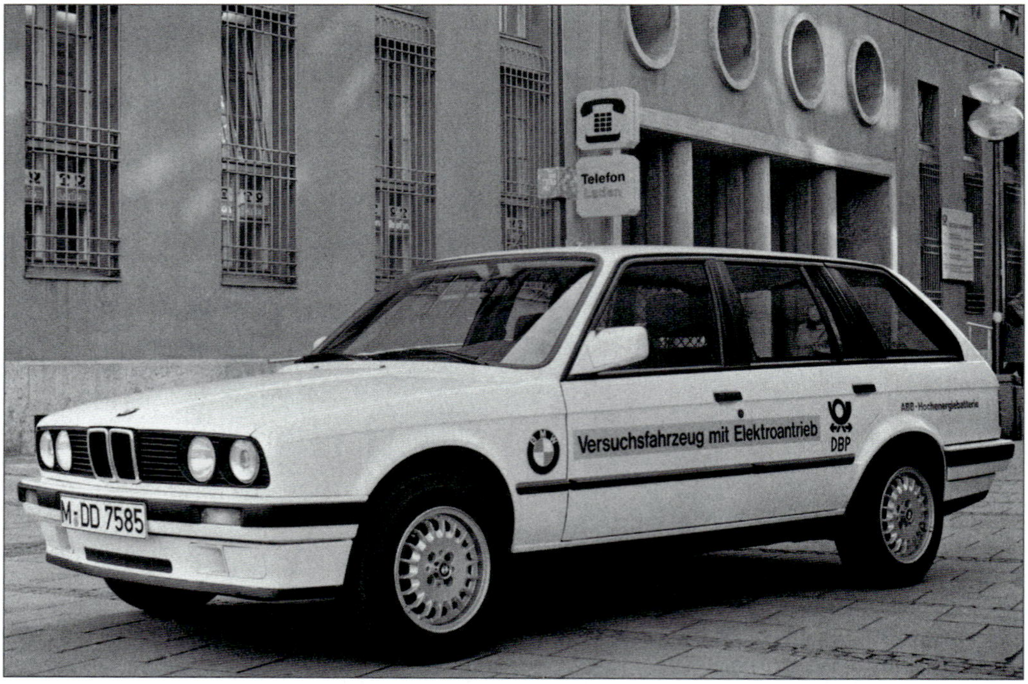

3 Series types provided the basis for electric city cars in the late 1980s, with a substantial 584 lb (265 kg) battery unit taking up much of the boot space and some of the rear passenger space. The four-wheel-drive 325 was adapted, with the drive to the front wheels used. The saloon had a range of some 90 miles (145 km) in 1989 and could exceed 60 mph (95 kmh), both 'city car' figures. The Touring type illustrated was also a front-wheel-drive vehicle. It had a better range, and was tested by the German Post Office in 1989.

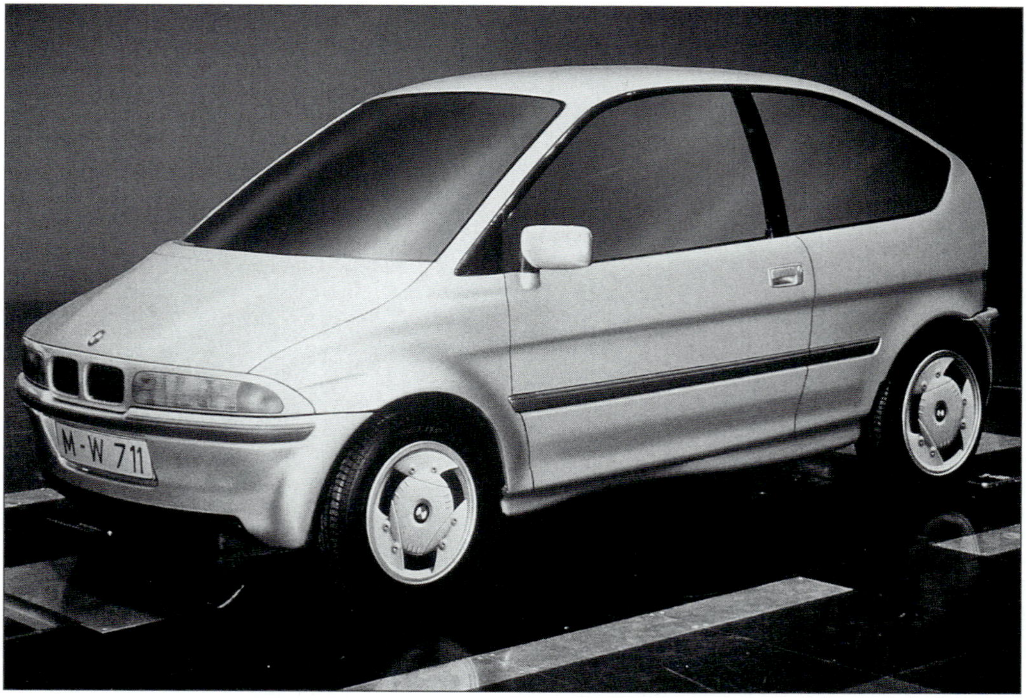

Moving on from adapted production cars, BMW Technik showed purpose-designed electric vehicles. E1 in 1991 was a compact four-seater with a range of some 150 miles. Oddly, E7 was an electric car project twenty-one years earlier – but that E stood for *Entwicklungs* . . .

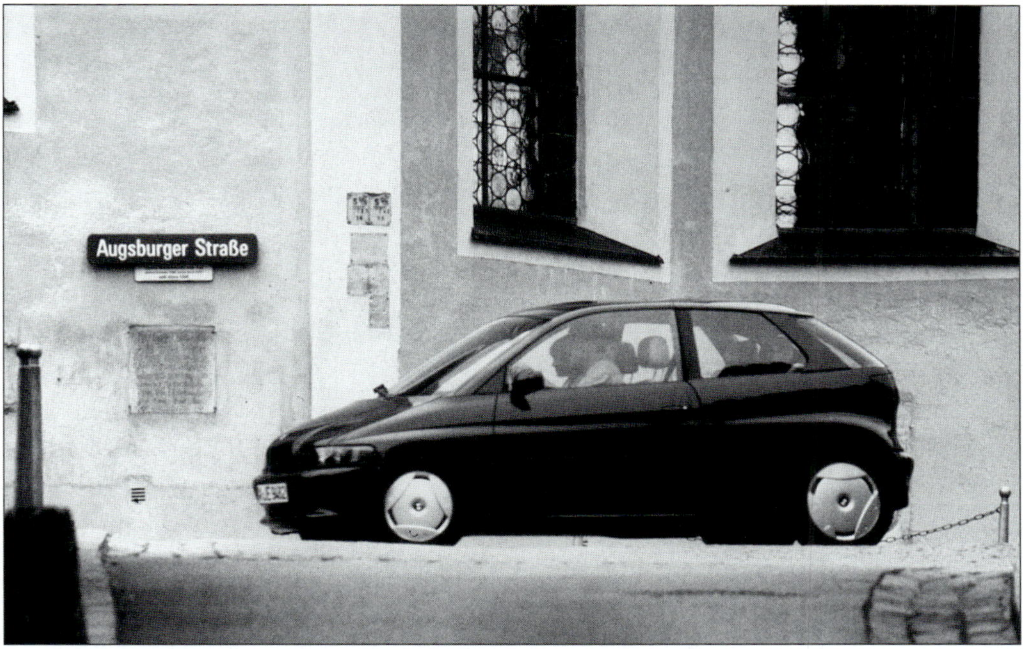

The more refined E1/2 (or E2) followed two years later. This was another rear-wheel-drive car with an aluminium chassis frame. The next stage was a small car with a motorcycle engine driving the front wheels and an electric motor the rear wheels (with no interconnection). Another small type was a motorcycle-engined three-seater, notable for its central driver's seat, unveiled in 1993.

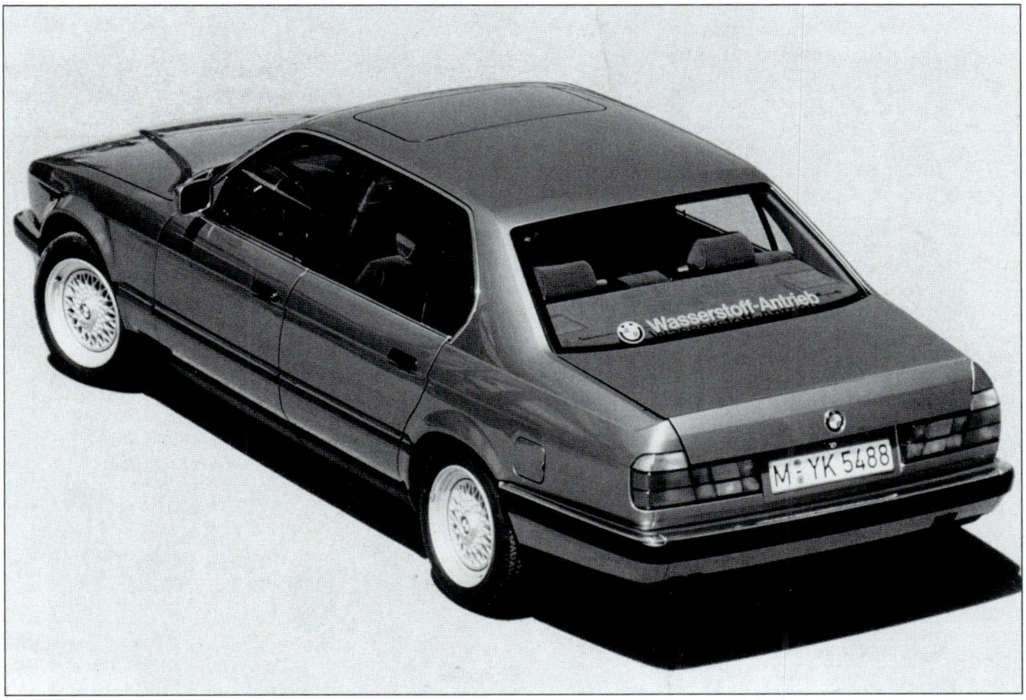

The label on the rear window of this 735 – '*Wasserstoff-Antrieb*' – indicates that the car is hydrogen-fuelled. Its engine had a ZF supercharger and produced 134 bhp, compared to just over 200 bhp when conventional fossil fuel was used. Hydrogen storage posed problems, and this 1989 test vehicle had a 20.5-gallon (93-litre) tank, which gave a range of only 180 miles (290 km). The fuel cell route should prove more viable . . .

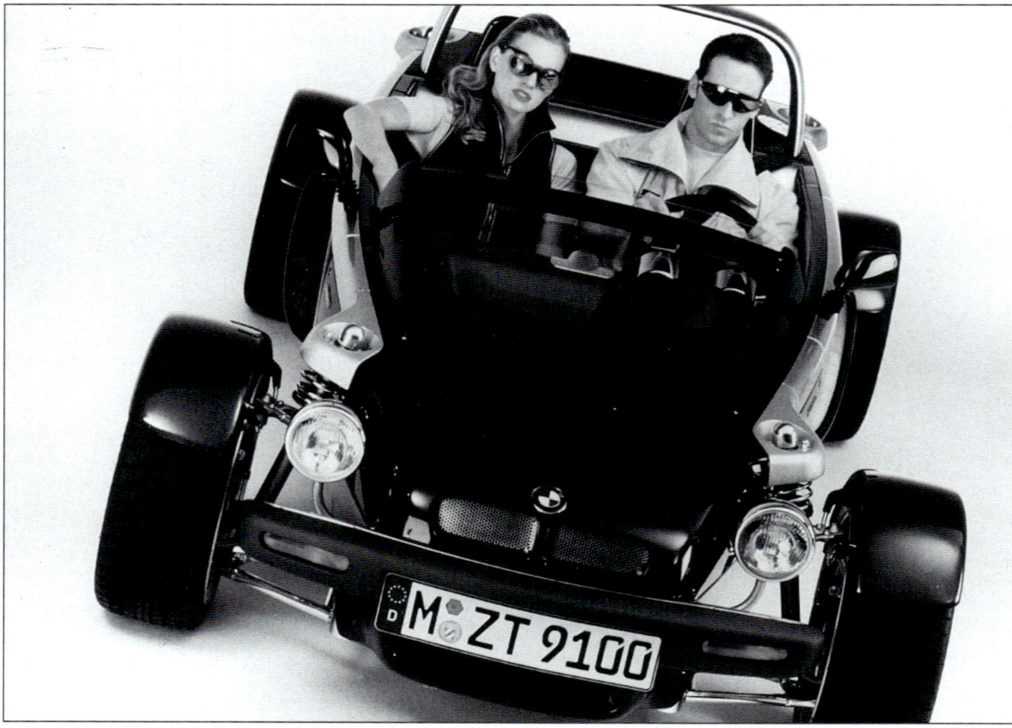

Just 4/2 was a fun machine, with only the nose to identify it as a BMW. A 'wrap-round frame' gave some protection against the weather, and BMW Technik gave assurances that it met crash protection requirements. A four-cylinder 987 cc motorcycle engine was mounted behind the seats and below the removable boot, which doubled as a suitcase.

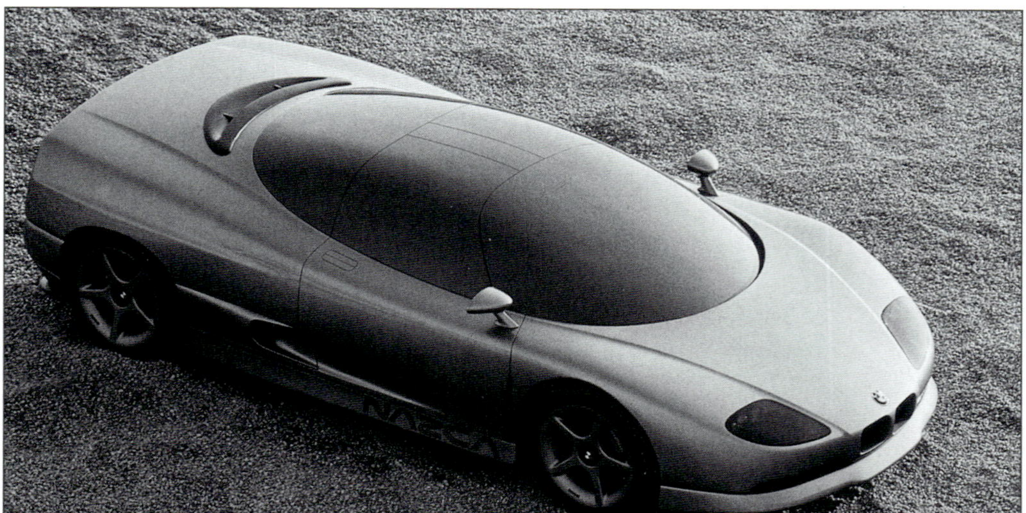

Many coachbuilders have worked their art around BMW mechanical components, but none has been more striking than this Nazca mid-engined coupé in 1991. The concept and execution were Italian, with mechanical design by Giorgio Stirano and body by Giugiaro. The chassis and body were in carbon fibre, and a mid-mounted BMW 5-litre V12 drove through a ZF five-speed gearbox. The cockpit was reached through a two-part door: the lower part opened conventionally, while the upper part opened like a gull-wing door – and was too cramped to be realistic for a production version. With the tops of the doors removed, the Nazca C2 was run as a spider. Estimates suggested a 185 mph (295 kmh) top speed, while wheelbase and wide track promised good handling qualities. This Italdesign project was exciting . . . but BMW went one better, with McLaren.

CHAPTER TWELVE

SPECIAL RELATIONSHIPS

BMW has been fortunate in its associations with specialist companies, their enthusiasm for its cars and their professionalism which has meant that responsibilities, particularly in racing, could be left in their hands in the knowledge that the reputation of the marque would be upheld. Foremost among these companies are Alpina and Schnitzer. Both started operating in the 1960s, and both have been committed to racing. Alpina moved on to become best known for its uprated and refined versions of production BMWs for the road, and while Schnitzer has continued to produce performance parts and conversions, its principal role through recent years has been running BMW teams. By coincidence, Alpina and Schnitzer have evolved so that they are virtually complementary.

Beyond these two there is a minor German industry focused on BMW tuning, and on cosmetic 'improvements' to give production models a touch of individuality, not always in the best taste and in any case an activity that has fallen away in the 1990s. This minor industry grew with the '*Neue Klasse*', and some companies have prospered. Hamann and Hartge, for example, built up portfolios of BMW work. Hamann has a solid reputation as a tuning wizard, for example with its Z3 coupé, while Hartge's cars have carried type numbers, and the company entered BMWs in top-level saloon-racing. GS Tuning has also converted cars, and raced them, and the list of specialists runs on, through Bamo, Koenig, Koepchen, Lorenz, Matthes, Mosselmann, Senni, Zender and others, and some working at less intense levels in Britain, France and Italy. Rafanelli's Bigazzi team has represented BMW in racing, enjoying quasi-works status until 1998.

Alpina has made very real contributions to BMW history, over a long period. Buckard Bovenstein first developed BMW engines in 1963, extracting an extra 10 bhp from the 1500 unit, and he set up Alpina in 1965, with a factory at Buchloe some 60 miles from Munich. Its first full conversions were to 02 cars at the end of the 1960s, and it soon proved its work in European touring car races. Alpina also supplied tuned engines to other racing teams.

Soon Bovenstein wanted Alpina to be recognized as a marque in its own right, feeling that its thoroughly engineered cars were complementary to regular BMWs, not simply filling niches in the conservative ranges. By the 1970s Alpina was a substantial operation, with its own model numbers prefixed 'B' or 'C', and in the 1990s it was producing around 500 cars a year. At the end of 1998, when it introduced its B3 version of the E46 3 Series saloon, it had built more than 1,200 earlier B3s.

The tendency was to develop variants with near-supercar potential, for example to slip in a straight six when the normal series model had a four-cylinder engine. That over-simplifies, for there was much more to the operation than following a big engine in a small car recipe, and for a quarter of a century Alpina's development facilities have been extensive. These contributed to BMW Motorsport's development of a turbocharged Formula 1 engine (just as Schnitzer's turbocharging experience made a useful contribution to this programme). And there is more to Alpina's 'enlarged' engines than the term suggests, for BMW's all-alloy blocks cannot be bored out, and so new blocks have to be cast to the same demanding specification. In 1999 Alpina's first diesel

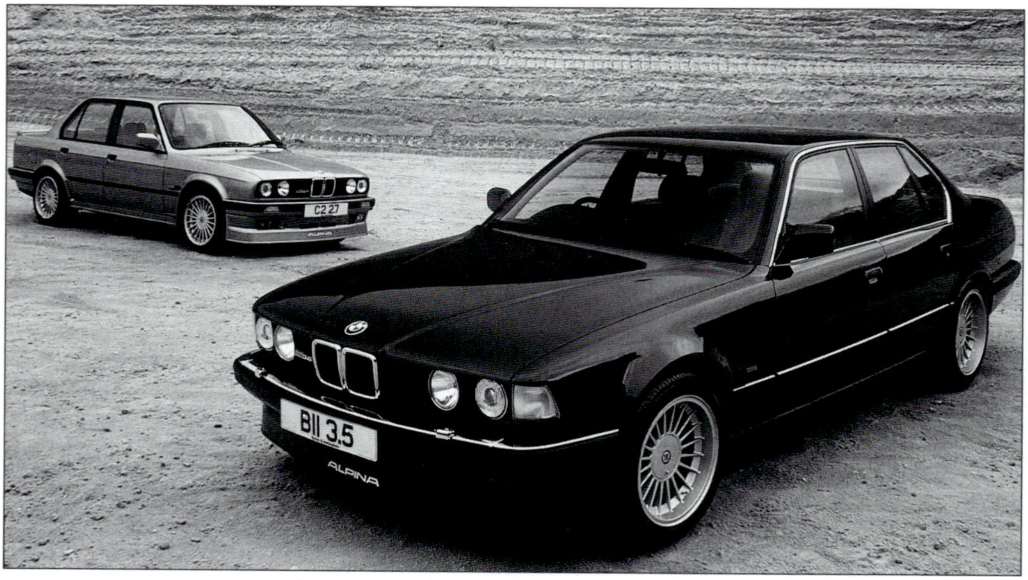

Variants of BMWs with the thin-spoked alloy wheels that have became an Alpina hallmark. The cars are a 3 Series-based 2.7 litre C2 of the second half of the 1980s and a slightly later 7 Series-based 3.5 litre B11.

appeared, the twin-turbo B10 Biturbo based on the 530d, and claimed to be capable of the near-universal German governed top speed of 250 kmh.

Alpina has also developed the Switchtronic transmission, on the basis of the BMW Steptronic automatic system. Switchtronic offers a manual option involving finger-tip changes via buttons on the steering wheel spokes, like in a modern Formula 1 car.

Schnitzer has been less concerned with innovative engineering, although like Alpina it has developed production BMWs, most notably 3 Series cars. Its origins are in German hill climbs and racing in the 1960s, when brothers Josef and Herbert Schnitzer formed the company, and Josef competed with BMW 1800s and 2000s. In 1966 the team won the German touring car championship, with the 2000TI.

The first Schnitzer development to attract wide attention was the 16-valve conversion of the 2002 engine, and this unit provided competitive power through to the mid-1970s, the decade when Schnitzer turned to turbochargers, and to the 3 Series. In single-seater racing, the French Martini team used Schnitzer dohc BMW engines in its MK 16 Formula 2 car, which Jacques Laffite drove to win the 1975 European title.

Meanwhile, with the 3.0CS/CSL coupés Schnitzer was in competition with Alpina, and both contributed to BMW's battles with Ford in the first half of the 1970s. With the 635CSi, the Schnitzer team won the European Touring Car Championship in 1983 and 1986. In 1988 Schnitzer drivers won the last European Touring Car championship for drivers, and several national championships. Attention then turned to the M3, which the team used to score a string of race successes in 1989–92, then to the 318i and 320i. By that time the immensely professional, and friendly, 'Charly' (Karl) Lamm was Schnitzer's team manager, and effectively BMW's team manager in series such as the British Touring Car Championship in the 1990s, and the FIA GT Championship, where McLarens were run. Schnitzer played a role in the development of the GTR version, and Lamm was BMW's racing director in this programme. Then as BMW Motorsport Team Schnitzer was withdrawn from touring car racing at the end of 1998, the management of BMW's open sports car racing programme was contracted to Schnitzer.

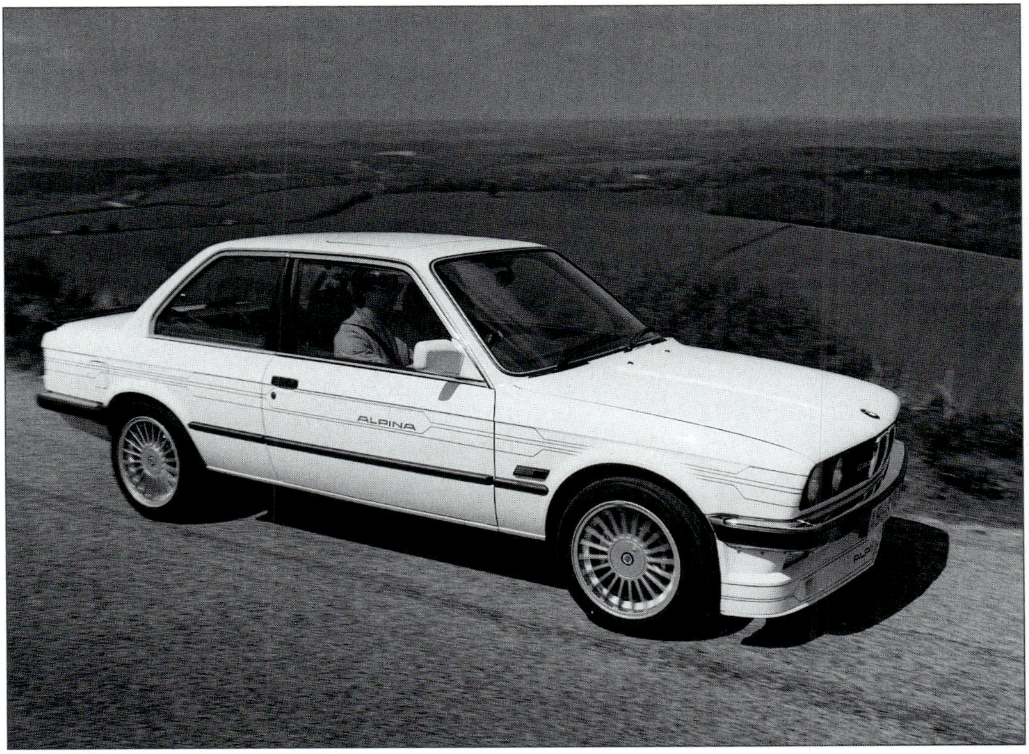

A 3 Series (E30) car in Alpina C1 guise, with a very prominent chin spoiler, the delicate Alpina lines on the flank and a small grille badge to distinguish it. It had a 2.3-litre 170 bhp straight six, and could reach 131 mph (210 kmh).

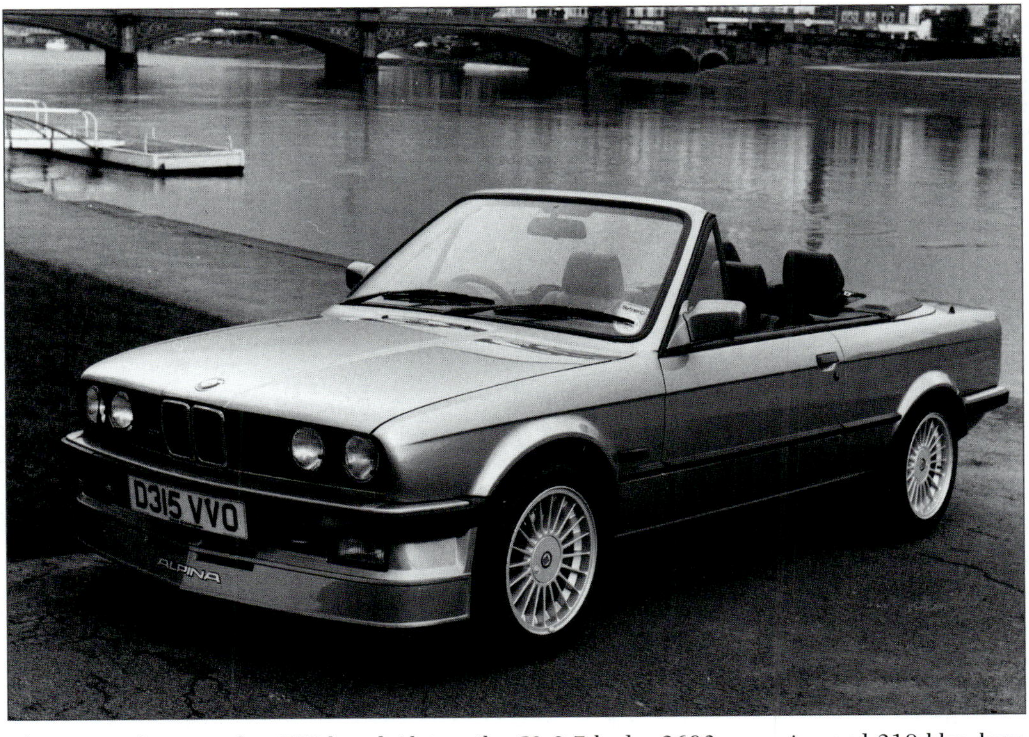

Three years later another E30-based Alpina, the C2 2.7 had a 2693 cc engine and 210 bhp, here in the clean-lined cabriolet body – a bewitching combination.

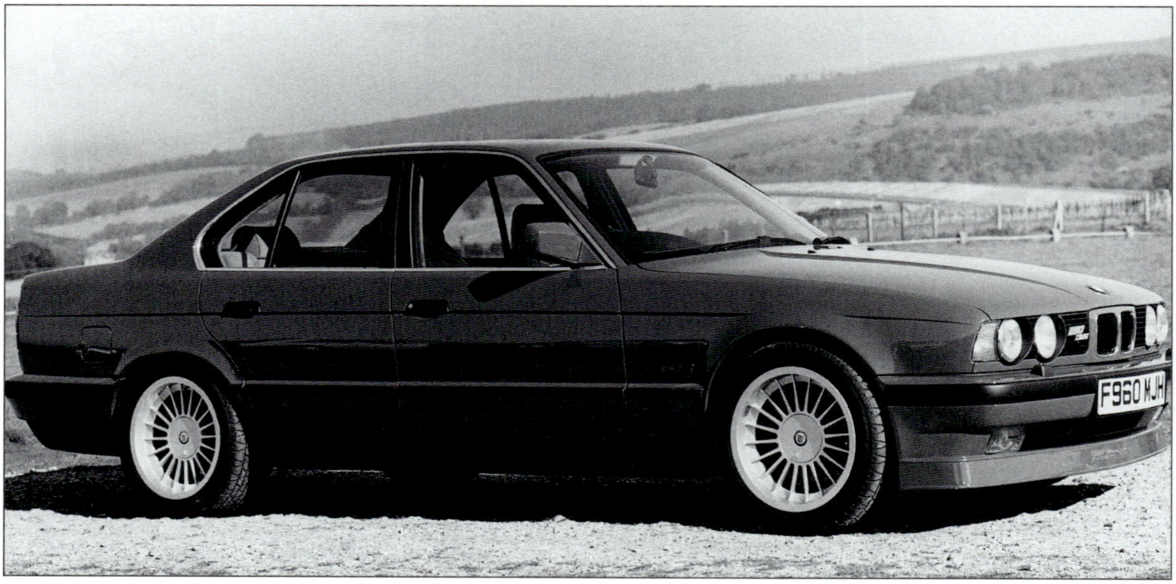

The B10 in this form was built 1988–95 (the B10 designation was used for several cars through to the 1990s, with straight six or V8 engines). This B10 was based on the 535i, and had nothing flamboyant in its body to suggest that it had 'supercar' performance, with a top speed over 150 mph (240 kmh). The B10 Biturbo which followed (1989–94) had a 360 bhp engine – 100 bhp more than this car – and could exceed 180 mph (290 kmh).

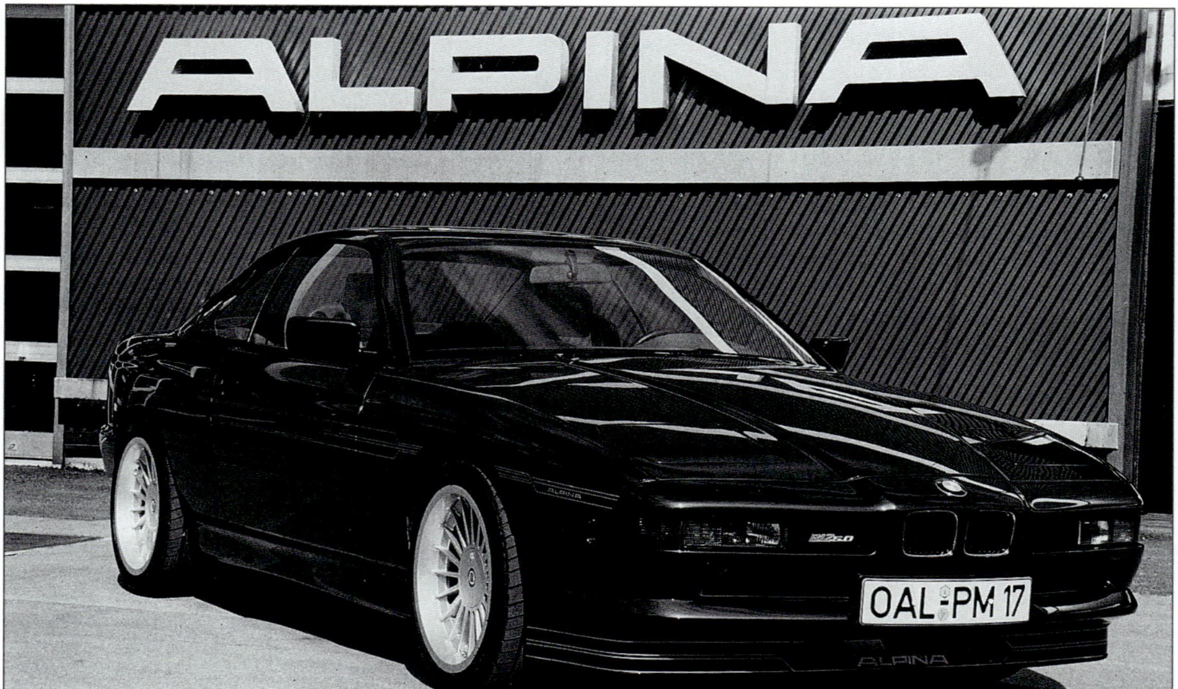

Supercar. BMW abandoned its M8 project in 1991, and that opened the door for Alpina's B12 coupé. This lost nothing of the 8 Series lines, and its appearance was enhanced by its wheels (which, incidentally, have valves concealed behind lockable covers). It also has a typical spoiler under the nose, and is lined out in Alpina's house style. The 5-litre V12 was rated at 350 bhp in this car of the mid-1990s, to give a maximum speed of the order of 175 mph (280 kmh). In quoted power output terms, this was bettered in Alpina road cars only by the B12s with 5.6-litre V12s.

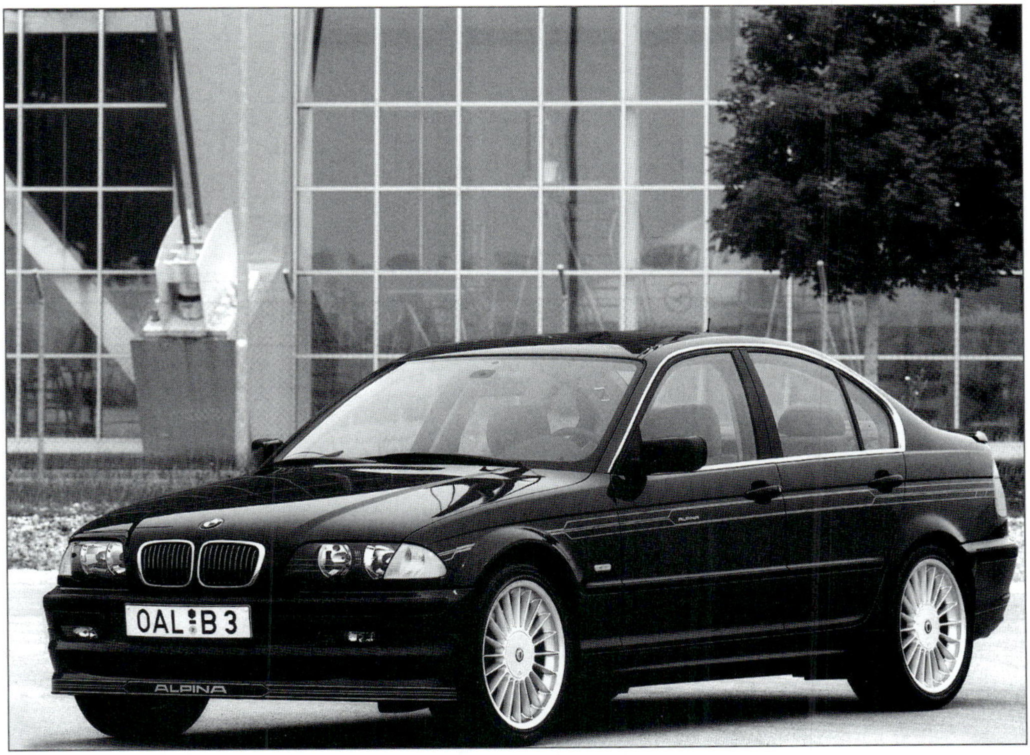

Late in 1998 Alpina introduced a new B3, its stunning version of the E46 3 Series saloon. Wide low-profile tyres meant that it looked lower than normal 3 Series cars, there were discreet front and rear spoilers, and immaculate Alpina paintwork. The 2.8-litre straight six was bored and stroked to give a capacity of 3152 cc, and further work introduced a new cylinder head and lightweight Mahle pistons. Output was increased from the normal 328i's 193 bhp to 265 bhp, and that gave a claimed top speed of 160 mph (255 kmh).

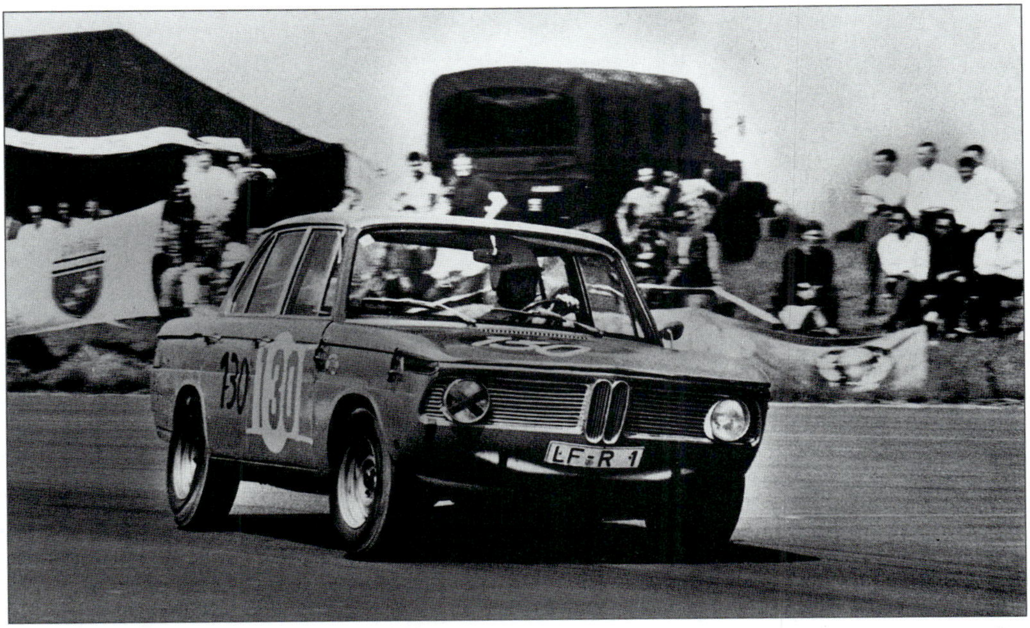

Josef Schnitzer forcing a 2000TI through a corner in a German championship race in 1966, the year the Schnitzer brothers' team won the national saloon car title. In those days Schnitzer cars were not always immaculate. . . .

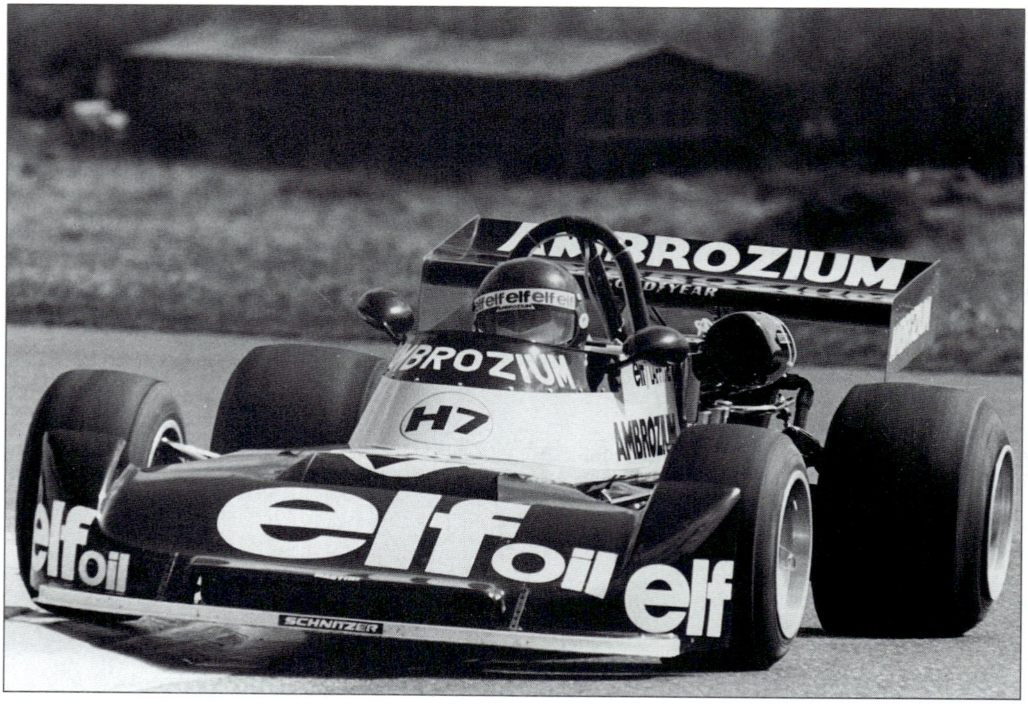

Schnitzer is seldom recalled for its involvement with single-seater racing, but it successfully developed Formula 2 engines on a BMW base in the mid-1970s. In 1975 Laffite drove the Schnitzer-BMW powered Martini MK 16 (here kerb-hopping at Thruxton) to six race victories, winning the European championship with 60 points to the runner-up score of 36. Schnitzer built a considerable number of these engines, first used in a winning car in 1974. They were rated at 290–300 bhp in 1975.

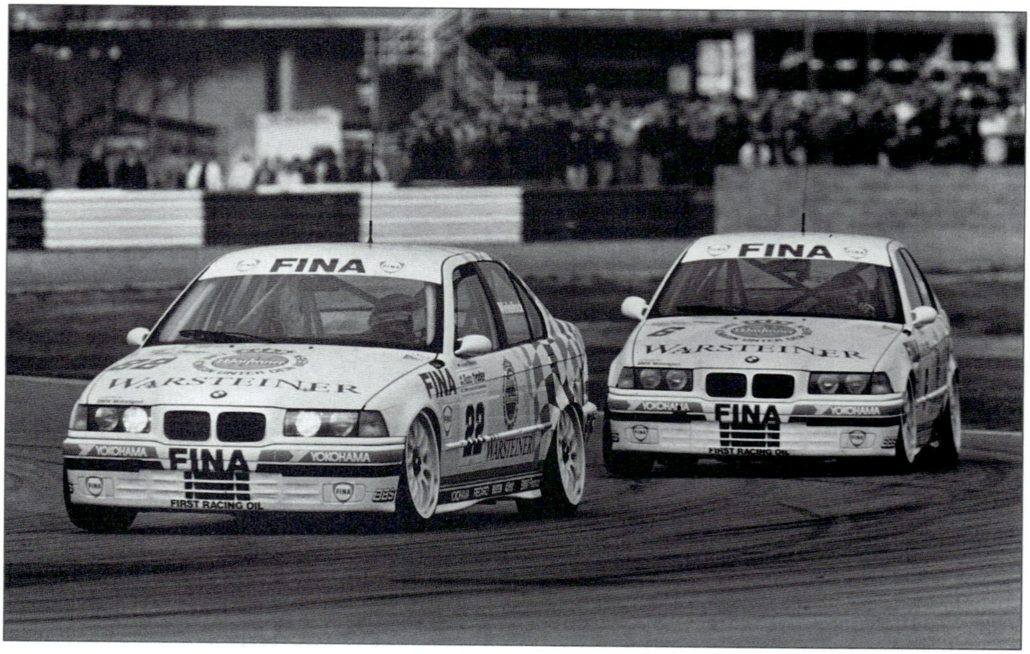

The Schnitzer team's assault on the 1993 British Touring Car Championship from a base at Silverstone was rewarded when Joachim Winkelhock and BMW took the titles, despite the weight handicap imposed on the four-door 318i as its front engine/rear drive layout gave it traction advantages over front-wheel drive cars. Steve Soper follows Winkelhock in this shot.

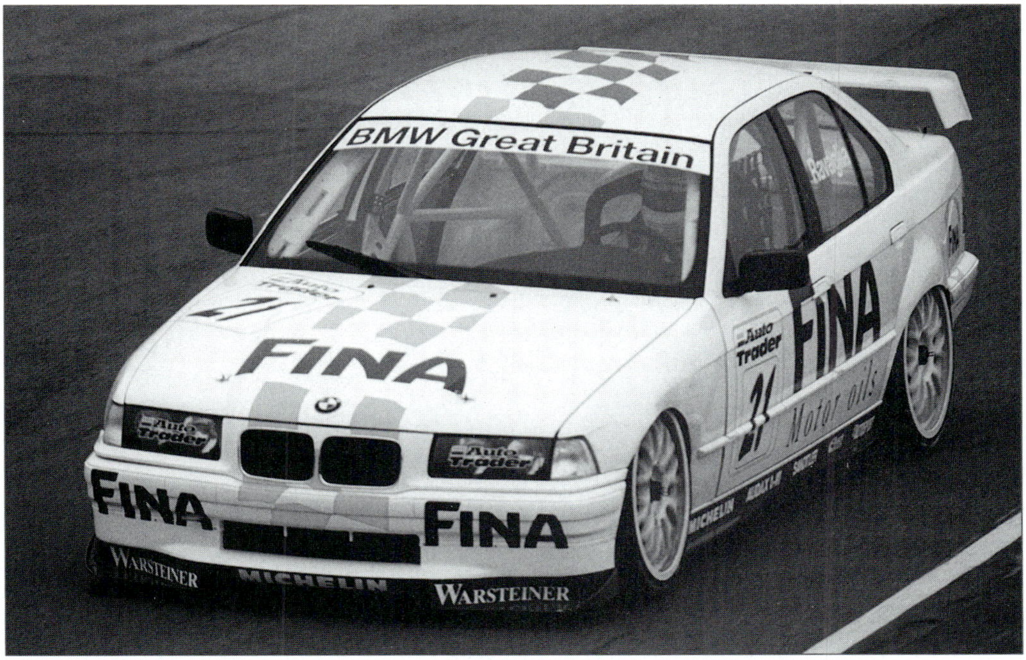

Major touring car championships in 1996 tended to fall to four-wheel drive cars, but in Britain Jo Winkelhock stoutly defended BMW honour with new 3 Series cars prepared and run by Schnitzer. New aerodynamic aids, front and rear, were developed and the cars sat even lower. The 16-valve four-cylinder S42 engine was based on the M42, and gave almost 300 bhp. Winkelhock won four British races in these cars, and his team-mate Roberto Ravaglia won one.

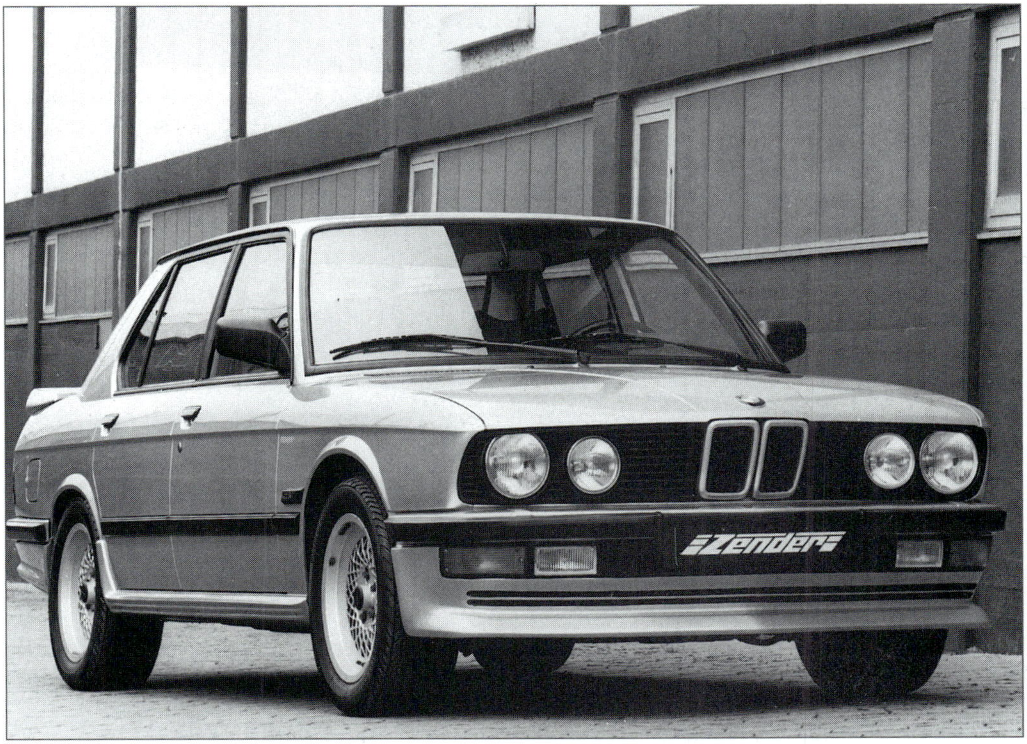

To complete an A–Z coverage in this chapter, here is a Zender. The conversion work was not as comprehensive as much carried out by Alpina, but this E28 528i-based car from the mid-1980s shows changes to the front and side panels, a rear wing, and intricate wide wheels.

ACKNOWLEDGEMENTS

In compiling this book the assistance of Heinrich Klebl and Walter Zeichner of BMW Mobile Tradition was invaluable, as was the help readily given by Chris Willows and Alun Parry of BMW (GB). David Carp's comments were most useful, and I am also grateful to Charles Armstrong-Wilson, editor of *BMW Car*, and Nick Godfrey of Sytner BMW. Where possible, performance figures have been taken from the most authoritative sources, the road tests published in *The Motor* and later in *Autocar*.

Many of the illustrations are from the BMW archives in Germany and Britain, others are from the author's collection, and the publishers are also grateful to Hugh Bishop, Keith Bluemel (p. 136), Bob Bull, Maurice Rowe and Gerry Stream.

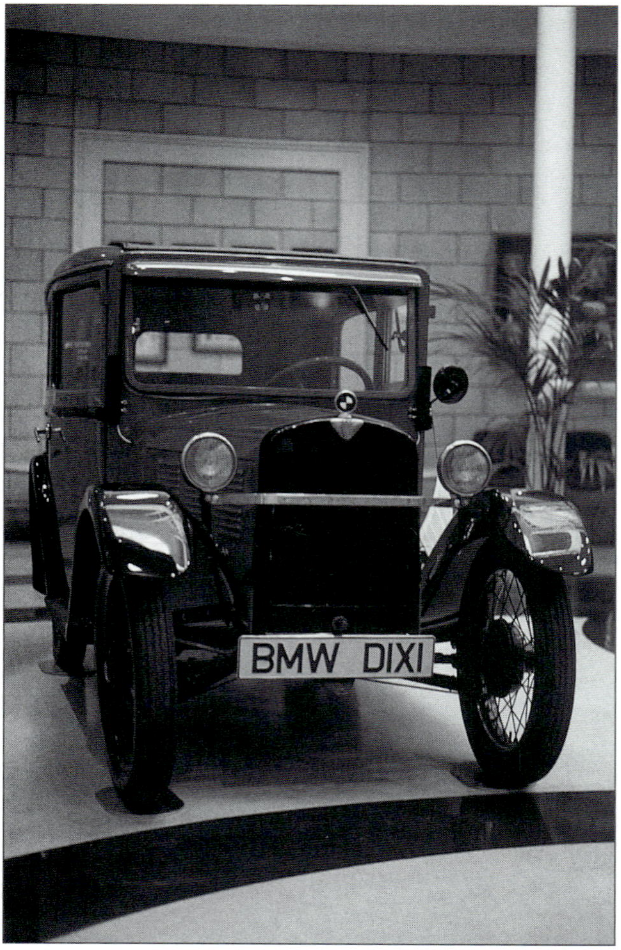

In a thoughtful gesture soon after BMW took over Rover, Bernd Pischetsrieder arranged for this BMW Dixi to be displayed at the British Motor Industry Heritage Trust museum at Gaydon.